Scott Perkins

DIGITAL LOGIC TESTING AND SIMULATION

DIGITAL LOGIC TESTING AND SIMULATION

ALEXANDER MICZO
Schlumberger Computer Aided Systems Research Center

HARPER & ROW, PUBLISHERS, New York
*Cambridge, Philadelphia, San Francisco,
London, Mexico City, São Paulo, Singapore, Sydney*

Sponsoring Editor: Peter Richardson
Project Editor: David Nickol
Cover Design: Wanda Lubelska Design
Text Art: Reproduction Drawings, Ltd.
Production: Delia Tedoff
Compositor: Waldman Graphics, Inc.
Printer and Binder: The Maple Press

Digital Logic Testing and Simulation
Copyright © 1986 by Harper & Row, Publishers, Inc.

All rights reserved. Printed in the United States of America. No part of this
book may be used or reproduced in any manner whatsoever without written permission,
except in the case of brief quotations embodied in critical articles and reviews.
For information address Harper & Row, Publishers, Inc., 10 East 53d Street, New
York, NY 10022.

Library of Congress Cataloging in Publication Data

Miczo, Alexander.
  Digital logic testing and simulation.

  Includes bibliographies and index.
  1. Digital electronics—Testing. I. Title.
TK7868.D5M49 1986   621.3815   85-21884
ISBN 0-06-044444-4

85 86 87 88 9 8 7 6 5 4 3 2 1

CONTENTS

PREFACE xi

CHAPTER 1 INTRODUCTION 1
- 1.1 Introduction 1
- 1.2 The Test 2
- 1.3 The Design Process 3
- 1.4 Design Automation 5
- 1.5 Economics of Testing 8
  - Problems 12
  - References 12

CHAPTER 2 COMBINATIONAL LOGIC TEST 14
- 2.1 Introduction 14
- 2.2 Approaches to Testing 14
- 2.3 Analysis of a Faulted Circuit 15
  - 2.3.1 Analysis at the Component Level 16
  - 2.3.2 Gate-Level Symbols 18
  - 2.3.3 Analysis at the Gate Level 20
- 2.4 The Stuck-at Fault Model 21
  - 2.4.1 The AND Gate Fault Model 22
  - 2.4.2 The OR Gate Fault Model 23
  - 2.4.3 The Inverter Fault Model 23
- 2.5 The Sensitized Path 23
  - 2.5.1 A Logic Analysis of the Faulted Circuit 23
  - 2.5.2 Analysis of the Sensitized Path Method 25

v

2.6 The *D*-Algorithm 28
  2.6.1 The D-*Algorithm: An Analysis* 29
  2.6.2 The Primitive D-*Cubes of Failure* 32
  2.6.3 Propagation D-*Cubes* 35
  2.6.4 Line Justification and Implication 37
  2.6.5 The D-*Intersection* 38
2.7 Other Path-Tracing Methods 43
  2.7.1 The Critical Path 43
  2.7.2 PODEM 47
2.8 Boolean Differences 50
2.9 Symbolic Path Tracing 55
  2.9.1 Poage's Method 55
  2.9.2 The Equivalent Normal Form 58
2.10 The Subscripted *D*-Algorithm 60
2.11 Summary 63
  Problems 65
  References 67

CHAPTER 3 SEQUENTIAL LOGIC TEST 68

3.1 Introduction 68
3.2 Definitions 68
3.3 Test Problems Caused by Sequential Logic 73
  3.3.1 The Effects of Memory 73
  3.3.2 Timing Considerations 77
  3.3.3 Hazards 79
3.4 Sequential Test Methods 80
  3.4.1 Seshu's Heuristics 80
  3.4.2 The Iterative Test Generator 82
  3.4.3 The Nine-Value Iterative Test Generator 87
  3.4.4 The Critical Path 88
  3.4.5 The Sequential Path Sensitizer 90
3.5 Sequential Logic Test Complexity 98
  3.5.1 Experiments with Sequential Machines 103
  3.5.2 A Theoretical Limit on Sequential Testability 108
3.6 Summary 114
  Problems 115
  References 116

CHAPTER 4 SIMULATION 118

4.1 Introduction 118
4.2 The Simulation Hierarchy 119
4.3 The Fault Simulator: An Overview 122
4.4 The Compiled Simulator 125

   4.4.1 Three-Valued Simulation 127
   4.4.2 Simulating Sequential Circuits 128
   4.4.3 Hazard Detection 129
   4.4.4 Parallel Fault Simulation 131
   4.4.5 Performance Enhancements 133
 4.5 The Event-Driven Simulator 134
   4.5.1 Zero Delay Simulation 135
   4.5.2 Nominal Delay Simulation 137
   4.5.3 Unit Delay Simulation 148
 4.6 Timing Verification 148
   4.6.1 Path Enumeration 149
   4.6.2 Block-Oriented Analysis 150
 4.7 Multiple-Valued Simulation 152
   4.7.1 Tri-State Logic 152
   4.7.2 MOS Simulation 154
 4.8 Fault Simulation 156
   4.8.1 Testdetect 156
   4.8.2 Deductive Fault Simulation 159
   4.8.3 Concurrent Fault Simulation 160
   4.8.4 Fault Simulation in Sequential Circuits 166
   4.8.5 Fault Simulation Performance 167
 4.9 Summary 169
   Problems 170
 The Simulator: A Description 172
 The Simulator (Listing) 174
   Problems Based on the Simulator 177
   References 182

CHAPTER 5 THE AUTOMATIC TEST PATTERN GENERATOR 184

 5.1 Introduction 184
 5.2 Overview of the ATPG 184
 5.3 Creating the Circuit Model 186
   5.3.1 Circuit Description by Components 186
   5.3.2 Circuit Description by Wire List 187
   5.3.3 Modeling the IC 189
 5.4 Fault Modeling 190
   5.4.1 Fault Equivalence and Dominance 190
   5.4.2 Checkpoint Faults 192
   5.4.3 Redundant Faults 194
   5.4.4 Technology-Related Faults 195
   5.4.5 Bridging Faults 199
   5.4.6 Selecting Fault Classes for Test 202
 5.5 The Pattern Generator 202
   5.5.1 Random Patterns 202

   5.5.2 *The Imply Operation* 204
   5.5.3 *Comprehension Versus Resolution* 205
   5.5.4 *Test Pattern Compaction* 206
   5.5.5 *The Asynchronous Environment* 208
 5.6 The Simulator 210
   5.6.1 *Fault Dictionaries* 210
   5.6.2 *Fault Dropping* 211
 5.7 Modeling Large Devices 212
   5.7.1 *Behaviorally Equivalent Circuits* 213
   5.7.2 *RAM Model* 213
   5.7.3 *Microprocessors* 215
   5.7.4 *Bidirectional Signals* 217
 5.8 User Features 217
   5.8.1 *User-Specified Inputs* 218
   5.8.2 *Test Pattern Languages* 219
   5.8.3 *The Simulator Output Display* 221
 5.9 Output Files 222
 5.10 Summary 223
   Problems 223
   References 226

CHAPTER 6 AUTOMATIC TEST EQUIPMENT 228

 6.1 Introduction 228
 6.2 The Manufacturing Test Environment 228
 6.3 The Functional Board Tester 232
   6.3.1 *The Reference Tester* 233
   6.3.2 *Architecture of a Stored Response Tester* 234
   6.3.3 *Diagnostic Tools* 237
 6.4 The Test Plan 240
 6.5 The Dynamic Functional Tester 243
 6.6 The In-Circuit Tester 247
 6.7 Developing a Board Test Strategy 250
 6.8 Summary 252
   References 253

CHAPTER 7 DESIGN-FOR-TEST 254

 7.1 Introduction 254
 7.2 Test Problems in Digital Circuits 254
 7.3 Ad Hoc Design-for-Testability Rules 256
   7.3.1 *The Solutions* 256
   7.3.2 *The Problems* 259
 7.4 Controllability/Observability Analysis 264
   7.4.1 *SCOAP* 265

CONTENTS

       7.4.2 *Other Testability Measures* 270
       7.4.3 *Test Measure Effectiveness* 271
       7.4.4 *Using the Test Pattern Generator* 272
7.5 The Scan Path 273
       7.5.1 *Level-Sensitive Scan Design* 276
       7.5.2 *Other Scan Methods* 280
7.6 The Partial Scan Path 282
7.7 Summary 285
     Problems 286
     References 287

CHAPTER 8 MEMORY SYSTEM DESIGN AND TEST 289

8.1 Introduction 289
8.2 Semiconductor Memory Organization 289
8.3 Memory Faults 291
8.4 Memory Test Patterns 292
8.5 Repairable Memories 295
8.6 Error-Correcting Codes 297
       8.6.1 *Vector Spaces* 297
       8.6.2 *The Hamming Codes* 299
       8.6.3 *ECC Implementation* 302
       8.6.4 *Reliability Improvements* 303
       8.6.5 *Iterated Codes* 305
8.7 Summary 305
     Problems 306
     References 307

CHAPTER 9 SELF-TEST AND FAULT TOLERANCE 308

9.1 Introduction 308
9.2 System Testing 309
       9.2.1 *The Ordering Relation* 309
       9.2.2 *The Microprocessor Matrix* 314
       9.2.3 *Graph Methods* 315
9.3 Self-Test 323
       9.3.1 *Signature Analysis* 323
       9.3.2 *Random Patterns* 329
9.4 Maintenance Processors 334
9.5 T-Fault Diagnosis 340
       9.5.1 *One-Step Diagnosis* 340
       9.5.2 *Sequential Diagnosis* 342
9.6 Fault Tolerance 343
       9.6.1 *Performance Monitoring* 343
       9.6.2 *Self-Checking Circuits* 345

        9.6.3 Burst Error Correction 347
        9.6.4 Triple Modular Redundancy 350
        9.6.5 Software-Implemented Fault Tolerance 351
9.7 Summary 352
     Problems 353
     References 354

CHAPTER 10 FUNCTIONAL TEST AND OTHER TOPICS 356

10.1 Introduction 356

10.2 Hardware Design Languages 357

10.3 Functional Simulation 361
       10.3.1 Functional Fault Simulation 361
       10.3.2 Functional Fault Modeling 362

10.4 The Functional ATPG 367

10.5 Artificial Intelligence Methods 372
       10.5.1 The State Machine Problem 372
       10.5.2 SCIRTSS 373
       10.5.3 Hitest 379
       10.5.4 Extracting Behavior from Structure 387
       10.5.5 The Expert System for Diagnosis 390

10.6 Hardware Simulation 394
       10.6.1 The Yorktown Simulation Engine 394
       10.6.2 The Metalogician 397
       10.6.3 The Physical Model 398

10.7 The SEM Noncontact Probe 399

10.8 Summary 402
     References 402

APPENDIX FINITE FIELDS: SOME DEFINITIONS 405

INDEX 409

PREFACE

Digital electronics has been the object of a major revolution. Circuits are shrinking in physical size while growing both in their speed and in their range of capabilities. Computers that once occupied a large room now reside on a single substrate about a quarter-inch on a side. This revolution is placing computing power in the hands of more and more users with applications ranging from home entertainment systems to large computer systems. The computer is no longer a mysterious back-room beast that is coddled by armies of field engineers. It is a commonplace instrument found in many homes and offices. It is used by people who expect it to work when they acquire it, who expect it to continue to work while they own it, and who expect it to be quickly repaired when something goes wrong.

Along with this revolution in hardware, there is a corresponding revolution in the software programs provided with the computer. Some of the most sophisticated of these programs are being used to design successors to the very host computers on which they reside. These programs are placing logic design capability into the hands of an ever-growing number of users. Furthermore, these development tools are themselves evolving. This continues to reduce the turnaround time from design of a logic circuit to receipt of fabricated parts.

This rapid advancement is not without its problems. A major problem, one which is growing in importance, is testing. Problems associated with testing of digital logic have been with us for as long as digital logic itself has existed. Unfortunately, the problems associated with the testing of digital logic are being exacerbated by the growing number of circuits placed on individual chips. One development group, designing a reduced instruction set computer (RISC), states, "the work required to . . . test a chip of this size approached the amount of effort required to design it. If we had started over, we would have used more resources on this tedious but important chore."[1]

[1]Foderaro, J. K., K. S. Van Dyke, and D. A. Patterson, "Running RISCs," *VLSI Design*, Sept.–Oct. 1982, pp. 27–32.

The increase in size and complexity of circuits placed on a chip, with little or no increase in the number of input/output (I/O) pins, effectively creates a bottleneck; more logic must be accessed with the same number of I/O pins, making it much more difficult to test the chip. Yet, the need for testing is becoming more important. The test must detect not only failures in individual units, but failures caused by defective manufacturing processes. A defect in a single unit may not significantly affect a company's balance sheet, but a defective manufacturing process in an extremely complex circuit could escape detection until after delivery of the product, resulting in a very expensive product recall.

Public safety is another factor that must be taken into account. Digital logic devices are being used more frequently in products that affect public safety. We must insure that these products are thoroughly tested and function as intended.

As a result of growing circuit complexity, testing is assuming an increasingly larger proportion of total product cost. Ironically, the very software design tools that make it possible to put more circuits on a chip at a reduced cost are effectively increasing the cost of circuit test. The difficulty in creating tests for new designs also contributes to delays in getting products into the marketplace. Product managers must balance the consequences of delaying shipment of a product for which adequate tests have not been developed against the consequences of shipping the product and facing the prospect of wholesale failure and return of large quantities of defective products.

Because of the testing problems arising from the increasing complexity of these devices, new test strategies are emerging. Increasing emphasis is being placed on finding a defect as early as possible in the manufacturing cycle, new algorithms are being devised to create tests for logic circuits, and more attention is being given to design-for-test techniques that require active participation by the logic designers. They must adhere to design rules intended to create designs that are more easily testable. This is simply a recognition of the fact that it is not possible for engineers to test untestable circuits.

Design-for-test seeks to modify hardware so that it is easier to test by conventional means, applying stimuli externally and monitoring response at physically accessible edge pins. Self-test or built-in-test modifies the design by embedding test mechanisms directly into the product being designed. The object is to place the test stimuli and the response evaluation mechanisms closer to the circuit that must be tested.

Fault tolerance also modifies the design, but the motive is to contain the effects of a fault. It is used when it is absolutely imperative that a logic device function correctly, even in the presence of defects. Passive fault tolerance incorporates redundancy into a device. Performance monitoring, or active fault tolerance, evaluates performance by means of self-testing or self-checking circuits, and this may include injecting test data into the device while it is operating in a functional mode. Errors in operation can be recognized, but recovery from error requires intervention by the processor or by an operator. An instruction may be retried or a unit may be removed from operation until it is repaired.

Digital processors employ some or all of the aforementioned techniques to deliver reliable computing. Successful designs carefully balance the use of these techniques for maximum benefit and require close cooperation between the test engineer and logic designer. It is clear, then, that there is no single solution to the test problem; there are solutions, and what works best in one product may not be applicable in another product. Furthermore, the best solution is often a combination of available solutions. This imposes a requirement that the designer understand the various concepts, their objectives, and their limitations.

This book is written for both the student and the practicing engineer. It assumes an understanding of logic design, sequential machine theory, and computer architecture. Although definitions are provided along the way, this is done primarily to insure that we share a common understanding of the use of the terms. Algorithms used to develop tests for digital circuits are explained and design practices which cause circuits to become untestable are examined. Therefore, the book should be useful to logic designers as well as to test engineers. Experience in industry indicates that logic designers are becoming more aware of the cost of testing (some industry studies indicate that manufacturing costs for testing logic boards can exceed 50% of the total cost of manufacturing that board). These designers frequently design untestable circuits because they do not understand why circuits are untestable, or, put another way, they do not know what intrinsic properties make a circuit testable. When the designer understands what design practices cause test problems, and why, he is in a better position to design testable circuits.

We begin, in Chapter 1, with an overview of the design and manufacturing processes. The growing importance of testing and its role in industry can be better appreciated when one is familiar with the design and manufacturing processes used to create digital products. In Chapter 2 we discuss combinational test methods. There has been a resurgence of interest in these with the growing use of scan in sequential circuits. The D-algorithm is given extensive treatment since it and derivatives of it are used extensively in industry.[2] Other topological or path-tracing methods covered include the critical path (also known as LASAR) and PODEM. The algebraic methods described next, while not extensively used for test generation, remain useful for analysis. Finally, the subscripted D-algorithm represents an example of symbolic propagation.

In Chapter 3 we examine methods for testing sequential circuits. We first examine the effects of stored states and circuit timing on the test problem. We then look at some of the methods for propagating tests through sequential logic, including the iterative test generator (ITG), the nine-value ITG, and the sequential path sensitizer (SPS) which extends the calculus employed in the D-algorithm to sequential elements. Finally, we examine some circuits which are testable but which cannot be tested without the aid of functional information. Simulation is discussed in Chapter 4, including both logic simulation and fault simulation. Topics discussed include compiled and table-driven simulators, event-driven simulation, and parallel, deductive, and concurrent fault simulation. Delay models for simulation include 0-delay, unit delay, and nominal delay elements. Timing verification in a synchronous design environment is also discussed.

---

[2]Breuer, M. A., et al., "A Survey of the State of the Art of Design Automation," *Computer*, vol. 14, no. 10, Oct. 1981, pp. 58–75.

Once we understand the algorithms for creating tests and simulating circuit models, we are then in a position, in Chapter 5, to examine the entire automatic test pattern generator (ATPG) environment, the software that implements the algorithms and creates test stimuli to be run on the test equipment. We look at fault modeling, fault collapsing, and techniques for enhancing the effectiveness and/or speed of operation of the ATPG. In Chapter 6 we look at the kinds of equipment used to apply tests to the circuits and monitor their response.

The first six chapters are concerned with traditional approaches for generating stimuli and applying them to circuits. In Chapter 7 we turn our attention to methods for designing testable logic. The task of testing logic can be substantially simplified if a logic design conforms to some rules. Memories have unique problems and solutions as a result of their regular or repetitive structure. In Chapter 8 we examine some algorithms designed to take advangage of this regularity.

In the final two chapters we examine trends and research from two perspectives. First, in Chapter 9, we look at testability from the standpoint of incorporating hardware into a design in order to gain greater access to internal circuits. One objective is to make testing and diagnosis easier by reducing the amount of logic through which the test must pass. Another objective is to detect faulty behavior during operation and thus to contain the effects of a fault. Then, in Chapter 10, we look at different ways to represent a design, such as hardware design languages and graphs, and we look at methods which use this information to create tests that are applied externally through the I/O pins.

The reader is cautioned to note that the field of digital logic testing is very dynamic, very volatile. It must be so in order to keep pace with the new circuits, new systems, and new architectures which continue to be introduced. New products provide faster, more efficient, and more economical computing. Their rapid strides imply a need for corresponding strides in test development if we hope to provide more reliable computing while maintaining reasonable costs. It is therefore incumbent on both the test engineer and the logic designer to remain abreast of trends in digital logic testing.

Alexander Miczo

DIGITAL LOGIC TESTING AND SIMULATION

Chapter 1

Introduction

1.1 INTRODUCTION

A test is an evaluation of a product or process. As such, it can take on different meanings for different people. It may involve a mere perusal of a product to determine whether it suits one's personal whims, or it may consist of a very long, exhaustive checkout of a complex system to insure compliance with numerous performance and safety requirements. Emphasis may be on speed of performance, accuracy, or reliability. The purchaser of an automobile may evaluate a purchase simply on the basis of color and styling. The automobile manufacturer must be concerned with two kinds of test. First, the design itself must be tested for such factors as performance, reliability, and serviceability. Second, individual units must each be tested to insure that they comply with the design specifications.

We shall consider testing as it relates to digital logic. We will concern ourselves primarily with technical issues but must not lose sight of the economic aspects of the problem. We will be concerned with both the cost of developing tests and the cost of applying tests to individual units. In some cases it becomes necessary to make trade-offs. For example, some algorithms for testing memories are very easy to create; a computer program to generate test vectors can be written in less than a half-day. However, the set of test vectors so created may be large enough to require several millennia to apply to an actual device. Obviously, we cannot keep a customer waiting that long. We must then accept the fact that it is sometimes necessary to invest more effort into initially creating the test set so as to reduce the cost of applying the test to individual units.

We begin this chapter by defining a test in a broad, generic sense. Then we put the subject of digital logic testing into perspective by briefly examining the overall design process. The problems related to the testing of digital components and assemblies can be better appreciated when viewed within the context of the overall design process. Within this process we note design stages where testing is required. We then look at design aids that have evolved over the years for designing and testing digital devices. Finally, we examine the economics of testing.

1.2 THE TEST

A test can be defined, in a rather general sense, as an experiment which is designed either to confirm or deny a hypothesis or to distinguish between two or more hypotheses. Figure 1.1 depicts a test configuration in which a stimulus is applied to a device under test (DUT) and the response is evaluated. If we know what response to expect from the correctly operating device, we can evaluate the response of the DUT to determine if it is responding correctly.

When the DUT is a digital logic device, the stimuli are called *test patterns* or *test vectors*. The response is usually observed at output pins of the device, although some test configurations permit monitoring of test points within the circuit that are not normally accessible during operation. Evaluation of the response is accomplished by comparing the device response to an expected response. The expected response can be determined by applying the stimuli to a known good device and recording the response or by creating a computer model of the device and simulating the input stimuli by means of the model.

If the response of the DUT differs from the expected response, then an *error* is said to have occurred. The error results from a *fault* in the circuit. Because a digital logic board may contain large numbers of potential faults, the error response is only the first step in a test. Further testing is usually necessary to distinguish which of several faults produced the erroneous response. This is accomplished by use of fault models. The process is essentially the same, that is, vectors are simulated against a model of the circuit, except that the computer model is modified to make it behave as though a fault were present in the model. By simulating the correct model and the model with the fault present, the responses can be compared, to determine how the faulted machine response differs from good machine response. Furthermore, by injecting several faults into the model, one at a time, and then simulating, it becomes possible to compare the response of a faulty DUT to various faulted models to determine which faulted model either duplicates or most closely approximates the behavior of the DUT.

If the DUT responds correctly to all applied stimuli our confidence in the DUT increases. However, we cannot conclude that the device is fault-free. We can only conclude that it does not contain any of the faults for which it was tested; it could contain other faults for which an effective test was not applied.

From the preceding paragraphs we see that there are two major aspects of the test problem:

Figure 1.1 Typical test configuration.

1.3 THE DESIGN PROCESS

1. Specification of test stimuli, and
2. Evaluation of stimuli to determine correct response.

Furthermore, this approach to testing is used both to detect the presence of faults and to distinguish between or diagnose faults for repair purposes.

In digital logic, the two phases of the test process listed above are referred to as test pattern generation and fault simulation. We shall have much more to say about both of these processes in later chapters. For the moment it is sufficient to say that they rank equally in importance; in fact, they complement one another. Stimuli capable of distinguishing between good circuits and faulted circuits do not become effective until they are simulated so that their effects can be determined. Conversely, extremely accurate simulation against very precise models without effective stimuli will not effectively uncover many defects.

1.3 THE DESIGN PROCESS

A digital device begins its life as a concept whose eventual goal is to fill a perceived need. The concept may originate from a completely new idea or it may result from a market research effort aimed at obtaining suggestions for enhancements to an existing product. In either case the concept often travels a long and arduous path before it reaches the marketplace—if, in fact, it ever reaches the marketplace.

For a new product the first step in the design process is an extensive requirements analysis. This results in a detailed product specification describing what the product must do. The object at this time is to maximize the likelihood that the final product will meet performance and functionality requirements at an acceptable price. Then, the behavioral description is prepared. It describes what the product will do. It may be rather brief, or it may be quite voluminous. For a complex product design, the product specification can be expected to be very formal and detailed, whereas the product specification for an enhancement to an existing product may consist of an engineering change document describing only the proposed changes.

After the product has been defined and a decision has been made to manufacture and market the device, a number of activities must occur, as illustrated in Figure 1.2. These activities are shown as occurring sequentially, but more often the activities overlap because, once a commitment to manufacture has been made, the objective is to get the product out the door and into the marketplace as quickly as possible. For a device with a given level of performance, time of announcement will frequently determine whether the product is competitive, that is, whether it falls above or below the performance-time plot illustrated in Figure 1.3.

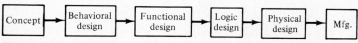

Figure 1.2 Design flow.

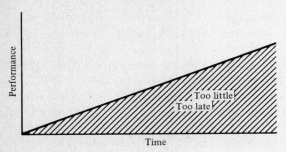
Figure 1.3 Performance-time plot.

During functional design, architects assemble together the functional units. At this stage of development the important decision must be made as to whether or not the product can meet the stated performance objectives, given the architecture and technology to be used. If not, alternatives must be examined. During this phase the logic is partitioned into physical units and assigned to specific units such as chips, boards, or cabinets. The partitioning process attempts to minimize input/output (I/O) pins and cabling between chips, boards, and units. Partitioning may also be used to advantage to simplify such things as test, component placement, and wire routing.

The functional design phase is followed by the logic design. Each of the functional units must be designed. This may require selecting or designing components, including the integrated circuits (ICs), and specifying the interconnection of these components.

Physical design specifies the physical position of components and the routing of wires to connect them. The placement may involve the assignment of circuits to specific areas on a piece of silicon, the placement of chips on a printed circuit board, or the assignment of boards to a cabinet. The routing task specifies the physical connection of devices after they have been placed. In some applications only one or two connection layers are permitted. Other applications may permit boards with 20 or more interconnection layers, with alternating layers of metal interconnections and insulating material.

The final design is sent to manufacturing, where it is fabricated. Engineering changes must frequently be accommodated due to logic errors or other unexpected problems such as noise, timing, heat buildup, electrical interference, or inability to mass-produce some critical parts.

In these various design stages there is a continuing need for testing. For requirements analysis the testing is essentially an evaluation of whether the product will fulfill its objectives, and testing techniques are frequently based on educated guesses concerning future needs. Attempts to introduce more rigor into this phase include the use of design languages such as PSL/PSA (Problem Statement Language/Problem Statement Analyzer)[1], which provides a way both to rigorously express the problem and to analyze the design. Another language useful in the initial design stages of a project is PMS (Processors, Memories, Switches)[2], which permits specification of a design via a set of consistent and systematic rules.

1.4 DESIGN AUTOMATION

The functional design can be more formally tested because the designer has a requirements document which tells him what he is expected to design. During this phase he has available numerous hardware design languages such as AHPL[3], CDL[4], and DDL[5], which permit him to express his design at a register transfer level and then simulate it with input vectors chosen to confirm correctness of the design or expose design errors. The design verification vectors must be sufficient to confirm that the design satisfies the behavior expressed in the product specification. Development of effective test stimuli at this state is highly iterative; a discrepancy between the designer's intent and his result signals the need for more stimuli to diagnose the reason for the discrepancy.

The logic design is tested in a similar fashion to the functional design. A major difference is that the simulators are more detailed or exhaustive in their analysis. Timing becomes a greater concern at the logic level and stimuli that were effective at the register transfer level may not be effective in ferreting out critical timing problems. On the other hand, stimuli that indicated correct or expected response at the register transfer level may, when simulated by a timing simulator, indicate incorrect response or marginal performance, or the simulator may simply indicate that it does not know, that is, cannot predict, the correct response.

The testing of physical structure is probably the most formal test level. The test engineer works from a very detailed design document to create tests which determine whether the behavior of the fabricated device corresponds to the design. Studies of fault behavior of the selected circuit family or technology permit the creation of fault models. These fault models are then used to create specific test stimuli which attempt to distinguish between the correctly operating device and a device with the fault.

This last category, which is the most highly developed of the design stages due to its more formal and well-defined environment, is where we will concentrate our attention. However, many of the techniques that have been developed for structural testing can be applied to design verification at the logic and functional levels.

1.4 DESIGN AUTOMATION

From virtually the beginning of the digital logic design era it has been recognized that many of the activities performed by architects and logic designers are tedious, repetitious, and error-prone and hence can and should be automated. The automation of these tasks is being accomplished by using the computer to perform repetitious tasks so that, in effect, each generation of computers assists in the design of its successors.

The mechanization of tedious design processes reduces the potential for errors caused by human fatigue, boredom, and inattention to mundane details. Early elimination of errors, which once was a desirable objective, has now become a sheer necessity. Formerly, discovery of an error in a design resulted in an engineering change which required modification to a board. This would frequently include connecting points together by means of color-coded wires. Yellow was a popular color for this task. Now, when a design is found to be in error, the problem is usually inside a chip containing many hundreds of gates and it can no longer be fixed with "yellow wires." Fixing it may require going through a complete design cycle to design and fabricate a new chip, which could delay delivery of a product by many weeks.

In addition to reducing errors, the elimination of tedious and time-consuming tasks provides an opportunity for designers to devote more time to creative endeavors. The designer can experiment with different solutions to a problem before a design becomes frozen in silicon. Various alternatives and trade-offs can be studied. This process of automating various aspects of the design process has come to be known as design automation (DA). It does not replace the designer, but rather enables him to be more productive and more creative. In addition, it is providing access to IC design for many logic designers who know very little about the intricacies of laying out an IC design. It is one of the major factors responsible for reducing the cost of digital products.

To support the kinds of design activities described in the previous section, a typical design automation system should provide the following capabilities:

Data management
   record data
   retrieve data
   define relationships
   perform rules checks

Design analysis/verification
   simulate
   check timing
   analyze performance

Design fabrication
   perform wire routing and placement
   create tests for structural defects

Documentation
   extract parts list
   create logic diagrams

The data management capability supports a data base which serves as a central repository for all design data. The data management programs accept data from the designer, format it, and store it in the data base. Some validity checks can be performed at this time to spot obvious errors. The data management programs must also be able to retrieve specified records from the data base. Different applications will require different records or combinations of records. As an example, one which we will elaborate on in a later chapter, a test program needs information concerning the specific ICs used in the design of a board; it needs information concerning their interconnections; and it needs information concerning their physical location on a board. A data base should be able to express hierarchical relationships[6]. This is

1.4 DESIGN AUTOMATION

especially true if a facility designs and fabricates both boards and ICs. The ICs are described in terms of logic gates and their interconnections, while the board is described in terms of ICs and their interconnections. A "where used" capability for a part number is useful if a vendor provides notice that he can no longer supply a particular part. Rules checks can include examination of fan-out from a logic gate to insure that it does not exceed some specified limit. The total resistive or capacitive loading on an output can be checked. Wire length may also be critical in some applications and the data management programs should be able to spot nets which exceed wire length maximums.

The data management system must be able to handle multiple revisions of a design or multiple physical implementations of a single architecture. This is true for manufacturers who build a range of machines all of which implement the same architecture. It may not be necessary to maintain an architectural level copy with each physical implementation. The system must be able to control access to and update of a design, both to protect proprietary design information from unauthorized disclosure and to protect the data base from inadvertent damage. A lockout mechanism is useful for preventing simultaneous updates, which could result in one or both of the updates being lost.

Design analysis and verification includes simulation of a design after it is recorded in the data base to verify that it is functionally correct. This may include simulation at a register transfer level with a hardware design language and/or simulation at a gate level with a logic simulator. Very definite relationships must be satisfied between clock and data paths. After a logic board with many components is built, it is usually still possible to alter the timing of critical paths by inserting delays on the board. On an IC there is no recourse but to redesign the chip. This evaluation of timing can be accomplished by simulating input vectors with a timing simulator, or it can be done by tracing specific paths and adding the delays of elements along the way. Other kinds of performance analysis that can be accomplished from the data base include computation of heat concentration on an IC or board and computation of the reliability of an assembly based on the reliability of individual components and manufacturing processes.

When the design has stabilized and has been entered into the data base and thoroughly evaluated to insure that it is correct, it can then be fabricated. Fabrication of a logic device requires placing chips on a board or circuits on a die and then interconnecting them. Because of the large numbers of chips or circuits that must be routed and connected, this is usually accomplished by placement and routing programs. Placement and routing can be a fully automated process for simple devices; for complex devices, it may require an interactive process whereby computer programs do most of the task and then require the assistance of an engineer to complete the task. Checking programs are often used after routing and placement to check that components have been placed in such a way that they do not occupy the same physical area. After a device has been placed and routed, other kinds of performance analysis can be accomplished. These include such things as computation of heat concentration on an IC or board and computation of the reliability of an assembly based on the reliability of individual components and manufacturing processes. Testing of the structure involves the creation of test stimuli that can be applied to the manufactured board or IC to determine whether it is fabricated correctly.

Documentation includes the extraction of parts lists and the creation of logic diagrams. The parts list is used to maintain an inventory of parts in order to fabricate assemblies. The parts list may be compared against a master list which includes information such as preferred vendors, second sources, or alternative parts that may be used if the original part is unavailable. Preferred vendors may be selected on the basis of an evaluation of their timeliness in delivering parts and the quality of parts received from them in the past. Logic diagrams are used by technicians and field engineers to debug faulty circuits as well as by the original designer or another designer who must modify or debug a logic design at some future date.

1.5 ECONOMICS OF TESTING

A question often confronted by the test engineer is: "How much testing is enough?" This may seem, at first glance, to be a frivolous question, since we would like to test our product sufficiently so that a customer never receives a defective product. When a product is under warranty or is covered by a service contract it represents an expense to the manufacturer when it fails because it must be repaired or replaced. In addition, there is an immeasurable cost in the loss of customer good will, an intangible but very real cost of defective products.

Unfortunately, we are faced with the inescapable fact that testing adds cost to a product. This cost comes from two sources: development of an initial test plan for a product and application of the test plan to individual units. We are therefore caught in a dilemma: testing adds cost to a product but failure to test also adds cost. We must therefore determine the right amount of testing. The right amount is that amount which minimizes the total cost of testing plus servicing or replacing defective components. In other words, we try to reach that point where the cost of additional testing exceeds the benefits derived. Exceptions, of course, must exist where public safety or national interests are involved.

There are two ways in which we can seek to minimize test cost. We can attempt to minimize the time spent in developing tests for the product, and we can attempt to minimize the amount of time required to apply the tests to individual units. The emphasis will be determined by the number of units to be manufactured. If the item is a consumer product with large volume, we can justify spending more time in developing a good test plan because the development cost will be amortized over many units. We can justify not only a more thorough test, but also a more efficient test, that is, one that reduces the amount of time spent in testing each individual unit. In low-volume products, testing can become a disproportionately large part of total product development cost and it may be impossible to justify the expense of making a test efficient. However, in critical applications it will still be necessary to prepare thorough tests.

We begin test development for a product by first noting that we want to test for faults most likely to occur. To do this we must become familiar with fault mechanisms and be able to model their behavior. We must know what faults occur

1.5 ECONOMICS OF TESTING

and how frequently they occur. Then, the effectiveness of a test can be measured as a function of the number of these faults that it can detect and their frequency of occurrence.

Once we have quantified test effectiveness, we can use the test effectiveness measure to help answer the question posed at the beginning of this section. For this we use the following equation[7]:

$$DL = 1 - Y^{(1-T)}$$

In this equation, DL represents defect level, that is, the fraction of devices shipped which are defective. The variable Y represents the yield of the manufacturing process and the variable T represents the test percentage; each of these is expressed as a fraction.

EXAMPLE ■

If it were possible to test for all defects, then

$$T = 1 \quad \text{and} \quad DL = 1 - Y^{(1-1)} = 0$$

On the other hand, if no defective units were manufactured, then

$$Y = 1 \quad \text{and} \quad DL = 1 - 1^{(1-T)} = 0$$

In either case, no defective units are shipped. ■ ■

This equation can be restated to express the test fraction T in terms of yield and defect level as follows:

$$T = 1 - \frac{\log(1 - DL)}{\log(Y)}$$

EXAMPLE ■

Integrated circuits (ICs) are manufactured on wafers—round, thin silicon substrates, usually ranging from 3 to 5 inches in diameter, as illustrated in Figure 1.4. After processing, the individual ICs are tested. The wafer is scribed and the dice that tested as bad are discarded. If the yield of good dice is 10%, and we want a defect level not to exceed 2%, what level of testing must we achieve? From the equation, we get

$$T = 1 - \frac{\log(1 - 0.02)}{\log(0.1)} = 1 - 0.0088 = 0.9912$$ ■ ■

The usefulness of this equation extends beyond its ability to compute testability. Testability is not a linear function. Experience indicates that testing follows the curve illustrated in Figure 1.5.

This curve tells us that we reach a point where substantial expenditures provide only marginal improvement in testability. At some point additional gains become exorbitantly expensive and may negate any hope for profitability of the product. However, looking again at the equation, we see that the defect level is a

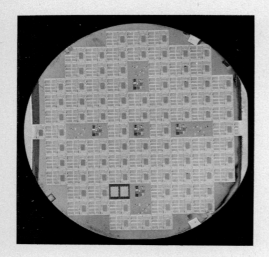

Figure 1.4 Wafer before scribing (*top*). Individual die (*center*). Close-up of die at 410X magnification (*bottom*). Can you spot the defect? (Courtesy of Fairchild Semiconductor)

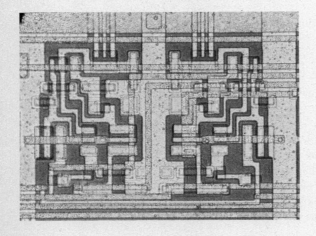

1.5 ECONOMICS OF TESTING

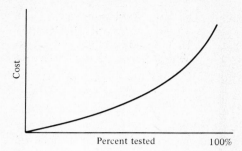

Figure 1.5 Cost curve for testing.

function of both testability and yield. Therefore, we may be able to achieve a desired defect level by improving yield.

EXAMPLE ■

We improve yield to $Y = 20\%$. What percentage of testing must be achieved to hold DL below 2%?

$$T = 1 - \frac{\log(1 - 0.02)}{\log(0.2)} = 1 - 0.0126 = 0.9874$$

■ ■

We see that raising yield reduces the testability requirements. Unfortunately, it is not quite that simple. The yield is a variable in another calculation and, strange as it may seem, a lower yield may improve profitability[8]. Given a wafer with N dies and a yield Y, there will be $Y \times N$ good dies on the wafer. These will be sold for Z dollars each, producing an income of $Y \times N \times Z$. This income must exceed the cost of producing, testing, and marketing the chips. If we shrink the size of the dies we get more of them on the wafer but the yield goes down. If shrinking the die size causes a sufficiently large increase in the number of good dice, then income increases. Given a fixed selling price, the object is to find the die size and yield which maximize the product term $Y \times N \times Z$. This calculation of die size and yield can have adverse secondary effects on testing requirements.

Although the shrinking die size reduces yield, the ideal situation is to minimize the loss of yield when the die is reduced in size. This can be done by using test information to learn more about failure mechanisms. We can record the frequency of occurrence of various faults and use this information to improve the manufacturing process, focusing our attention on those areas where frequency of occurrence of defects is greatest.

The importance of being able to achieve a low defect level in digital components can be appreciated when we look at a typical printed circuit board. Suppose that on a given board there are 70 components, each of which has a defect level $DL = 0.01$. Then, assuming that no defects are introduced in the manufacturing process, the likelihood of getting a defect-free board is $(0.99)^{70} = 0.49$; that is, there is less than a 50% chance of producing a defect-free board—and this assumes that no manufacturing defects were introduced.

The ability to compute defect level is also important because it may tell us that, given certain reasonable levels of testability and yield beyond which we cannot hope to improve further, we will have to expect a certain percentage of defective units to be shipped and we must plan our business strategy accordingly, whether it be to stock more spare parts or to improve our service department.

Another important aspect of test economics is the cost of locating and replacing a defective part. Detecting a defective unit is usually only part of the job. Consider again the board with 70 integrated circuits. If we discover during operation of the board that it is defective, then it is necessary to locate the exact part that has failed. This can be a time-consuming and error-prone task. Replacing suspect components that have been soldered into a printed circuit board can introduce new defects. Each replacement of a component must be followed by retest to insure that the component replaced was the one that actually failed; hence it is time-consuming, tying up both technician and expensive test equipment. Consequently, a goal of test development must be to create tests that are capable not only of detecting faulty operation but also of pinpointing, whenever possible, the faulty component. In practice, a test often produces a list of suspected components and the objective must be to shorten the list as much as possible.

One solution to the problem of locating faults during the manufacturing process is to detect faulty devices as early as possible. This strategy is an acknowledgment of the so-called power-of-ten rule. This rule or guideline asserts that the cost of locating a defect increases by an order of magnitude at every level of integration. For example, if it cost N dollars to detect a faulty chip at incoming inspection, it may cost ten times as much if the component escapes detection at incoming inspection and is soldered onto a printed circuit board. If the component is not detected at board test, it may cost 100 times as much if the board with the faulty component is placed into a complete system. If the defective system is shipped to a customer and a field engineer must make a trip to a customer site, the cost increases by another power of ten. The obvious implication is that there is a tremendous economic incentive to find defects as early as possible.

PROBLEMS

1.1 Assume that the relative cost of repairing defects, C_d, expressed as a function of the percentage t of faults tested, is $C_d = (100 - 0.7t)m$, where m is the number of units to be manufactured. Further, assume that the cost C_P of achieving a particular test percentage t is $C_P = t/(100 - t)$. What value of t will minimize total cost?

1.2 Draw a graph of defect level versus fault coverage, using each of the following values of yield as a parameter: $Y = \{0.01, 0.10, 0.30, 0.50, 0.70, 0.90, 0.99\}$.

REFERENCES

1. Teichrow, D., and E. A. Hershey, III, "PSL/PSA: A Computer-Aided Technique for Structured Documentation and Analysis of Information Processing Systems," *IEEE Trans. Software Eng.*, vol. SE-3, no. 1, Jan. 1977, pp. 41–48.

REFERENCES

2. Bell, C. G., and A. Newell, *Computer Structures: Readings and Examples*, McGraw-Hill, New York, 1971.
3. Hill, F. J., and G. R. Peterson, *Digital Systems: Hardware Organization and Design*, 2nd ed., Wiley, New York, 1978.
4. Chu, Y., *Introduction to Computer Organization*, Prentice-Hall, Englewood Cliffs, N.J., 1970.
5. Duley, J. R., and D. L. Dietmeyer, "A Digital System Design Language (DDL)," *IEEE Trans. Comput.*, vol. C-17, Sept. 1968, pp. 850–861.
6. Sanborn, J. L., "Evolution of the Engineering Design System Data Base," *Proc. 19th Design Automation Conf.*, 1982, pp. 214–218.
7. Williams, T. W., and N. C. Brown, "Defect Level as a Function of Fault Coverage," *IEEE Trans. Comput.*, vol. C-30, no. 12, Dec. 1981, pp. 987–988.
8. Oldham, W. G., "The Fabrication of Microelectronic Circuits," *Sci. Am.* vol. 237, no. 3, Sept. 1977, pp. 111–128.

Chapter 2
Combinational Logic Test

2.1 INTRODUCTION

In the first chapter we saw that it was important to determine when to test and also how much to test. In this chapter we begin to investigate the questions of what to test and how to test.

In electronics the two most common defects are shorts and opens. A short is an electrical connection where one should not exist, and an open is lack of a connection between two points where a connection should exist. We will look at the effects that these and other defects have on circuit behavior. We will then examine the algorithms that have emerged over the years to test for the presence of these defects. We will discuss the strengths and weaknesses of these algorithms and attempt to put the subject matter into historical perspective.

2.2 APPROACHES TO TESTING

Testing of digital logic involves the application of stimuli to a device under test (DUT) and evaluation of the response to determine whether the device is functioning correctly. An important part of the test is the creation of effective stimuli. Early approaches to creation of stimuli, circa 1950s, involved either the application of all possible input combinations to device inputs or the application of stimuli intended to verify specific functional capabilities of the device.

The application of 2^n test vectors to a device with n inputs was effective if n was small and if there were no sequential circuits on the board. Because the number of tests, 2^n, grows exponentially with n, the number of tests required increases rapidly. For example, given a circuit with 32 inputs, if we apply patterns and evaluate response at the rate of 10 million test vectors per second, it would take almost seven minutes to test the circuit. For a circuit with 40 inputs this number increases to 30 hours. Worse still, as we shall see in the next chapter, if the circuit has sequential logic there is no assurance that this technique will completely test the circuit.

Another approach to testing was to exercise the functionality of a device. The logic designer or a test engineer would write sequences of inputs that were intended to exercise the device in its normal functional modes of operation. The data transformation devices, such as the arithmetic logic unit (ALU), would be used to perform arithmetic and logic operations on arguments provided by the engineer. These and other test sequences would exercise storage devices such as registers and flip-flops and data routing devices such as multiplexers. If the circuit produces all the right answers, it is tempting to conclude that the circuit is free of defects. That, however, is the wrong conclusion because the circuit may have a fault which simply was not detected by the applied stimuli. This lack of accountability is a major problem with the approach. There is no practical way to evaluate the effectiveness of the tests. The effectiveness of the test stimuli can be estimated by observing the number of products that fail after being shipped, but that is a costly solution. Furthermore, we are still left with the problem of diagnosing the cause of the malfunction.

In 1959, R. D. Eldred[1] advocated testing hardware rather than function. This was to be done by creating tests for specific faults. The most commonly occurring faults would be modeled, where the *fault model* is a computer model of the circuit that has been modified to conform to some premise or conjecture about real physical defects. Then, input stimuli would be created which could distinguish between the fault-free and the faulted models. The advantages of this approach, as we shall see, are:

> We can create specific tests for faults most likely to occur.
>
> We can compute the effectiveness of our test set by determining how many of these commonly occurring faults will be tested by the set of test vectors created.
>
> We can associate specific defects with specific test patterns so that if a DUT responds incorrectly to a test pattern, there is information pointing to a faulty component or set of components.

This method advocated by Eldred has become a standard approach to developing tests for digital logic failures.

2.3 ANALYSIS OF A FAULTED CIRCUIT

A prerequisite for being able to test for faults in a digital circuit is an understanding of the kinds of faults that can occur and the consequences of those faults. To that end, we will analyze the circuit of Figure 2.1. We will hypothesize the presence

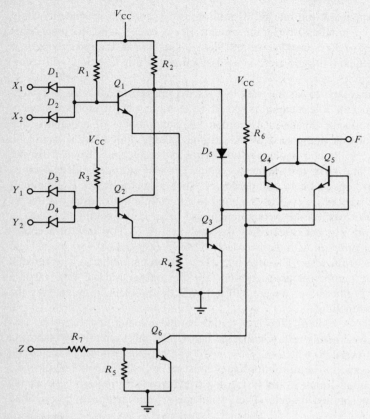

Figure 2.1 Component level circuit.

of a fault in the circuit, namely a short across resistor R_4. Then we will devise a test capable of detecting the presence of that fault.

2.3.1 Analysis at the Component Level

Before we begin our analysis, we first make some remarks about the components used in the circuit. Unless otherwise explicitly stated, we shall use the positive logic convention. Any voltage between ground (GND) and $+0.8$ volts represents a logic 0. A voltage between $+2.4$ and $+5.0$ volts (V_{CC}) represents a logic 1. A voltage between $+0.8$ and $+2.4$ volts represents an indeterminate state. The NPN transistors Q_1 through Q_6 behave like on/off switches when used in digital circuits. A low voltage on the base cuts off the transistor so that it cannot conduct. In effect, it behaves as though an open circuit exists between the emitter and the collector. A high voltage on the base causes the transistor to conduct and it then behaves as though there were a direct connection between the emitter and the collector.

We are now ready to analyze the fault and its effects on the circuit. We begin

2.3 ANALYSIS OF A FAULTED CIRCUIT

by noting that with the resistor shorted, the base of Q_3 is held at ground. It will not conduct and therefore behaves like an open switch. This causes the voltage at the collector to go high, a logic 1. With the collector of Q_3 high, the base of Q_5 and the emitter of Q_4 are high. Q_4 will not be able to conduct because its base cannot be made more positive than its emitter. However, Q_5 is capable of conducting, depending on the voltage applied to its emitter by Q_6.

If Z is high ($Z = 1$), then the positive voltage on the base of Q_6 causes it to conduct; hence it is, in effect, shorted to ground. Therefore, the base of Q_5 is more positive than the emitter, the transistor Q_5 conducts, and the output goes to logic 0. If Z is low ($Z = 0$), Q_6 is cut off. Since it does not conduct, the base and emitter of Q_5 are at the same potential, and it is cut off. Therefore, the output of Q_5 goes high and the output of F is at logic 1. We see that as a result of the fault, the output F is the complement of input Z, and is totally independent of any signals appearing on X_1, X_2, Y_1, and Y_2.

We now know how the circuit behaves with the fault present. What we wish to determine is how to devise a test that will tell us if the fault is present. We assume that the output F is the only point in the circuit that we can observe, that is, we cannot probe inside the circuit. This restriction tells us that the only way to detect the fault is to create input stimuli for which the output response is a function of the presence or absence of the fault. The response of the faulted circuit will then be opposite to the response of the fault-free circuit.

We begin by considering what happens if there is no fault in the circuit. If the fault is not present, the output is dependent not only on Z but also on X_1, X_2, Y_1, and Y_2. If these inputs are at values which cause the Q_3 output to go high, then the faulted circuit cannot be distinguished from the fault-free circuit because the circuits produce identical signals at the output of Q_3 and hence identical signals at the output F. However, if the output of Q_3 is low, then an analysis of the circuit as done previously reveals that when $Z = 0$ the output F goes to 0 and when $Z = 1$ the output goes to 1. Therefore, when Q_3 is low the resulting signal at the output F is opposite to what it would be if the fault were present. We conclude that we want to apply a signal to the base of Q_3 that can cause the collector to go low. A positive signal on the base will produce the desired result. Now, how do we get a high signal on the base of Q_3? To determine that, we have to analyze the circuits preceding Q_3.

Consider the circuit made up of Q_1, R_1, D_1, and D_2. If X_1 or X_2 or both of them are at logic 0, then the base of Q_1 is at ground potential and hence Q_1 acts like an open switch. Likewise, if Y_1 or Y_2 or both are at logic 0, then Q_2 acts like an open switch. If both Q_1 and Q_2 are open, then the base of Q_3 is at ground. But we wanted a high signal on the base of Q_3. If either Q_1 or Q_2 conducts, then there is a complete path from ground through R_4, through Q_1, or Q_2, through R_2 to V_{CC}. Then, with the proper resistance values on R_1, R_2, and R_4, a high voltage signal appears at the base of Q_3. Therefore, we conclude that we must have a high signal on X_1 and X_2 or Y_1 and Y_2 (or both) in order to distinguish whether or not the fault is present. Note that we must also know what signal is present on input Z. With $X_1 = X_2 = 1$, or $Y_1 = Y_2 = 1$, the output F assumes the same value as Z if the fault is not present and the opposite value if the fault is present.

2.3.2 Gate-Level Symbols

One of the difficulties with analyzing circuits as done in the preceding section is the tedious task of analyzing each individual component in the circuit in order to calculate the input signals required to distinguish between the good circuit and a circuit with the fault under consideration. It requires circuit engineers capable of analyzing complex circuits because, within a given technology, there are many ways to design circuits at the component level to accomplish the same end result, logically speaking. It is not obvious, in a large circuit with thousands of individual components, exactly what logic function is being performed by a particular group of components. Further complicating the task is the fact that there are several technologies in use and each has its own unique way to perform digital logic operations. For instance, the circuit made up of D_1, D_2, Q_1, R_1, and R_2 constitutes a two-input AND function. The output, taken at the emitter of Q_1, is high only if X_1 and X_2 are high. The same operation is accomplished in a MOS technology by means of the circuit shown in Figure 2.2. If X_1 and X_2 are high, then the output F is connected to V_{CC} and the output is high. The two circuits perform the same logic operation but bear absolutely no physical resemblance to one another.

Designers have long used logic symbols to represent their designs. These symbols reduce the complexity of the logic circuit drawings and have the advantage of being technology-independent, in the sense that circuits which are logically identical in behavior and conform to the same logic diagrams are often implemented in more than one technology. The diagrams have the advantage that they can be understood by individuals with little or no circuit engineering knowledge. We introduce here, in Figure 2.3, the symbols that will be used in this text.

Included with each logic symbol is a truth table describing the behavior of the circuit for all input combinations. The AND circuit is shown with three inputs. Its behavior can be summarized in the observation that its output is high (logic 1) if all of its inputs are high, otherwise its output is low (logic 0). Alternatively, we may say that the output is true only if all inputs are true. The OR circuit output is true if any of the inputs are true, and false only if all inputs are false. The inverter causes the output to be the complement of the input. The NAND first performs an AND on the inputs, then inverts the result. The inversion process will be repre-

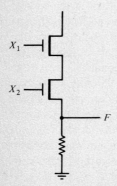

Figure 2.2 MOS AND gate.

2.3 ANALYSIS OF A FAULTED CIRCUIT

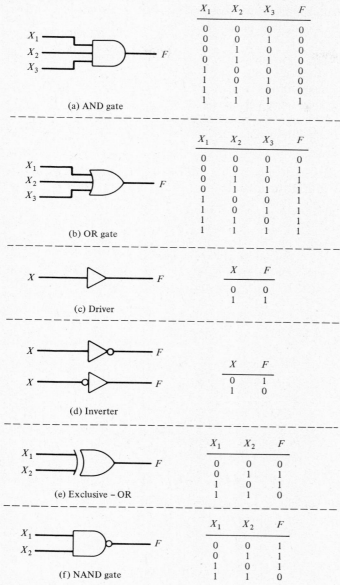

Figure 2.3 Basic switching elements.

sented by a circle or "bubble" on an input or output of a circuit. A NOR (not pictured) is represented by an OR symbol with a bubble on its output. Its output is true if all inputs are false, otherwise its output is false.

The AND circuit and the OR circuit are frequently called *gates*. A low on any input to the AND circuit inhibits or gates out signals applied to the other inputs, thus preventing them from getting through to the output. An input signal, when high, permits the other signals to pass through the AND circuit. In similar fashion, an input to the OR inhibits other signals when high and permits them to pass when low.

The inputs and outputs of logic functions will be called *terminals*. Any wire which connects two or more terminals will be called a *net*. The term net will also apply to any set or collection of interconnected terminals. An input terminal which is physically accessible at an integrated circuit pin or logic board pin will be called a *primary input*. An output terminal which is physically accessible will be called a *primary output*. An output terminal of a logic function will also sometimes be called a *node*.

2.3.3 Analysis at the Gate Level

With these symbols and definitions we now go back and determine how to logically represent the circuit of Figure 2.1. We already stated that Q_1, D_1, D_2, and R_1 constitute a two-input AND gate. The components Q_2, D_3, D_4, and R_3 form another two-input AND gate. The common connection of the emitters of Q_1 and Q_2 form an OR gate, transistors Q_3 and Q_6 each act as an inverter, and Q_4, Q_5 together form an exclusive-NOR (EXNOR, an exclusive-OR with its output complemented). Therefore, the circuit of Figure 2.1 can be represented by the logic diagram of Figure 2.4.

Now reconsider the fault that we examined previously. We saw that the output of Q_3 could not be driven to a low state when R_4 was shorted. This is equivalent to the NOR gate output in the circuit of Figure 2.4 being stuck at a logic 1. Consequently, we want to assign inputs which will cause the output of the NOR gate, when fault-free, to be driven low. This requires a 1 on one of the two inputs to the NOR gate. If we arbitrarily pick the upper input and require that it be at logic 1, then the upper AND gate is required to produce a logic 1 and this requires that inputs X_1 and X_2 both be at logic 1. As before, we must assign a known value to input Z so that we know what value is to be expected at the circuit primary output F for the fault-free and faulted circuits. The reader will (we hope) agree that the circuit representation of Figure 2.4 is much easier to analyze.

The circuit representation of Figure 2.4, in addition to being easier to work with and requiring fewer details to keep track of, has the advantage that it can be understood by people who are familiar with logic but not familiar with the behavior

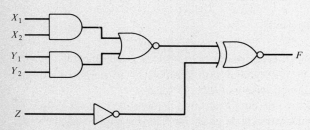

Figure 2.4 The gate equivalent circuit.

of electronic circuits. Furthermore, it is universal, in the sense that the circuit can be represented in terms of these symbols regardless of whether it is implemented in MOS, TTL, ECL, or some other technology. As long as the circuit can be logically modeled, it can be represented by these symbols. Another important advantage of this representation, as we shall see, is the fact that computer algorithms can be defined upon these logic operators, which are, for the most part, independent of the particular technology chosen to implement the device. If the circuit can be expressed in terms of these symbols, then the circuit description can be processed by the computer algorithms.

2.4 THE STUCK-AT FAULT MODEL

We have seen that we can represent a circuit composed of resistors, diodes, and transistors as an interconnection of logic elements. By altering the model of a particular logic gate so as to represent a faulted logic gate, we can analyze the behavior of the logic circuit containing that gate and develop tests to distinguish between the fault-free circuit and the faulted circuit. But, for what kind of faults should we test? We can create an extremely difficult problem if we come up with the wrong answer.

An n-input combinational circuit can implement any of 2^{2^n} functions. To verify with *absolute* certainty that the circuit implements the correct function we would have to apply all 2^n input combinations and confirm that the circuit responded correctly to each stimulus. As we saw previously, that could take an inordinate amount of time.

We represented the fault in the circuit of Figure 2.1 as a NOR gate output stuck-at-1 (S-A-1). What happens if diode D_1 is open? Under that condition it is not possible to pull the base of Q_1 to ground potential from input X_1. Therefore input 1 of the AND gate, represented by D_1, D_2, R_1, and Q_1, is S-A-1. What happens if there is an open from the common connection of the emitters of Q_1 and Q_2 to the emitter of Q_1? Then there is no way that Q_1 can provide a path from ground, through R_4, Q_1, and R_2 to V_{CC}. The base of Q_3 is unaffected by any changes in the AND gate. Since the common connection of Q_1 and Q_2 represents an OR operation (called a wired-OR or DOT-OR), the fault is equivalent to an OR gate input stuck-at-0 (S-A-0). At it turns out, most component faults can be represented as inputs or outputs of simple logic gates S-A-0 or S-A-1. The S-A-x, $x \in \{0, 1\}$, fault model has become nearly universal. It has the attraction that it permits enumeration of faults. For an n-input logic gate, it is possible to assign a specific number of faults. Furthermore, it is possible to specifically define the faults and their effect on gate behavior. This permits implementation of computer algorithms, as we shall see, and furthermore, by knowing the exact number of faults in a circuit, it is possible to keep track of those for which tests have been generated and those for which tests have not been generated and thus create an effectiveness measure for the algorithm.

Another important consideration related to using the S-A-x model is the fact that it is basically impossible to model all faults that could conceivably occur in a circuit. Consider a circuit containing m nets which interconnect the various components. Each of these nets may individually be classified at any given time as

Fault-free
Stuck-at-1
Stuck-at-0

If we consider all possible combinations, then there are 3^m machines described by the m nets and the three possible states of each net. Of these possibilities, only one corresponds to a completely fault-free machine.

If we also consider all possible combinations of shorts between nets, then there are

$$\sum_{i=2}^{m} \binom{m}{i} = 2^m - m - 1$$

shorts that could occur in an actual circuit. The reader will note that we keep getting back to the problem of "combinatorial explosion"; that is, the number of choices or the number of problems to solve figuratively explodes. To attempt to test for every stuck-at or short fault combination is clearly impractical. In such cases, we must be pragmatists, not purists.

The impracticality of trying to test for every conceivable fault in a circuit has led to adoption of the *single-fault assumption*. When attempting to create a test, it is assumed that a single fault exists. Most frequently, it is assumed that an input or output of a gate is stuck-at-1 or stuck-at-0. Physical defects that occur in actual practice frequently produce symptoms that behave as though they were inputs or outputs S-A-0 or S-A-1. Many years of experience by many digital electronics companies with the stuck-at fault model has demonstrated that the stuck-at model is effective. A good stuck-at test which detects all or nearly all single stuck-at faults in a circuit will also detect all or nearly all multiple stuck-at faults and short faults. There are technology-dependent faults for which the stuck-at fault model must be modified or augmented; these will be discussed in a later chapter.

We take note of the fact that tests are usually created on the assumption that solid failures exist; intermittent faults dependent on environmental or other external factors such as temperature, humidity, or line voltage are assumed, when creating tests, to be solid failures. In the following paragraphs we describe fault models for AND, OR, and the inverter. Fault models for other basic circuits can be deduced from these.

2.4.1 The AND Gate Fault Model

The AND gate is fault-modeled for inputs stuck-at-1 and the output stuck-at-1 and stuck-at-0. This provides a total of $n + 2$ tests for an n-input AND gate. The test for an input S-A-1 consists of putting a logic 0 on the input being tested and logic 1s on all other inputs so that the input being tested is the controlling input; it alone determines the value on the output. If the fault is not present the output goes to a logic 0 and if the fault is present the output goes to a logic 1. An input pattern of

2.5 THE SENSITIZED PATH

all 1s will test for the output S-A-0. Any of the tests for an input S-A-1 will also test for the output S-A-1, hence it is not necessary to explicitly test for the output S-A-1. However, it is possible to detect an output S-A-1 without detecting any inputs S-A-1 if two or more inputs have logic 0s on their inputs; therefore it can be useful to retain the output S-A-1 as a separate fault. When tabulating faults tested by a sequence of test vectors, counting the output as tested when none of the inputs is tested provides a more accurate estimate of fault coverage. Note also that we do not test for inputs S-A-0 on the AND gate because the all-1s input vector used to test for the output S-A-0 will test for all possible input S-A-0 faults.

2.4.2 The OR Gate Fault Model

An n-input OR gate, like the AND gate, requires $n + 2$ tests. However, when testing input stuck-at faults, the input values are opposite to what they would be for an AND gate. The input being tested is set to 1 and all other inputs are set to 0. The all-0s input tests for the output S-A-1 and any input S-A-1.

2.4.3 The Inverter Fault Model

The inverter is tested for output S-A-0 and S-A-1. If it fails to invert, possibly due to a short across a transistor, this will be detected by one of the stuck-at tests.

2.5 THE SENSITIZED PATH

We have defined the stuck-at fault as the fault model of interest for basic logic gates. We have also defined tests for detecting stuck-at faults on these gates. However, individual logic gates do not usually occur in actual practice; rather, they are interconnected with hundreds or thousands of other similar gates to form complex circuits. When the gate we wish to test is embedded in a much larger circuit and there is no immediate access to the gate, it becomes necessary to use the surrounding circuitry to set up the inputs to the gate under test and to make the effects of a fault visible at an observable output.

2.5.1 A Logic Analysis of the Faulted Circuit

We consider again the circuit of Figure 2.4. We wanted to test for the output of the NOR gate S-A-1. It was necessary to assign values to the inputs X_1, X_2, Y_1, Y_2 in such a way as to cause the signal on the output of the NOR gate to go low. Then, if the fault was actually present, the value at the output of the NOR gate in the faulty circuit would be different from the value present in the fault-free circuit. It was also necessary to cause the effects of the fault to become visible at the output F. This presented no problem because the only gate between the point of the fault and the output was an EXNOR and either a logic 0 or logic 1 on input Z would permit the fault to produce visible effects at output F.

We now examine the more complex circuit of Figure 2.5. We want to create

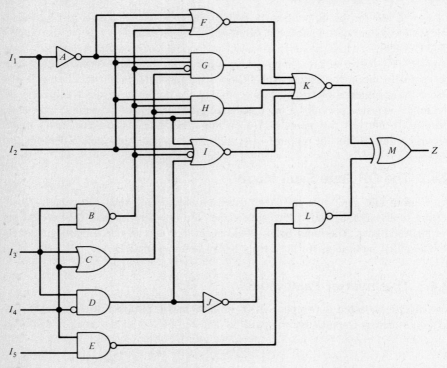

Figure 2.5 A more complex example.

a test for input 3 of gate H S-A-1 (i.e., the input driven by gate C. On schematic drawings we will number inputs from top to bottom in ascending order). Since gate H is an AND gate, the test for input 3 S-A-1 requires that input 3 be set to logic 0 and the other inputs be set to logic 1. We are immediately confronted with two problems: we must compute logic values on primary inputs I_1 through I_5 which will cause the required test values to appear on the inputs of H, and we must compute values on I_1 through I_5 which make the fault effect visible at the output. Furthermore, the values computed on I_1 through I_5 in these two operations must not conflict.

We first attempt to create a sensitized path from the fault origin to the output. A *sensitized path* of a fault f is a signal path originating at the fault origin f whose value all along the path is functionally dependent on the presence or absence of the fault. If the sensitized path terminates at a net that is observable by test equipment, then the fault is *detectable*. The process of creating a sensitized path is called *propagation*.

We start by noticing that H goes to a NOR gate labeled K. In order to sensitize a path through K, all other inputs to K must be at logic 0. If any of them is at logic 1, the output of K will go to 0, irrespective of the value on H. Therefore, gates F, G, and I must all produce 0s on their outputs. We record these requirements in an assignment table and continue forward.

2.5 THE SENSITIZED PATH

We next come to the EXOR gate labeled M. A choice exists at this point because either a 0 or a 1 on the bottom input of M will permit the sensitive path to be extended through M to the output Z. We arbitrarily assign a 0 to the bottom input of M. This requires that gate L produce a 0. We record this requirement in the assignment table and also record, in a decision table, the fact that a choice existed at gate M.

We have now created a sensitized path from the point of the fault to the output. However, we have not yet created a test. Logic values were assigned to the outputs of gates F, G, I, and L. We must now satisfy those assignments. We begin by selecting an element from the assignment table and determining what must be done to satisfy the value assigned to its output.

If the assignment table is viewed as a push-down stack, then the first element selected is gate L. The value 0 was assigned to its outputs, requiring that both of its inputs be set to 1. Gates J and E are therefore pushed onto the stack with the notation that they must produce logic 1s. We then pick E from the stack and determine what must be done to produce a logic 1 on its output. (The reader may find it instructive to keep track of the contents of the assignment stack and decision table with pencil and paper, adding to the stack and decision table and crossing off as assignments are first made and then satisfied.) For gate E, a choice exists. A logic 0 can be assigned to either input I_4 or I_5. We choose I_4 and record this choice in the decision table, and then select J from the stack. To get a 1 on its output requires a 0 from D, which in turn requires a 0 from I_3. Note that no choice exists because I_4 was already assigned a logic 0 and it becomes inverted at the input to D.

We return to the stack and find that a 0 is required from gate I, so one of its inputs must be set to 1. Its top input, which is processed first, was already assigned a 1. Gate G is selected next; it must produce a 0. Again processing the top input first, we find that gate A must provide a 0, which requires that I_1 produce a 1. Since I_1 has already been assigned that value, it requires no further processing. Continuing in this fashion with the remaining elements on the stack, we arrive at a set of inputs $(I_1, I_2, I_3, I_4, I_5) = (1, 1, 0, 0, x)$ which satisfies the requirements, where x denotes a "don't care," meaning that no value is required.

The fault-free machine produces a logic 1 in response to the set of inputs assigned. If input 3 of gate H is stuck at 1, then the output of H goes to 1, causing the output of K to go to 0, which in turn causes Z to go to 0, hence the fault is detectable.

2.5.2 Analysis of the Sensitized Path Method

We will now analyze the process which just took place and make some observations. The process of satisfying assignments by backing up to the inputs is called *justification*; it is also sometimes called the *consistency* operation. The two processes, propagation and justification, can be used to find a test for almost any fault in the circuit (redundant logic, as we shall eventually see, presents testing problems). Furthermore, propagation and justification can be applied in either order. We started by propagating from the point of fault to an output. It would be possible to first justify the assignments on the four inputs of gate H, then propagate forward to the output, one gate at a time, each time justifying all assignments made in that step of the propagation. If the first approach is chosen the propagation and justification phases are each entered only once. During the propagation phase all required assignments are placed on the assignment stack. Then, in the justification phase, the assignment stack expands and contracts. When the stack is finally empty, the justification phase is complete. In the second approach, processing begins with the justification process, attempting to satisfy initial assignments on the gate whose input or output is being tested. Each time the assignment stack empties, control reverts to the propagation mode and the sensitive path is extended one gate closer to the outputs. Then, control again reverts to the justification routine until the assignment table is again empty. Control passes back and forth in this fashion until the sensitized path reaches an output and all assignments are satisfied.

Implication When assignments are made to individual gates, these assignments sometimes possess implications beyond the immediate area of the gate being processed. Consider the assignment of a logic 0 to I_3. This implies that the output of gate B is a logic 1 regardless of what value is on the bottom input of B. That, in turn, implies that the output of F is a 0 and the output of G is a 0. Furthermore, gate D produces an output of 0 as a result of the assignment on I_3 and this implies that the inverter J produces a 1. The value of this observation lies in the fact that these implications can be determined at the time when an assignment is made and can save computation time. For instance, if the assignment $I_3 = 0$ has already been made, and we are attempting to propagate the fault from H through K, it is not necessary to do anything to satisfy the assignments $F = G = 0$.

The Decision Table During the propagation and justification processes, we occasionally arrived at gates where choices existed. For example, when we required a 0 from the three-input NOR gate F, we first tried to assign a 1 to the upper input. However, the choice caused a problem because it resulted in an assignment $I_1 = 0$, which conflicted with a previous assignment $I_1 = 1$. Because a choice existed, we were able to back up and make an alternative choice which eventually proved successful. In large circuits with much fan-out, complex multilevel decisions must often be made. If all decisions at a given gate have been tried without success, then the decision stack must be popped and a decision made at the next available decision junction. Furthermore, assignments to all gates following the junction at which the decision was made must be erased and any mechanism used to keep track of decisions for the gate that was popped off the decision stack must be reset.

The implication operation is of value here because it can often eliminate a number of decisions. For example, the initial test for gate H assigned a logic 1 to input I_2. But the assignment $I_2 = 1$ implies that the output of gate F has the value 0. Therefore, if implication is performed, there is no need to justify $F = 0$, and that in turn eliminates the need to make a decision at gate F.

The Fault List The fault, input 3 of gate H, was selected arbitrarily for the purpose of demonstrating the propagation and justification techniques. In practice, the entire

2.5 THE SENSITIZED PATH

set of potential faults would be compiled into a fault list. The ultimate objective is to have a test for every fault on the list. One way to accomplish this is to simply go down the list and create a test vector for each fault in the fault list. For the circuit of Figure 2.5, assuming that there are $n + 2$ faults for each gate, except for the inverters, which have two faults, the fault list would contain 57 entries. Since there are only five inputs it is obvious that there would be many duplicate entries in a list of test vectors if one were created for each of these 57 faults. If the previously described procedure were exercised for every fault, the resulting 57 test vectors could be ordered numerically by converting the input vectors into decimal equivalents and then discarding duplicates.

It is clear that having a test for every fault does not imply the existence of a unique test vector for every fault. In fact, a set of stimuli developed for one fault very frequently detects several other faults. For example, the test for input 3 of H S-A-1 causes the fault-free circuit to assume the value $Z = 1$. If input 3 of H were actually S-A-1 the output would go to 0. But several other faults would also cause Z to assume the value 0, the most obvious being the output of M S-A-0. Other faults causing a 0 output include input 1 of gate M S-A-0 and the output of gate K S-A-0. In fact, any fault along the sensitive path which causes the value on the sensitive path to assume a value other than the correct value will be detected by the test vector.

The importance of this observation lies in the fact that if we can determine which previously undetected faults are detected by each new test vector, then we can check them off in the fault list and do not need to develop test vectors to specifically test for those faults. We will describe several techniques for accomplishing this later.

Making Choices The sensitive path method for generating tests was used during the early 1960s[2]. One of the problems with this method was the fact that it attempted to drive a test to a single output. When the propagation routine reached a net with fan-out it would arbitrarily select a single path and continue to pursue its objective of reaching an output. Unfortunately, this blind pursuit of an output would occasionally ignore easy solutions.

In Figure 2.6 the EXOR goes to two AND gates. The AND gate labeled P is not difficult to control; its input from Q can be set to 1 with a 0 on the primary input. Gate N may be more difficult to control because its upper input comes from other logic and a test at its output must be driven through more combinational

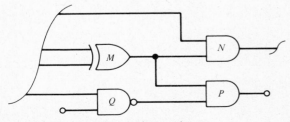

Figure 2.6 Choosing the best path.

logic. However, an arbitrary choice could result in an attempt to drive a test through the upper gate. In fact, if the program did not examine the function associated with the second fan-out, it could go right past a primary output and attempt to propagate a test through complex sequential logic.

By ordering the inputs and the fan-out list for each gate, the program can be forced to favor inputs which are easiest to set to controlling values and the propagation path which most easily reaches a primary output whenever a decision must be made. An algorithm for performing this ordering will be described in detail in Chap. 7. The algorithm, called SCOAP, very methodically computes this ordering for all gates in a circuit.

The Reconvergent Path A difficulty inherent in the sensitive path method is the fact that it might not be able to create a test for a fault when, in fact, a test does exist. This can be illustrated by means of the circuit in Figure 2.7[3].

Consider the output of NOR gate 6 S-A-0. We require inputs 2 and 3 to be 0 so as to get a 1 on the output of 6 in the fault-free circuit. If we elect to propagate through gate 9 we require that input 1 be 0. Then the output of 9 is 0 for the fault-free circuit and it is 1 if the output of 6 is S-A-0. In order for 9 to be the controlling input to gate 12, the other three inputs to gate 12 must be at 0.

To get a 0 at the output of gate 10, one of its inputs must be set to 1. Since the output of gate 6 is S-A-0, we get a 1 by setting input 4 to 1. The output of gate 7 then assumes the value 0 and that, together with the 0 on input 3, causes the output of gate 11 to assume the value 1. The sensitive path is now inhibited, so there does not appear to be a test for the fault. But a test does exist. The input assignment $(0,0,0,0)$ will detect a S-A-0 fault at the output of gate 6.

2.6 THE *D*-ALGORITHM

Failure to generate a test for the circuit of Figure 2.7 occurred because the sensitive path method attempts to propagate fault symptoms through only a single path. In our example, fan-out occurred and we were required to make a choice. Rather than make a choice, we will now propagate the sensitive path through all paths when we arrive at a net with fan-out.

We start by introducing the *D* notation of Roth[4]. It will simplify explanations and keep the logic diagrams less cluttered. The *D* (as in discrepancy) is a composite signal. It simultaneously denotes the signal value on the good machine (GM) and the faulted machine (FM) according to the following table:

GM \ FM	0	1
0	0	$\overline{D}$
1	D	1

Conceptually, the *D* can be thought of as representing two circuits superimposed. Where the good machine and the faulted machine have the same value,

2.6 THE D-ALGORITHM

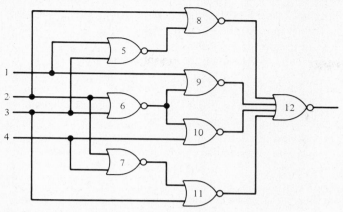

Figure 2.7 Effect of reconvergent fan-out.

the composite machine will have a 0 or 1. Where they have different values, the composite machine will have a D, indicating logic 1 on the good machine and 0 on the faulted machine, or $\overline{D}$, indicating a 0 on the good machine and 1 on the faulted machine.

At the output of gate 6 in Figure 2.7, where we hypothesized the presence of the S-A-0, we want the good machine to have the value 1; therefore we assign a D to that net. We want to propagate this D to a primary output. Since the output of 6 fans out to two NOR gates, we want to propagate this D through both NOR gates. We therefore have the signal D present on one input of gate 9 and also on one input of gate 10. We know from previous sections that the other inputs of these two NOR gates must be 0s. Therefore, we assign the value 0 to primary inputs 1 and 4. But what value appears at the outputs of 9 and 10? Bearing in mind that the inputs are 0 and D on both NOR gates, and that D represents a 1 on the fault-free circuit and 0 on the faulted circuit, we have a fault-free circuit in which NOR gate inputs 0 and 1 are ORed together and inverted to give a 0 on the output, and a faulted circuit in which NOR gate inputs, both 0, are ORed and inverted to give a 1 on the output. Hence, the outputs of gates 9 and 10 are $\overline{D}$.

We now have two sensitive paths converging on gate 12, both of which have the value $\overline{D}$. If NOR gates 8 and 11 both have output 0, then the inputs to 12 are $(0, 0, 0, 0)$ for the good machine and $(0, 1, 1, 0)$ for the faulted machine. Since 12 is also a NOR gate, its output is 1 for the good machine and 0 for the faulted machine, that is, its output is a D. Of course, we are not yet done. We need to obtain 0 from gates 8 and 11. Since the inputs are completely assigned, all we can do is inspect the circuit to determine whether the input assignments to 8 and 11 cause the necessary values to occur on their outputs. Luckily, that turns out to be the case.

2.6.1 The D-Algorithm: An Analysis

We have gone through a small example rather quickly and have been able to deduce with little difficulty what must be done at each step. Unfortunately, such intuitive descriptions, while appealing, cannot easily be programmed into a computer. We must provide a more rigorous framework. We will begin with a brief description of cube theory, which Roth used to describe the *D*-algorithm.

We define a *singular cube* of a function as an assignment

$$(x_1, \ldots, x_n, y_1, \ldots, y_m) = (e_1, e_2, \ldots, e_{m+n})$$

where the x_i are inputs, the y_j are outputs, and $e_i \in \{0, 1, x\}$. A singular cube in which all input coordinates are 0 or 1 is called a *vertex*. A vertex can be obtained from a singular cube by converting all *x*s on input coordinates to 0s and 1s.

A singular cube *a contains* the singular cube *b* if *b* can be obtained from *a* by changing some of the *x*s in *a* to 1s and 0s. Alternatively, *a* contains *b* if it contains all of the vertices of *b*. The *intersection* of two singular cubes is the smallest singular cube containing all of their common vertices. It is obtained through use of the intersection operator, which operates on corresponding coordinates of two singular cubes according to the following table:

I	0	1	x
0	0	—	0
1	—	1	1
x	0	1	x

The dash (—) denotes a conflict. If one singular cube has a 0 in a given position and the other has a 1, then they are in conflict; the intersection does not exist. Two singular cubes are *consistent* if a conflict in their output intersections implies a conflict on their input intersections. In terms of digital logic, this simply says that a stimulus applied to a combinational logic circuit cannot produce both a 1 and a 0 on an output. The term "singular" is used to denote the fact that there is a one-to-one mapping between input and output parts of the cube. We will henceforth drop the term singular; it will be understood that we are talking about

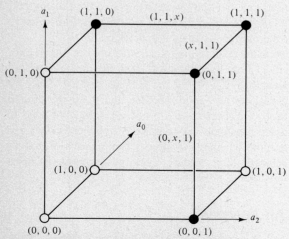

Figure 2.8 Cube representation of a function.

2.6 THE D-ALGORITHM

singular cubes. Furthermore, to simplify notation, we will restrict our attention in what follows to single-output cubes, the definitions being easily generalized to the multiple-output case.

A *cover* C is a set of pairwise consistent, nondegenerate cubes, all referring to the same input and output variables. Given a function F, a *cover of F* is a cover C such that each vertex $v \in F$ is contained in some $c \in C$. A *prime* cube of a cover is one which is not contained in any other $c \in C$. If the output part of a cube has the value 0, we will call the cube a 0-point; if it has value 1, we will call it a 1-point; and if it has value x (don't care), we will call it an x-point. An *extremal* is a prime cube which covers a 0-point or 1-point that no other prime cube covers.

EXAMPLE ■

The function $F = a_0 a_1 + \bar{a}_0 a_2$ can be represented by the cube of Figure 2.8. The set of vertices for this cube is:

a_0	a_1	a_2	F
0	0	0	0
0	0	1	1
0	1	0	0
0	1	1	1
1	0	0	0
1	0	1	0
1	1	0	1
1	1	1	1

■ ■

The following is a covering for the function which consists of prime cubes (asterisks denote extremals):

$$\begin{array}{cccc} * & 1 & 1 & x & 1 \\ & x & 1 & 1 & 1 \\ * & 0 & x & 1 & 1 \end{array} \Big\} p_1$$

$$\begin{array}{cccc} * & 1 & 0 & x & 0 \\ & x & 0 & 0 & 0 \\ * & 0 & x & 0 & 0 \end{array} \Big\} p_0$$

We shall denote the set of cubes for which the output is a 1 as p_1. Likewise, p_0 denotes the set of cubes whose output is 0. The reader should verify that each vertex of F is contained in at least one extremal. Two intersections follow:

```
x  1  1          1  0  x  0
0  x  1  1       0  x  0  0
---------        -----------
0  1  1  1            —
```

In the first intersection, the cube $(0, 1, 1, 1)$ is the smallest cube which contains all points common to the two vectors intersected. The second intersection is null. From Figure 2.8 it can be seen that the two cubes have no points in common. The set of extremals contains all of the vertices, hence it completely specifies the function for all defined outputs.

The reader familiar with the terms "implicant" and "prime implicant" may note a similarity between them and the cubes and extremals of cube theory. An *implicant* is a product term which covers at least one 1-point of a function F and does not cover any 0-points. An implicant is *prime* if

1. For any other implicant there exists a 1-point covered by the first implicant which is not covered by the second, and
2. When any literal is deleted the resulting product term is no longer an implicant of the function.

Implicants and prime implicants deal with product terms that cover 1-points, whereas cubes deal with both 1-points and 0-points. The cover corresponds to the set of implicants for both the function F and its complement $\overline{F}$. The collection of extremals corresponds to the collection of prime implicants for both the function F and its complement $\overline{F}$.

2.6.2 The Primitive *D*-Cubes of Failure

A *primitive* is an element which exists in its smallest basic form; it cannot be further subdivided. Up to this point we have regarded the basic switching gates as primitives. As we shall see, the D-algorithm permits primitives which are composites of several basic switching gates. Fault models for primitives used in the D-algorithm are called primitive D-cubes of failure (PDCF). We will use the two-input AND gate to describe the procedure for generating a PDCF. We start with a cover for the AND gate which contains all of the vertices

1	2	3	
0	0	0	
0	1	0	p_0
1	0	0	
1	1	1	p_1

If input 1 is S-A-1, then the output is completely dependent on input 2. The cover then becomes

1	2	3	
0	0	0	
1	0	0	f_0
0	1	1	
1	1	1	f_1

(When referring to the faulted machine, we denote the set of 0-points as f_0 and the set of 1-points as f_1.) We now have two distinct circuits. The first one produces

2.6 THE D-ALGORITHM

an output of 1 only when both inputs are at 1. The second circuit produces an output of 1 whenever the second input is a 1, regardless of the value applied to the first output. A cursory examination of the two sets of vertices reveals an input combination, (0, 1), that causes a 0 output from the fault-free circuit and a 1 from the faulted circuit. The vector (0, 1) is clearly, then, a test for the presence of the S-A-1 fault on input 1. Are there any other tests for input 1 S-A-1? We can determine the answer by performing a point-by-point comparison of vertices from the two sets of vertices. In this case, we find that there is only one test for input 1 S-A-1. This test is the primitive D-cube of failure for the AND primitive faulted with input 1 S-A-1. The comparison of vertices from the two sets can be performed using the intersection table of the previous section. When we get to the output we do not flag it as a conflict; rather, we assign a $\overline{D}$, where D and $\overline{D}$ have the meanings described previously.

If the two-input AND gate is faulted with the output S-A-1, then the cover for this two-input AND gate becomes

1	2	3	
0	0	1	
0	1	1	f_1
1	0	1	
1	1	1	

Now it turns out that there are three tests for the output S-A-1 and we may choose any of these three tests for the fault. However, if we examine the three tests, we notice that all three tests have at least one 0 on an input. That is reasonable since a 0 on any input of an AND gate should cause its output to go to 0—unless it is S-A-1. The value of this observation lies in the fact that we do not have to set values on both inputs. This usually simplifies the creation of tests. We conclude, then, that $(0, x)$ and $(x, 0)$ are tests for the output of the two-input AND gate S-A-1. Is there a way in which this can be computed algorithmically?

Consider again the input S-A-1 fault for the two-input AND gate. The cover for the good circuit can be described in terms of the extremals. For the good circuit, this cover is

1	2	3	
0	x	0	p_0
x	0	0	
1	1	1	p_1

For the faulted gate, this cover is

1	2	3	
x	0	0	f_0
x	1	1	f_1

The vertex $(0, 1)$ is contained in the input parts of the cubes $(0, x, 0) \in p_0$ and $(x, 1, 1) \in f_1$. We notice that the input parts of these two cubes can be intersected to give us the original vertex $(0, 1)$. The intersection of an element from p_0 with an element from f_1 has produced a test for input 1 of the AND gate S-A-1. This, then, suggests the following general method for finding test(s) for a particular fault:

1. Construct a cover consisting of extremals for both the fault-free and the faulted circuits.
2. Intersect members of f_0 with members of p_1.
3. Intersect members of f_1 with members of p_0.

Since there must be at least one vertex which produces different outputs for the good circuit and the faulted circuit (why?), either step 2 or step 3 (or both) must result in a nonempty intersection. Note that the intersections do not necessarily result in a vertex.

EXAMPLE ■

Consider the output of the two-input AND gate S-A-1. The cover f_1 consists of the single cube $(x, x, 1)$. Intersecting it with the extremals in p_0 results in the two tests $(0, x, \overline{D})$ and $(x, 0, \overline{D})$. (When performing steps 2 and 3 above, we intersect only the input parts.)   ■ ■

We have developed PDCFs for a rather elementary circuit, namely an AND gate. We leave it as an exercise for the reader to develop PDCFs for other elementary gates such as OR, NAND, NOR, and Invert. We point out here that the technique for creating PDCFs is quite general. If we have a cover for a circuit G and its faulted counterpart, we can use the method just described to create a test for the circuit. As an example, consider the circuit of Figure 2.9.

We denote by G^* the circuit with input 1 S-A-1. The Karnaugh maps for G and G^* are:

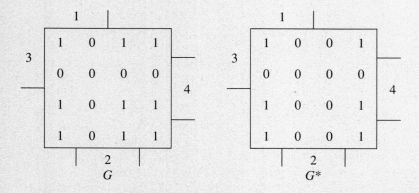

2.6 THE D-ALGORITHM

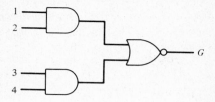

Figure 2.9 AND-OR-Invert circuit.

The extremals for G and G^* are:

1	2	3	4	G		1	2	3	4	G^*	
1	1	x	x	0	$\left.\begin{array}{c}\\\\\end{array}\right\} p_0$	x	1	x	x	0	$\left.\begin{array}{c}\\\\\end{array}\right\} f_0$
x	x	1	1	0		x	x	1	1	0	
0	x	0	x	1	$\left.\begin{array}{c}\\\\\\\\\end{array}\right\} p_1$	x	0	0	x	1	$\left.\begin{array}{c}\\\\\end{array}\right\} f_1$
0	x	x	0	1		x	0	x	0	1	
x	0	0	x	1							
x	0	x	0	1							

The complete set of intersections $p_0 \cap f_1$ and $p_1 \cap f_0$ yields:

$$0 \; 1 \; 0 \; x \; D$$
$$0 \; 1 \; x \; 0 \; D$$

Either of these two vectors, when applied to the circuit, will distinguish between the fault-free circuit and the circuit with input 1 S-A-1.

2.6.3 Propagation D-Cubes

We can, theoretically at least, develop a test for a combinational circuit of any size by starting with the covers for the fault-free circuit and a circuit with some specified fault. Furthermore, the method is not restricted to stuck-at faults. Since it starts with either a Karnaugh map or a set of covering cubes, a test can be developed to distinguish between any two circuits, faulted or not, through the simple expedient of intersecting their covers.

This observation is primarily of academic interest; there is a practical problem with the approach. Circuits that are large enough to perform useful functions are generally too large to be easily represented by Karnaugh maps and covers. Recognizing this, the D-algorithm provides for partitioning of a circuit into an interconnection of primitives for processing. Associated with each primitive is a set of rules for propagating tests through the primitive and justifying tests back to its inputs.

In propagation we have a sensitized signal, a D or $\overline{D}$, on one or more inputs of a circuit, and we want to assign the remaining inputs in such a way that the output is totally dependent on the sensitized signal. We also assume, in keeping with our single-fault assumption, that the circuit through which we are propagating is fault-free, that is, the fault of interest has occurred elsewhere and we are now trying to drive it to an observable output.

Since we are trying to drive a test through a circuit, we must create a situation in which the response at the output of the fault-free circuit is a 0 and the response at the output of the faulted circuit is a 1, or conversely. This tells us that if the input part of the cube for the fault-free circuit is in p_0, then the input part of the cube for the faulted circuit must be in p_1, and vice versa. This suggests that we again want to perform intersections. We will perform intersections but we cannot use the previous intersection table because it prohibited conflicts. We are now actually looking for conflicts so we use the following table:

	0	1	x
0	0	$\bar{D}$	0
1	D	1	1
x	0	1	x

The row and column labels represent the values on input i of the first and second cubes, respectively. Since we are intersecting elements from p_0 with elements from p_1 we will always get a conflict on the output. We will also get a conflict on at least one input coordinate position. If we perform all possible intersections we can create a table of entries called the *propagation D-cubes*. Then, when we must propagate through a circuit, we search through the table for an entry which has D and $\bar{D}$ entries matching the signals on the input position(s) of the circuit through which we are attempting to propagate a signal. That entry tells us what values must be on the other inputs of the circuit.

EXAMPLE ■

Using the cover for the AND-OR-Invert of Figure 2.9, and intersecting p_0 with p_1, we obtain the following propagation D-cubes for the AND-OR-Invert:

1	2	3	4	G
D	1	0	x	$\bar{D}$
1	D	0	x	$\bar{D}$
D	1	x	0	$\bar{D}$
1	D	x	0	$\bar{D}$
0	x	D	1	$\bar{D}$
0	x	1	D	$\bar{D}$
x	0	D	1	$\bar{D}$
x	0	1	D	$\bar{D}$

■ ■

There are actually 16 propagation D-cubes. We get the other eight by intersecting p_1 with p_0. We can also get them simply by exchanging D and $\bar{D}$ signals on both the inputs and outputs. In practice, it is often necessary to restrict the propagation D-cube tables to contain only those propagation D-cubes having a single D or $\bar{D}$ among the inputs. This is because it is possible to have as many as 2^{2n-1} propagation D-cubes for a function with n inputs (see Problem 2.13). For a function with six inputs, this could result in a table of 2048 entries if all single and multiple

2.6 THE D-ALGORITHM

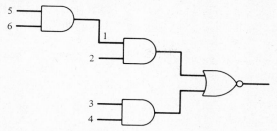

Figure 2.10 AOI with AND gate input.

D and $\overline{D}$ signals were maintained on the inputs. Multiple D and $\overline{D}$ values on the inputs are needed much less frequently than single D or $\overline{D}$ signals and can be created from the cover when needed.

2.6.4 Line Justification and Implication

We created a set of inputs for a primitive circuit and saw how to propagate the resulting test through other logic in order to make the test visible at an output. We also need to justify signal assignments made to the outputs of primitives during the propagation phase. Consider the circuit of Figure 2.10. It is the AND-OR-Invert (AOI) with input 1 now sourced by an AND gate. We want to again test input 1 for the S-A-1 condition. Therefore we need a 0 on input 1 of the AOI. Because we are familiar with the behavior of the AND gate we can easily deduce that either input 5 or 6 must be at 0 to get the required 0 at the output of the AND. Alternatively, we can go to the cover for the AND gate and select an entry from p_0. The selected entry will tell us what values must be applied to the inputs in order to get the required 0 on the output.

The selected entry may not always be acceptable. In Figure 2.11 we again consider the AOI as a primitive. It is configured as a 2-to-1 multiplexer by virtue of the inverter. If we are trying to obtain a test for a S-A-1 fault on input 2 of the AOI, we start by applying $(1, 0, 0, x)$ to nets 1, 2, 3 and 4. We must then justify our assignments. Assuming we can justify the 1 on net 1, we must then justify the 0 assigned to net 2. When we examine the cover for the inverter, we find that we need a 1 on the input. This requires a 1 on the output of the AND gate. We then

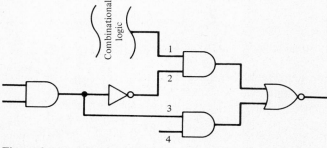

Figure 2.11 AOI as a multiplexer.

seek to justify the 0 on net 3, but it requires a 0 from the AND gate. We have created a conflict. We obviously cannot get a 0 and 1 simultaneously from the AND gate.

To resolve this conflict we must make an alternative decision. Fortunately another PDCF, $(1, 0, x, 0)$, exists for the fault. With this alternative PDCF net 3 no longer requires an assignment. The original PDCF $(1, 0, 0, x)$ implied a 0 at the output of the AND gate and hence to the input of the inverter. That in turn implied a 1 on the output of the inverter and produced a conflict. Had the implications of the test $(1, 0, 0, x)$ been extended, the computations required to justify the assignment on net 1 could have been avoided.

2.6.5 The *D*-Intersection

We have developed covers, primitive *D*-cubes of failure, and propagation *D*-cubes. We now use these to create tests for circuits composed of numerous interconnected primitives. This will be accomplished by means of the *D*-intersection, which we define with the help of another of our ubiquitous intersection tables.

	0	1	x	D	$\overline{D}$
0	0	—	0	—	—
1	—	1	1	—	—
x	0	1	x	D	$\overline{D}$
D	—	—	D	μ	λ
$\overline{D}$	—	—	$\overline{D}$	λ	μ

The *D*-intersection table defines the results of a pairwise intersection of corresponding elements of two vectors whose elements are members of the set $\{0, 1, D, \overline{D}, x\}$. The elements represent the values on the inputs of a circuit as well as the values on the outputs of individual primitives in the circuit. The dash (—) indicates a conflict, in which case the intersection does not exist. We postpone discussion of λ and μ until later.

The *D*-intersections will be used to extend a sensitized path from the point of a fault to the inputs and outputs of the circuit. We begin by selecting a fault and assigning a PDCF. The propagation *D*-cubes and the cover are then used in conjunction with the *D*-intersection table to form subsets of connected nets, where we say that two nets are *connected* if the values assigned to them are the direct result of

1. The assignment of a PDCF, or
2. A succession of one or more nontrivial *D*-intersections.

A nontrivial intersection requires that the vectors being intersected have at least one common coordinate position in which neither of them has an x value.

The set of all connected nets forms a subcircuit called the *test cube*, also sometimes called a *D-chain*. A test cube will have associated with it an *activity vector* and a *D-frontier*. The activity vector consists of those nets of the test cube which

2.6 THE D-ALGORITHM

1. Are outputs of the test cube, and
2. Have a value D or $\overline{D}$ assigned.

The D-frontier is the set of gates with outputs not yet assigned which have one or more input nets contained in the activity vector. The objective is to start with the PDCF and form an expanding test cube via D-intersections between an existing test cube and the propagation D-cubes and members of the primitive covers until the test cube reaches the circuit inputs and outputs.

EXAMPLE ■

We will generate a test for the output of NOR gate number 9 in Figure 2.12 S-A-0.

We assign a PDCF and use a column header to help keep track of assignments as we proceed.

1	2	3	4	5	6	7	8	9
x	x	x	x	x	x	0	0	D

There is no propagation required so we proceed to justify the values assigned to gates 7 and 8. We start with gate 7 and assign an extremal from p_0.

1	2	3	4	5	6	7	8	9
x	x	x	x	x	x	0	0	D
1		x	x			0		
1	x	x	x	x	x	0	0	D

Since input 1 has fan-out, we extend the implication of the value assigned to input 1.

1	2	3	4	5	6	7	8	9
1	x	x	x	x	x	0	0	D
1				0				
1	x	x	x	0	x	0	0	D

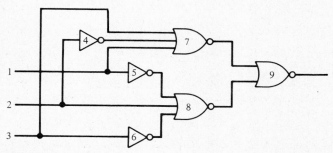

Figure 2.12 NOR circuit.

We next justify the value assigned to gate 8. We again assign an extremal from p_0. However, because of the implication step, we know that we cannot select an extremal which requires a 1 from gate 5, so we assign the value 1 to input 2.

1	2	3	4	5	6	7	8	9
1	x	x	x	0	x	0	0	D
	1						0	
1	1	x	x	0	x	0	0	D

One final implication produces a 0 on the output of gate 4.

1	2	3	4	5	6	7	8	9
1	1	x	x	0	x	0	0	D
			0					
1	1	x	0	0	x	0	0	D

■ ■

We have successfully computed the test; no entries remain in the assignment table. In this example, input 3 and, subsequently, gate 6 do not require assignments. At each point where fan-out was present, we extended implications.

EXAMPLE ■

We now use the D-algorithm to create a test for the circuit of Figure 2.7. We will list all operations in tabular form, assign numbers to relevant steps as we proceed, and refer to the step numbers as we explain the operations.

	1	2	3	4	5	6	7	8	9	10	11	12
1.	x	0	0	x	x	D	x	x	x	x	x	x
2.	0					D			$\overline{D}$			
	0	0	0	x	x	μ	x	x	$\overline{D}$	x	x	x
3.	0	0	0	x	x	D	x	x	$\overline{D}$	x	x	x
4.				0		D				$\overline{D}$		
	0	0	0	0	x	μ	x	x	$\overline{D}$	$\overline{D}$	x	x
5.	0	0	0	0	x	D	x	x	$\overline{D}$	$\overline{D}$	x	x
6.								0	D	D	0	$\overline{D}$
	0	0	0	0	x	D	x	0	λ	λ	0	$\overline{D}$
5.	0	0	0	0	x	D	x	x	$\overline{D}$	$\overline{D}$	x	x
6.								0	$\overline{D}$	$\overline{D}$	0	D
	0	0	0	0	x	D	x	0	μ	μ	0	D
7.	0	0	0	0	x	D	x	0	$\overline{D}$	$\overline{D}$	0	D
8.				x			1				0	
	0	0	0	0	x	D	1	0	$\overline{D}$	$\overline{D}$	0	D
9.		x			1					0		
10.	0	0	0	0	1	D	1	0	$\overline{D}$	$\overline{D}$	0	D

2.6 THE D-ALGORITHM

In the first step we assigned a PDCF for the output of gate 6 S-A-0. We then propagated it through gate 9. The intersection produced the result μ on the output of gate 6. We now give the rules for processing the μ and λ symbols.

1. If one or more μs occur, convert them to the corresponding D or $\overline{D}$ signals that appear in the test cube and propagation D-cube.
2. If one or more λs occur, complement all D and $\overline{D}$ signals in the propagation D-cube, perform the intersection again, and convert the resulting μs according to rule 1.
3. If μs and λs both occur, the intersection is null.

In accordance with rule 1, we convert the μ on the output of gate 6 to a D. Because gate 6 fans out to two gates, the activity vector now consists of gates 6 and 9 and the D-frontier consists of gates 10 and 12. We will refrain from implying signals in this example. We choose to propagate through gate 10 in step 4. We again produce a μ which is converted to a D.

We next propagate, in step 6, through gate 12. This time we obtain λ on gates 9 and 10. We therefore go back and complement the D and $\overline{D}$ signals in the propagation D-cube, and for convenience we relabel it as step 6'. This results in μ appearing on gates 9 and 10. These are then both converted to $\overline{D}$ in step 7. In this step we propagated a multiple path through gate 12. The inputs to gate 12 have the value $(0, 1, 1, 0)$ for the fault-free machine and $(0, 0, 0, 0)$ for the faulted machine. If propagation D-cubes with multiple D and $\overline{D}$ signals are not stored, it would be necessary to create the required propagation D-cube, using the cover consisting of vertices.

Finally, having propagated the signal to the output, we must now justify our assignments to internal gates. In step 8 we justify the assignment of a 0 to gate 11 by assigning a 1 to gate 7 and an x to input 3. In step 9 we do the same for gate 8. We must, of course, also justify the values assigned to gates 7 and 5, but at this stage it merely requires confirming that the values on their inputs will satisfy the requirements on the outputs since there are no more assignments that can be made. The final test cube is shown in line 10.

■ ■

We were fortunate that we did not have to invoke rule 3; μ and λ did not occur simultaneously. If they had, it would indicate that the test cube and the propagation D-cube had D and $\overline{D}$ signals in more than one common position. Furthermore, some of the signals were in agreement and some conflicted. Therefore, complementing all D and $\overline{D}$ signals in the propagation D-cube would not have resolved the conflict.

A flowchart of the D-algorithm is included as Figure 2.13. The D-algorithm is sometimes referred to as a two-dimensional algorithm, in contrast to path sensitization, which has been characterized as one-dimensional. Strictly speaking, the path sensitization method is not even an algorithm, but rather a *procedure*. The distinction lies in the fact that an algorithm can always find a solution *if a solution exists*. In other respects they are similar, since both an algorithm and a procedure

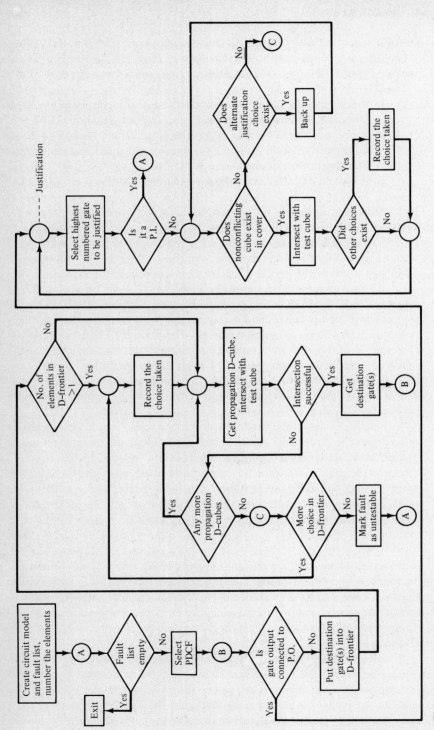

Figure 2.13 *D*-Algorithm flowchart.

can be programmed, so that a next step or a criterion for termination always exists. The reader is cautioned to note that authors are not consistent in usage of these terms, some calling an algorithm that which is more accurately called a procedure. While we may not always strictly adhere to this distinction, the reader should be aware that when an author sets out to demonstrate that his method is an algorithm, he must show that it will find a solution whenever a solution exists.

The D-algorithm was introduced by J. Paul Roth in 1966. It has come to be the most popular technique[5] used in the implementation of automatic test pattern generation systems. One of Roth's major contributions is the fact that, when he published his results in 1966, he also provided a formal proof that his technique was indeed an algorithm—that, in contrast to the sensitized path, it would find a test for a fault if a test existed.

The proof that the D-algorithm is actually an algorithm is somewhat involved. We outline the proof here; the interested reader can find the detailed proof in Roth's original paper. The proof consists of showing that if a test cube $c(T, F)$ exists for failure F, the test cube $c(T, F)$ must be contained in a primitive D-cube of failure of F. Also, a test cube must contain a connected chain of coordinates having values D or $\overline{D}$ linking the output of the faulted gate to a primary output. Given a particular gate through which the test passes on its way to an output, the test cube $c(T, F)$ must be contained in some propagation cube of the gate in question since the propagation D-cubes are constructed so as to define all possible combinations by which a test can be propagated through the gate. Finally, the manner in which cubes are intersected during creation of a test assures that all possible chains can be constructed, implying that, given a particular test, the D-algorithm will find that test if it does not first find some other test.

2.7 OTHER PATH-TRACING METHODS

A number of methods have been developed for the purpose of creating tests for digital logic. In all of the methods the ultimate goal is to find, for each fault of interest, an input pattern that will create a path originating at the hypothesized fault and terminating at an output that can be observed with the proper test equipment. The signal all along the path between these points, up to and including the output, is dependent on whether or not the fault is present. In this section we will look at the critical path method, which attempts to create a path by starting at output pins and working its way back to primary inputs, and we will look at path-oriented decision making (PODEM), which starts at the inputs and works toward output pins.

2.7.1 The Critical Path

The *critical path*, known commercially as LASAR (logic automated stimulus and response)[6], is interesting for several reasons, not the least of which is the fact that it has enjoyed considerable commercial success, having been implemented on several different computers by several companies. Of interest technically is the fact that the LASAR implementation only recognizes the NAND gate. It was known well before the advent of digital logic circuits that the NAND is a universal building block; logic functions could be expressed using only the NAND function[7]. In order for LASAR to create tests for a circuit designed with other types of gates, those circuit elements must be modeled in terms of the NAND gate.

Processing rules for a circuit to be processed by LASAR are defined in terms of "forcing" values and "critical" values as they apply to the NAND gate. The forcing rules for an n-input NAND gate are:

1. If the output of a NAND gate is 0, then the inputs are all forced to 1.
2. If the inputs are all 1, the output is forced to 0.
3. If the output is 1, and all inputs except input i are at 1, then input i is forced to 0.

A value on a node is *critical* if its existence is required for establishment of a test. The rules are:

1. If the output of a NAND gate is a 0, and it is critical, then the inputs are all critical 1s.
2. If the output is a critical 1, and if all inputs except input i are 1s, then input i is a critical 0.

If a NAND gate has a critical 0 on its input, then the other inputs are all *necessary* 1s; that is, it is necessary that they be 1s in order for input i to be critical. In order for a NAND gate to provide a necessary 1 on its output, at least one of its inputs must have a 0 assigned. That input is always *arbitrary* or noncritical.

The creation of a test starts with selection of an output pin and assignment of a 0 or 1 state to that pin. From that pin, an attempt is made to extend critical values as far back as possible toward the inputs, using the rules for establishing critical values. Then, after the path is extended as far back as possible, the necessary states are established. When complete, a critical path extends from an output pin back either to some internal net(s) or to one or more input pins (or both). The critical paths define a series of nets or signal paths along which any gate input or output will, if it fails, cause the selected output to change from its correct value. Since the establishment of a 0 on an output pin requires 1s on all the inputs to the NAND gate connected to that output, it is possible to have several critical paths converging simultaneously on an output pin.

Upon successful creation of a test, the next test begins by permuting the critical 0 on the lowest level NAND gate which has one or more inputs not yet tested, that is, the critical 0 closest to the primary inputs. The 0 is assigned to one of the other inputs to that NAND gate and the input that was 0 is now assigned the value 1. The test process then backs up again from the critical 0 to primary inputs, attempting to satisfy these new assignments. A successful test at any level may result in a critical 0 at a lower level becoming a candidate for permutation before another critical 0 on the NAND gate that was just processed. However, once selected, a NAND gate will be completely processed before another one is selected closer to the output. Eventually this process reaches the output pin, which

2.7 OTHER PATH-TRACING METHODS

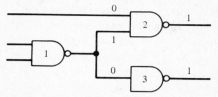

Figure 2.14 Critical assignments.

is then complemented, if the complement value has not already been processed, and the process is repeated.

The practice of postponing necessary assignments until the critical path(s) has been extended as far back as possible can help to minimize the number of conflicts that occur. Figure 2.14 illustrates a situation where a net fans out to two NAND gates (one of which is actually an inverter). Assuming that the outputs of gates 2 and 3 are both critical, if we first establish the upper input of gate 2 as far back as possible, and then extend the necessary 1 on the lower input to gate 2, we will later have to reverse the assignments on gate 2 in order to get a 0 on the input to gate 3. Since the 1 on the output of gate 3 is critical, by the rules for critical assignments, the input to gate 3 is also critical, hence it will be processed before the necessary 1 on the input to gate 2. This avoids having to undo some assignments.

Conflicts can occur despite postponement of necessary assignments. When this occurs, the rule is to permute the lowest arbitrary assignment that will affect the conflict. This is continued until a self-consistent set of assignments is achieved.

EXAMPLE ■

We will use the critical path method to develop a test for the circuit of Figure 2.15.

We first assign a 0 to the output of NAND gate 8. This forces 1s on all of its inputs. We then select gate 5 and attempt to extend the critical path through one of its inputs. To do this, we assign $(1, 2, 3) = (0, 1, 1)$. Hence, input 1 is critical and inputs 2 and 3 are necessary. We must then get a 1 on the output of gate 6. We try to extend another critical path. Since the middle input of gate 6 is the complement of the value on input 3, we cannot extend a second critical path back through gate 6 without disturbing the critical path

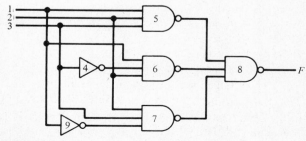

Figure 2.15 Creating a critical path.

already set up through gate 5. However, the values already assigned on 1, 2, and 3 do satisfy the critical 1 value needed at the output of gate 6. We then try to extend the critical path through gate 7. This also fails. Worse, still, the values already assigned to the inputs cause a conflict with the critical 1 assigned to the output of gate 7 because they force gate 7 to produce a logic 0. We therefore go back to gate 5 and permute the assignments on its inputs. We assign a critical 0 to the middle input and we see that we now have an assignment $(1, 2, 3) = (1, 0, 1)$ that will produce 1s on the outputs of 5, 6, and 7. We have a critical path from input 2, through gates 5 and 8, to the output F. We also have critical paths from the outputs of gates 6 and 7 to the output F. ■■

As previously mentioned, LASAR recognizes only NAND gates. Consequently, when processing a circuit, the components used in the circuit must be modeled in terms of NAND gates. Some simple transformation rules exist which permit conversion of other logic gate types into a NAND-equivalent form. These are shown in Figure 2.16.

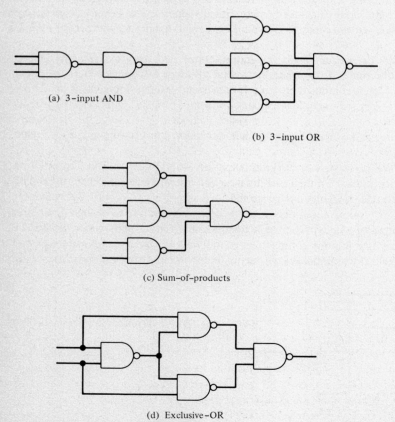

(a) 3-input AND

(b) 3-input OR

(c) Sum-of-products

(d) Exclusive-OR

Figure 2.16 Some simple transformations.

2.7 OTHER PATH-TRACING METHODS

The substitution of gates on an individual basis can be quite costly in terms of the number of gates in the resulting circuit. The three-input NOR gate requires five NAND gates and three levels of logic. Some circuits, such as two-level sums of products, can be implemented at a more global level with two levels of NAND gates. It takes fewer gates, hence data entry is less error-prone and it is more realistic in terms of number of gates required to model the original circuit. A global modeling process is, however, more error-prone, since groups of gates now, rather than individual gates, must be converted to their equivalent form, a process which is susceptible to transformation errors. An alternative approach is to simply go ahead and substitute on an individual gate basis and then go through the remodeled circuit and purge all occurrences of consecutive pairs of one-input NAND gates[8].

2.7.2 PODEM

The D-algorithm selects a fault from a list and works from the point of the fault toward inputs and outputs, employing propagation and justification techniques as it proceeds. In circuits that rely heavily on reconvergent fan-out, such as parity checking and error detection and correction (EDAC) circuits, the D-algorithm can become mired in large numbers of conflicts. These conflicts result from the fact that the justification process, proceeding back toward primary inputs along two or more paths, frequently arrives at a point where signals converge with conflicting requirements. As a result, the D-algorithm must find a node where an arbitrary decision was made and choose an alternative assignment. This can be a very time-consuming and/or memory-intensive process, depending on how the conflicts are handled. In EDAC circuits, the amount of time can become prohibitively large.

The PODEM[9] test generation system attempts to address the problem of uncontrollably large numbers of remade decisions by selecting a fault and then working directly at the inputs to create a test. PODEM begins by assigning xs to all inputs. It then starts assigning arbitrary values to primary inputs. Implications of the assignments are propagated forward. If either of the following two propositions is true, the assignment is rejected.

1. The signal net for the stuck fault being tested has the same logic level as the stuck level.
2. There is no signal path from an internal signal net to a primary output such that the internal signal net has value D or $\overline{D}$ and all other nets on the signal path are at x.

Proposition 1 excludes input combinations which cause the good machine and the faulted machine to assume identical values at the site of the fault, irrespective of the fault's presence or absence. Proposition 2 excludes input combinations that cause all possible paths from the sensitized path to outputs to become blocked. If the test is not complete, and if there is no path to an output that is free to be assigned, then there is no way to propagate the test to an output.

When PODEM makes assignments to primary inputs, it employs a "branch-and-bound"[10] method. In this method, the process can be viewed as a tree, as

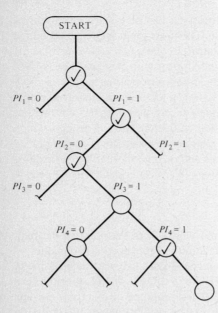

Figure 2.17 Branch-and-bound tree.

shown in Figure 2.17. An assignment is made to a primary input (PI) and, if it does not violate either proposition 1 or 2, the assignment is retained and a branch is added to the tree. If a violation results, the assignment is rejected and the node is flagged; the tree is thus bounded. If the node had been previously flagged, then the node is completely rejected and it is necessary to back up in the tree until an unflagged node is encountered. The process continues until a successful test is created or the process returns to the start node, at which time it is concluded that a test does not exist. The criterion for a successful test is the same as that employed by the D-algorithm, namely that a D or $\overline{D}$ has been propagated from the point of a fault to a primary output.

If PODEM rejects the initial assignment to the ith input selected, and there are n inputs, then it in effect eliminates 2^{n-i} combinations from further consideration when creating a test for a particular fault. If the initial assignment to the first primary input is rejected, this in effect cuts the number of combinations to be considered in half. We say, therefore, that PODEM examines all input combinations implicitly. It does not have to examine them explicitly in order to determine whether a test exists. Since it will consider all possible input combinations if necessary to find a test, it can be concluded that if PODEM does not find a test, a test does not exist; hence it follows that PODEM is an algorithm.

The PODEM algorithm can be implemented by using a last-in, first-out (LIFO) stack. As PIs are selected, they are placed on the stack. A node is flagged if the initial assignment has been rejected and the alternative is being tried. If a node violates one of the two propositions, and it is flagged, it is popped off the stack, hence bounding the graph. Nodes continue to be popped off until an unflagged node is encountered. The process terminates when a test is found or the stack becomes empty.

2.7 OTHER PATH-TRACING METHODS

EXAMPLE ■

We illustrate the PODEM algorithm by means of the circuit in Figure 2.18.

In this circuit we will take the primary inputs in the order in which they are numbered and our first choice of trial assignment at any input will, for the purpose of explanation, be the logic value 0. We will create a test for the output of gate 8 S-A-0. In this case, the assignment of a 0 to input 1 results in a violation of proposition 2. Therefore, the assignment is rejected and we assign a 1 to input 1. We follow that by assigning a 0 to input 2. With a 0 on the output of gate 6 and an x on the output of gate 7, there is no violation of the propositions, so the assignment is retained. Likewise, a 0 on input 3 causes no violation, so it is retained. However, when we assign a 0 to input 4 we find that the output of gate 8 goes to 0; this is a violation of proposition 1, so we back up and assign a 1 on input 4. Similarly, assigning a 0 to input 5 produces another violation of proposition 1, so we assign a 1. We now have a test, $(1, 0, 0, 1, 1)$, for the output of gate 8 S-A-0. The fault-free circuit takes on the value 0 and the faulted circuit takes the value 1.

■ ■

Selection of inputs and assignment of initial values can influence the success of the PODEM algorithm. Therefore, the algorithm, as implemented, contains additional capability which permits selection of initial primary inputs and logic values that have a high probability of helping PODEM to create a test. The capability consists essentially of a backtrace process, which starts at the gate under test or some other gate where assignments must be made to propagate a test and traces back to the PIs. Along the way it is noted which binary value is most likely to help meet the test objective. Consider the example above. If the upper input to gate 9 is to be tested for a stuck-at-1, then we would like to start by assigning a 0 to input 1. We now must develop a logic 1 on the lower input. If we back up from the lower input of gate 9 to a primary input (any primary input), we see that we need a 1 on one of the inputs to gate 8. We arbitrarily choose the upper input. This implies a 1 on the output of gate 6, which implies that we need a 1 on both inputs. Therefore, we set input 2 to a 1 and propagate its implications forward. There is no rule violation so we retain the assignment. We then select the other input to gate 6 and back up again to another input, which in this case turns out to be input 3. This backtrace operation appears quite similar to the justification process

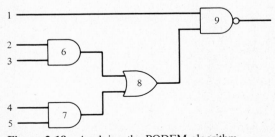

Figure 2.18 Applying the PODEM algorithm.

used in the *D*-algorithm. However, there is a difference. This backtrace does not require that assignments be justified. It is informal in that it looks for inputs that have a high probability of success. It is not actually part of the PODEM algorithm, but rather is an aid whose sole purpose is to speed up the algorithm by helping it to select inputs and logic values with a high probability of success. The input and value chosen may, in fact, be rejected by the algorithm.

2.8 BOOLEAN DIFFERENCES

Some of the methods developed to create tests for digital logic employed algebraic or symbol manipulation techniques. These appeared in the 1960s and early 1970s but never achieved the popularity of the path-tracing methods. Here we will describe the method of Boolean differences.

Given a function F which describes the behavior of a digital circuit, if a fault occurs that transforms the circuit into another circuit whose behavior is expressed by F^*, then the 1-points of the function T,

$$T = F \oplus F^*$$

define the complete set of tests capable of distinguishing between the two functions.

EXAMPLE ■

We will create a test for a shorted inverter (gate 5) in the circuit of Figure 2.19.

We begin with the equation for its behavior:

$$F = x_4 \cdot (\overline{x_1} + x_2) \cdot (x_1 + x_3)$$

With a shorted inverter, the equation becomes:

$$F^* = x_4 \cdot (x_1 + x_2) \cdot (x_1 + x_3)$$

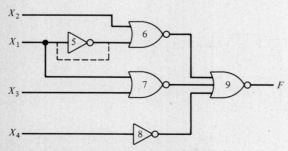

Figure 2.19 Circuit with shorted inverter.

2.8 BOOLEAN DIFFERENCES

Then

$$\begin{aligned}
F \oplus F^* &= \overline{F} \cdot F^* + F \cdot \overline{F^*} \\
&= (\overline{x_4} + x_1 \cdot \overline{x_2} + \overline{x_1} \cdot \overline{x_3}) \\
&\quad \cdot (x_1 \cdot x_4 + x_1 \cdot x_3 \cdot x_4 + x_1 \cdot x_2 \cdot x_4 + x_2 \cdot x_3 \cdot x_4) \\
&\quad + (\overline{x_1} \cdot x_3 \cdot x_4 + x_1 \cdot x_2 \cdot x_4 + x_2 \cdot x_3 \cdot x_4) \\
&\quad \cdot (\overline{x_4} + \overline{x_1} \cdot \overline{x_2} + \overline{x_1} \cdot \overline{x_3}) \\
&= x_1 \cdot \overline{x_2} \cdot x_4 + \overline{x_1} \cdot \overline{x_2} \cdot x_3 \cdot x_4 \\
&= \overline{x_2} \cdot x_4 \cdot (x_1 + \overline{x_1} \cdot x_3) \\
&= \overline{x_2} \cdot x_4 \cdot (x_1 + x_3)
\end{aligned}$$

From this equation it is evident that if we choose $x_2 = 0$ and $x_4 = 1$, then a logic 1 on either x_1 or x_3 will cause the fault-free machine and the faulted machine to produce different outputs (verify this); hence we have a test capable of detecting the short in the inverter. ■ ■

We restrict our attention, for the moment, to input faults. Given a function $F(x_1, x_2, \ldots, x_n)$, we define the Boolean difference[11] of F with respect to its ith input variable as

$$D_i(F) = F(x_1, \ldots, x_i, \ldots, x_n) \oplus F(x_1, \ldots, \overline{x_i}, \ldots, x_n)$$

The following properties[12] hold for the difference operator (in what follows, the AND operation takes precedence over the exclusive-OR):

1. $D_i(F) = D_i(\overline{F})$
2. $D_i(F(x_1, \ldots, x_i, \ldots, x_n)) = D_i(F(x_1, \ldots, \overline{x_i}, \ldots, x_n))$
3. $D_i(D_j(F)) = D_j(D_i(F))$
4. $D_i(F \cdot G) = F \cdot D_i(G) \oplus G \cdot D_i(F) \oplus D_i(F) \cdot D_i(G)$
5. $D_i(F + G) = \overline{F} \cdot D_i(G) \oplus \overline{G} \cdot D_i(F) \oplus D_i(F) \cdot D_i(G)$
6. $D_i(F \oplus G) = D_i(F) \oplus D_i(G)$

We outline the proof for property 4, but first we state some properties of the Exclusive-OR operator.

(a) $F \oplus F = 0$
(b) $F \oplus 0 = F$
(c) $F \oplus G = G \oplus F$
(d) $G = F \oplus F \oplus G$
(e) $F + G = F \oplus G \oplus F \cdot G$
(f) $F \cdot G \oplus F \cdot H = F \cdot (G \oplus H)$
(g) $F(x) = x_i \cdot F(x_1, \ldots, 1, \ldots, x_n) \oplus \overline{x_i} \cdot F(x_1, \ldots, 0, \ldots, x_n)$

We now sketch the proof. For notational convenience we omit the subscript associated with the variable x_i and the functions F and G. It is understood that the functions are differenced with respect to the ith variable, x_i and F_e, $e \in \{0, 1\}$, denotes $F(x_1, \ldots, e, \ldots, x_n)$. We first use property (g) to expand the left-hand side:

$$D_i(F \cdot G) = D_i[(x \cdot F_1 \oplus \bar{x} \cdot F_0) \cdot (x \cdot G_1 \oplus \bar{x} \cdot G_0)]$$
$$= [(x \cdot F_1 \oplus \bar{x} F_0) \cdot (x \cdot G_1 \oplus \bar{x} \cdot G_0)]$$
$$\oplus [(\bar{x} \cdot F_1 \oplus x \cdot F_0) \cdot (\bar{x} \cdot G_1 \oplus x \cdot G_0)]$$
$$= x \cdot F_1 \cdot G_1 \oplus \bar{x} \cdot F_0 \cdot G_0 \oplus \bar{x} \cdot F_1 \cdot G_1 \oplus x \cdot F_0 \cdot G_0$$

We take note of the first two terms in the expansion and use properties (a) and (b) to add the terms indicated in braces:

$$= G_1 \cdot x \cdot F_1 \oplus \{G_1 \cdot \bar{x} \cdot F_0 \oplus G_1 \cdot \bar{x} \cdot F_0\}$$
$$\oplus G_0 \cdot \bar{x} \cdot F_0 \oplus \{G_0 \cdot x \cdot F_1 \oplus G_0 \cdot x \cdot F_1\}$$
$$\oplus \bar{x} \cdot F_1 \cdot G_1 \oplus x \cdot F_0 \cdot G_0$$

We then drop the braces, group terms 1 and 2 and also 4 and 5, and then use properties (c) and (f):

$$= \{[(x \cdot F_1 \oplus \bar{x} \cdot F_0) \cdot G_1] \oplus [(x \cdot F_1 \oplus \bar{x} \cdot F_0) \cdot G_0]\}$$
$$\oplus \bar{x} \cdot F_0 \cdot G_1 \oplus x \cdot F_1 \cdot G_0 \oplus \bar{x} \cdot F_1 \cdot G_1 \oplus x \cdot F_0 \cdot G_0$$

The term in braces is recognized as $F \cdot D_i(G)$. We now have:

$$D_i(F \cdot G) = F \cdot D_i(G) \oplus x \cdot G_0 \cdot D_i(F) \oplus \bar{x} \cdot G_1 \cdot D_i(F)$$

where the second and third terms were obtained by grouping product terms with a common x or $\bar{x}$ variable and factoring. We factor once again to get:

$$D_i(F \cdot G) = F \cdot D_i(G) \oplus D_i(F) \cdot [\bar{x} \cdot G_1 \oplus x \cdot G_0]$$
$$= F \cdot D_i(G) \oplus D_i(F) \cdot [G \oplus G \oplus \bar{x} \cdot G_1 \oplus x \cdot G_0]$$
$$= F \cdot D_i(G) \oplus G \cdot D_i(F) \oplus D_i(F) \cdot [G \oplus \bar{x} \cdot G_1 \oplus x \cdot G_0]$$

When G is expanded to $x \cdot G_1 \oplus \bar{x} \cdot G_0$, the expression in square brackets is recognized as $D_i(G)$. We leave the details as an exercise.

We now consider again the circuit of Figure 2.19. We will attempt to create a test for input 3. However, rather than try to solve the problem by brute force as we did previously, this time we attempt to exploit the difference relationships. We start by defining the following functions:

$$g = x_4$$
$$h = (\bar{x}_1 + x_2)(x_1 + x_3)$$

We then use property 4 to compute the difference relative to input 3:

$$D_3(g \cdot h) = g \cdot D_3(h) \oplus h \cdot D_3(g) \oplus D_3(g) \cdot D_3(h)$$

A cursory glance at the expression tells us that we still have a rather imposing amount of work to perform. Are there any shortcuts? Fortunately, the answer is yes. We digress briefly to define the concept of independence. A function $F(X)$, $X = (x_1, \ldots, x_i, \ldots, x_n)$ is *independent* of x_i if $F(X)$ is logically invariant under complementation of x_i. This definition leads to the following theorem.

2.8 BOOLEAN DIFFERENCES

Theorem. A necessary and sufficient condition that a function $F(X)$ be independent of x_i is that $D_i(F) = 0$.

If the function $F(X)$ is independent of x_i, then the difference operator possesses the following properties:

7. $D_i(F) = 0$
8. $D_i(F \cdot G) = F \cdot D_i(G)$
9. $D_i(F + G) = \overline{F} \cdot D_i(G)$

Alternatively, if $F(X)$ is a function only of x_i, then

10. $D_i(F) = 1$

With these additional properties, we now return to our problem. We notice that $g = x_4$ is independent of x_3; therefore

$$D_3(g) = 0$$

hence
$$D_3(g \cdot h) = g \cdot D_3(h)$$

If we now define two new functions,

$$u = \overline{x_1} + x_2$$
$$v = x_1 + x_3$$

then we can apply property 4 to $D_3(h)$ to get

$$g \cdot D_3(h) = g \cdot D_3(u \cdot v)$$
$$= g \cdot u \cdot D_3(v) \quad \text{(from property 9)}$$

We can now use property 5 to get

$$D_3(x_1 + x_3) = \overline{x_1} \cdot D_3(x_3) \oplus \overline{x_3} \cdot D_3(x_1) \oplus D_3(x_3) \cdot D_3(x_1)$$

Again, by virtue of the independence theorem, we can discard the last two terms and we now have:

$$D_3(F) = x_4 \cdot (\overline{x_1} + x_2) \cdot \overline{x_1} \cdot D_3(x_3)$$
$$= \overline{x_1} \cdot x_4(\overline{x_1} + x_2)$$
$$= \overline{x_1} \cdot x_4$$

The circuit of Figure 2.19 is a multiplexer with an enable input. The select line is x_1, the enable is x_4, and the data inputs are x_2 and x_3. The final equation says that an error in input x_3 will be visible at the output if the multiplexer is enabled and if input x_3 is selected ($x_1 = 0$). The Boolean difference method has, in effect, created a sensitized path from input x_3 to an output. It now remains but to apply a 1 and a 0 to x_3 in order to exercise and test the complete path from the input to the output.

Up to this point in the discussion we have limited our attention to primary inputs. It is also possible to detect specific faults internal to a circuit by using the

Boolean difference. First, consider the internal node to be just another input x_{n+1}. Then express the behavior of the circuit as a function of the original inputs and the new input. The internal node will, in general, be some function G of the same set of inputs. If we want the internal node to be tested for an S-A-1 (S-A-0), then we want to create a path from the newly created "input" to the output and, in addition, we want that "input" to assume the value 0 (1). Hence, for an S-A-1 fault, we want to compute the solution for

$$\bar{x}_{n+1} \cdot D_{n+1}(F) = 1$$

For the S-A-0 fault, we want the solution for

$$x_{n+1} \cdot D_{n+1}(F) = 1$$

EXAMPLE ■

In order to contrast the amounts of computation required, we will again create a test for the shorted inverter, this time using the Boolean difference. The output of gate 5 is now treated as an input. We express F as

$$F = x_4 \cdot (x_2 + x_5) \cdot (x_1 + x_3)$$

In this case, the function G is simply $\bar{x}_1$.

Now applying the difference operator and the given properties to the function F, we get

$$\begin{aligned} G \cdot D_{n+1}(F) &= G \cdot [x_4 \cdot (x_1 + x_3) \cdot D_5(x_2 + x_5)] & \text{(properties 4 and 7)} \\ &= G \cdot [x_4 \cdot (x_1 + x_3) \cdot (\bar{x}_2 \cdot D_5(x_5))] & \text{(property 5)} \\ &= G \cdot [x_4 \cdot (x_1 + x_3) \cdot \bar{x}_2] & \text{(property 10)} \end{aligned}$$

The expression within the square brackets specifies necessary conditions on the inputs in order to propagate the fault to the output. Since the fault is a shorted inverter, either value of x_1 will distinguish the faulty circuit from the fault-free circuit. ■ ■

The Boolean difference has been developed quite thoroughly; for instance, if G is a function $G(u, v)$ of u and v, and $u = u(x_1, \ldots, x_n), v = v(x_{n+1}, \ldots, x_{n+m})$, where u and v share no variables in common, then the following *chain rule* holds:

$$D_i(G) = D_1(G) \cdot D_i(u)$$

where $D_1(G)$ is the difference of G with respect to u and $D_i(u)$ is the difference of u with respect to its ith variable. With the chain rule, the Boolean difference behaves much like the path sensitization approaches.

EXAMPLE ■

We apply the chain rule to input 3 of the circuit of Figure 2.19. We first separate the expression for the circuit into subexpressions which have no variables in common.

if
$$F = \overline{(\overline{x}_2 \cdot x_1 + \overline{x}_1 \cdot \overline{x}_3 + \overline{x}_4)}$$
$$u = \overline{x}_1 \cdot \overline{x}_3$$
$$v = \overline{x}_2 \cdot x_1 + \overline{x}_4$$

then
$$F = \overline{(u + v)}$$

and
$$D_3(F) = D_1(F) \cdot D_3(u)$$

From this point it is simple to compute the final result, which we leave as an exercise. ∎

2.9 SYMBOLIC PATH TRACING

The two methods described in this section trace signal paths in logic circuits by manipulating symbols which represent the nets and logic elements. Logic expressions are formed and tests for faults in the circuit are derived from the expressions.

2.9.1 Poage's Method

In 1963, J. P. Poage described a method for analyzing digital circuits, making use of what he called *fault parameters*[13]. For each net in a circuit he associated three such parameters, k_0, k_1, and k_n. These parameters are characterized by the fact that, at any given time, one and only one of these parameters is true. The following equations define the conditions under which they are true:

$$k_0 = 1 \quad \text{if net } k \text{ is S-A-0}$$
$$k_1 = 1 \quad \text{if net } k \text{ is S-A-1}$$
$$k_n = 1 \quad \text{if net } k \text{ is fault-free}$$

From these conditions, it follows that

$$k_n \cdot k_1 = k_n \cdot k_0 = k_1 \cdot k_0 = 0$$

Given a net G in a circuit, if we represent it as

$$G^* = G \cdot k_n + k_1$$

then the net will have value 1 if G is true and fault-free or if net k is S-A-1. If we define

$$\overline{G^*} = \overline{G} \cdot k_n + k_0$$

then the net will have value 0 if G is both false and fault-free or if the net is S-A-0. We can express a function in terms of these parameters:

$$F^*(x_1, \ldots, x_m, a_0, b_0, \ldots, a_m, b_m, \ldots)$$

By proper substitution of nonconflicting values for the fault parameters we can obtain a specific expression for any single or multiple fault in the circuit. Consider the expression $F = A \cdot B$ for the AND gate. If we associate nets 1, 2, and 3 with inputs A and B and output F, respectively, then the corresponding function F^* becomes:

$$F^* = (A^* \cdot B^*) \cdot 3_n + 3_1$$
$$= ((A \cdot 1_n + 1_1) \cdot (B \cdot 2_n + 2_1)) \cdot 3_n + 3_1$$
$$= [(A \cdot 1_n + 1_1) \cdot 3_n + 3_1] \cdot [(B \cdot 2_n + 2_1) \cdot 3_n + 3_1]$$

The expressions in square brackets, each containing a single input literal, are called *literal propositions*. Poage has shown that functions realized by combinational circuits can always be written in terms of literal propositions. This permits use of a compact form

$$F^* = A(1_1, 3_1) \cdot B(2_1, 3_1)$$

When $\overline{F^*}$ is true, the output has value 0. In that case,

$$\overline{F^*} = (\overline{A^*} + \overline{B^*}) \cdot 3_n + 3_0$$
$$= (\overline{A} \cdot 1_n + 1_0) \cdot (\overline{B} \cdot 2_n + 2_0) \cdot 3_n + 3_0$$
$$= [(\overline{A} \cdot 1_n + 1_0) \cdot 3_n + 3_0] + [(\overline{B} \cdot 2_n + 2_0) \cdot 3_n + 3_0]$$
$$= \overline{A}(1_0, 3_0) + \overline{B}(2_0, 3_0)$$

In general, the complement of a literal proposition can be obtained by complementing the input literal and the subscripts of the fault parameters.

EXAMPLE ■

We apply Poage's method to the circuit of Figure 2.20.

$$F^* = F(7_1) = 5^*(7_1) + 6^*(7_1)$$
$$= \overline{1^*}(5_1, 7_1) + \overline{2^*}(5_1, 7_1) + 3^*(6_1, 7_1) \cdot 4^*(6_1, 7_1)$$
$$= \overline{A}(1_0, 5_1, 7_1) + \overline{B}(2_0, 5_1, 7_1) + C(3_1, 6_1, 7_1) \cdot D(4_1, 6_1, 7_1)$$

This equation identifies conditions for which the output will be a 1. Bear in mind that the quantities in parentheses are a convenient shorthand notation representing all conditions under which the term will compute to the value 1. For example, the first term will be true if A is false or if net 1 is S-A-0 or if net 5 is S-A-1 or if net 7 is S-A-1.

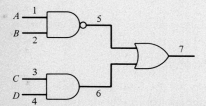

Figure 2.20 Applying Poage's method.

2.9 SYMBOLIC PATH TRACING

To determine the conditions under which $F = 0$, we complement F^*.

$$\overline{F^*} = A(1_1, 5_0, 7_0) \cdot B(2_1, 5_0, 7_0) \cdot [\overline{C}(3_0, 6_0, 7_0) + \overline{D}(4_0, 6_0, 7_0)]$$

Then, $F = 0$ if net 7 is S-A-0
                            or if A and B are true and C is false
                            or if 5 and 6 are S-A-0
                            or if 5 is S-A-0 and C or $D = 0$
                            or etc. ∎∎

In effect, by assigning specific values to the fault parameters, a function is created which represents the behavior of the circuit under single- or multiple-fault conditions. To derive the equation for circuit behavior corresponding to a particular fault or set of faults, simply retain the appropriate fault parameter(s) and delete all others.

EXAMPLE ∎

For the single fault corresponding to net 1 S-A-0, we get

$$F^* = \overline{A}(1_0) + \overline{B} + C \cdot D$$

Recalling that $\overline{A}(1_0)$ actually means

$$\overline{A}(1_0) = \overline{A} \cdot 1_n + 1_0$$

we get $F^* = \overline{A} \cdot 1_n + 1_0 + \overline{B} + C \cdot D$

Now, if the fault is not present, then

$$F^* = \overline{A} \cdot 1_n + \overline{B} + C \cdot D = \overline{A} + \overline{B} + C \cdot D$$

If the fault is present, then

$$F^* = \overline{A} \cdot 0 + 1 + \overline{B} + C \cdot D = 1$$

From this, we conclude that net 1 S-A-0 always produces a 1 on the output and a successful test for that fault must cause the fault-free circuit to assume the value 0. Inspection of the equation for F leads to the test $A = B = 1$ and $C = 0$ or $D = 0$. ∎∎

We now consider a converse problem. Given a particular input vector, list all single and multiple faults detected by this vector. The first step in this process is to determine the response of the fault-free machine. Suppose we have the test $(A, B, C, D) = 1, 0, 1, 1$. Then $F(1, 0, 1, 1) = 1$. Therefore, we are interested in those fault parameters which cause F to assume the value 0. From the equation

$$F = \overline{A} + \overline{B} + C \cdot D$$

we see that the first term is a 0 and the other two are both 1. Therefore, we look for fault parameters which cause the second and third terms to change without altering the value of the first term. Since we are attempting to create an output of 0, we use

$$F^* = A(1_1, 5_0, 7_0) \cdot B(2_1, 5_0, 7_0) \cdot (\overline{C}(3_0, 6_0, 7_0) + \overline{D}(4_0, 6_0, 7_0))$$

We want values which cause $\overline{F^*}$ to assume the value 1. Since A is already a 1, we can ignore it (but it must not conflict). B is a 0, so if net 2 is S-A-1, or net 6 is S-A-0, or net 7 is S-A-0, then the second term will be a 1. Both C and D are at 1, hence we require that net 3 or 4 or 6 or 7 be S-A-0 in order for the third term to be a 1. We finally conclude that the test (1, 0, 1, 1) will detect the following faults:

single faults—

$$\text{net 7 S-A-0}$$

double faults—

net 5 S-A-0 and net 3 or 4 or 6 S-A-0
net 2 S-A-1 and net 3 or 4 or 6 S-A-0

2.9.2 The Equivalent Normal Form

In 1966 D. B. Armstrong published a method called the *equivalent normal form* (ENF)[14]. It is similar to the method developed by Poage but is somewhat more direct. Numbers are first assigned to the gates and letters to the nets. Then, starting at an output and working back toward primary inputs, the ENF replaces individual literals by products of literals or sums of literals. When an AND gate is encountered during backtracing, a product term is created in which the literals are the names of nets connected to the inputs of the AND gate. Encountering an OR gate causes a sum of literals to be formed, while encountering an inverter causes a literal to be complemented.

EXAMPLE ■

We illustrate the procedure by reducing the circuit in Figure 2.21 to its equivalent normal form.

$$M = (I + L)_9$$
$$= ((B \cdot H)_5 + (J \cdot K)_8)_9$$
$$= ((B \cdot (E + G)_4)_5 + ((\overline{H})_6 \cdot (\overline{B} + \overline{D})_7)_8)_9$$
$$= ((B \cdot ((\overline{A})_1 + (F \cdot D)_3)_4)_5 + (((\overline{E} \cdot \overline{G})_4)_6 \cdot (\overline{B} + \overline{D})_7)_8)_9$$
$$= ((B \cdot ((\overline{A})_1 + ((\overline{C})_2 \cdot D)_3)_4)_5 + ((((A)_1 \cdot ((C)_2 + \overline{D})_3)_4)_6 \cdot (\overline{B} + \overline{D})_7)_8)_9$$
$$M = B_{59} \cdot \overline{A}_{1459} + B_{59} \cdot \overline{C}_{23459} \cdot D_{3459} + A_{14689} \cdot C_{234689} \cdot \overline{B}_{789}$$
$$+ A_{14689} \cdot C_{234689} \cdot \overline{D}_{789} + A_{14689} \cdot \overline{D}_{34689} \cdot \overline{B}_{789}$$
$$+ A_{14689} \cdot \overline{D}_{34689} \cdot \overline{D}_{789}$$

■ ■

The final equation expresses the function in a standard sum-of-products form, hence the term equivalent normal form. In creating the equation, it is required that

1. The identity of every signal path from the inputs to the outputs must be retained.
2. No redundant terms or literals resulting from the expansion may be discarded.

2.9 SYMBOLIC PATH TRACING

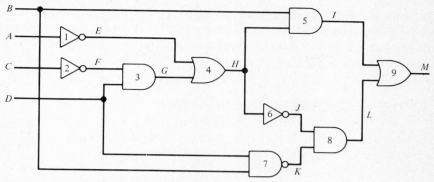

Figure 2.21 Circuit for computing ENF.

Each literal in an ENF consists of a subscripted input variable or its complement which identifies a path from the variable to an output. The ENF may contain redundant terms and/or redundant literals, even in circuits which themselves are irredundant. From the manner in which the ENF is constructed, it can be seen that there will be at least one subscripted literal for every path from each input variable to an output. It is also easy to see that the complemented literals correspond to paths which contain an odd number of inversions.

Testing a literal for S-A-1 requires that the literal have the value 0 and all other literals in that same product term have the value 1. In addition, at least one literal in each of the other product terms must be 0 in order that the value of the expression be functionally dependent on both the literal presently being tested and the signal path from that literal to the output. If all literals in a given term are at 1, and if at least one literal in each of the other terms is a 0, then the literals in the given term are all tested for S-A-0. Armstrong proved the following theorem:

Theorem. A test devised for a literal appearance in the ENF sensitizes the path in the original circuit associated with the literal appearance.

This next corollary follows from the theorem and preceding discussion.

Corollary. If a set of literals $\{L\}$ is selected whose paths contain every connection in the circuit, and if a set of tests $\{T\}$ can be found which tests at least one appearance of each of these literals for S-A-1 and S-A-0, then the set of tests detects every S-A-1 and S-A-0 fault in the circuit.

Armstrong provided a method for selecting terms and literals in order to create a minimal or near-minimal test. The test selection process consists of finding a test which will test as many as yet untested literals as possible. Start by making the smallest possible number of assignments necessary to sensitize an as yet untested literal, then continue assigning values to the literals in order to sensitize additional paths, if possible. Following this, a matrix of tests versus faults is created. Each row corresponds to a test and each column corresponds to a fault.

If test i detects fault j, then a 1 is placed in row i, column j, otherwise that entry in the matrix contains a 0. The matrix permits selection of a subset of tests in the manner of the Quine-McCluskey algorithm for selecting a cover of prime implicants.

2.10 THE SUBSCRIPTED D-ALGORITHM

It has been observed that the D-algorithm must recompute a propagation path from the output of a gate to a primary output for each input fault to be tested. This effort becomes increasingly redundant for circuits which have many gates with a large number of inputs. Elimination of these redundant computations is one of the objectives of the subscripted D-algorithm, or A-algorithm (AALG)[15].

The AALG goes farther, however. Whereas the D-algorithm selects a single fault and justifies fixed binary values on the inputs of the corresponding gate, AALG simultaneously justifies symbolic assignments on all inputs in a process called *back-propagation*. The first step is to select a gate to be processed and assign the symbol D_0 to its output. If the gate has m inputs, then a symbol D_i, $1 \le i \le m$, is assigned to each of its inputs. The D_i are called *flexible signals*; they may represent 0 or 1, depending on which values are required for a particular test. The symbol D_0 is propagated to a primary output, using the same forward-propagation techniques employed in the D-algorithm. Logic assignments are made to gate inputs so as to enable the sensitized signal to propagate to the gate outputs, and these enabling assignments are justified using the same techniques employed with the standard D-algorithm.

After the D_0 signal has been successfully propagated to an output, all of the D_i are back-propagated to primary inputs. If the back-propagation is completely successful, then tests for the output fault and all of the gate input faults can be computed simply by inspecting values at the primary inputs. This is illustrated in the circuit of Figure 2.22, where the input vector I has value

$$I = (x, 0, D_1, D_2, 0)$$

This vector is interpreted by referring back to the gate where the D_i originated. We note that a test for the output of gate 14 S-A-0 requires that both of its inputs be 1, that is, $D_1 = D_2 = 1$. Tests for S-A-1 on inputs 1 and 2 of gate 14 require that $D_1, D_2 = 0, 1$ and $1, 0$, respectively. Therefore, the tests for these three faults are

$$(x, 0, 1, 1, 0)$$
$$(x, 0, 0, 1, 0)$$
$$(x, 0, 1, 0, 0)$$

The assignments on the inputs are not unique. The input vector I could also have been assigned the values $(D_1, 1, x, D_2, 0)$, or $(D_1, 1, x, 0, D_2)$, or $(x, 0, D_1, 0, D_2)$, depending on choices made at gates where decisions were required.

We now discuss the rules for back-propagating. Basically, each D_i is back-propagated toward the inputs along as many paths as possible. This is done through

2.10 THE SUBSCRIPTED D-ALGORITHM

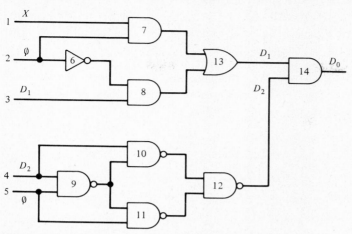

Figure 2.22 Illustrating the subscripted D-algorithm.

replication. When symbolically propagating back through an element, the symbol D_i at the output is replicated at the inputs, according to the following rules:

> If the output is assigned D_i, and the gate inverts the signal, then the inputs are assigned $\overline{D}_i$.
>
> The signal D_i (or $\overline{D}_i$) is replicated at all inputs if none of the inputs has been previously assigned.
>
> The signal D_i can be replicated at some inputs if the others are assigned noncontrolling values.

EXAMPLE ■

On a three-input NAND gate, with one of the inputs assigned a logic 1, and D_j assigned to its output during back-propagation, the remaining two inputs are assigned $\overline{D}_j$. ■ ■

This proliferation of D_i signals enhances the likelihood of establishing a sensitized path from one or more primary inputs to input i of the gate presently being tested, in contrast to propagation of a single replica, which may require considerable backtracking in response to conflicts. However, it is still possible to encounter conflicts. In fact, with flexible signals increasing exponentially in number as progress continues toward the inputs, conflicts are virtually inevitable in any realistic circuit. Efficient handling of conflicts is imperative if performance is to be realized.

A conflict can occur during back-propagation as a result of a signal D_i and a conflicting value of that same signal attempting to control a gate, or as a result of two different signals D_i and D_j attempting to control a gate, or a conflict may occur at a gate with fan-out if two or more signal paths reconverge at the gate and one of the paths has a flexible signal while another has a fixed binary value. The situation where conflicting values of the same flexible signal try to control a gate is illustrated in the upper path of Figure 2.22. The assignment of D_1 on the output of gate 13 during back-propagation initially results in the replication of D_1 on each of the inputs, hence on the outputs of gates 7 and 8. Back-propagation then produces replicas of D_1 on both inputs of gate 7 and on both inputs of gate 8. However, we are now faced with the prospect of flexible signal D_1 on both the input and output of the inverter. This conflict can be resolved by assigning a 0 or 1 to the output of gate 6. If we choose a 1, then the input to gate 6 is forced to a 0, and the lower input to gate 7 is a 0. That forces a 0 on the output of gate 7 and also causes the upper input to gate 7 to be reassigned to x.

The conflict between flexible signals D_j and D_k can be illustrated by assigning D_0 to gate 12. Forward propagation to the output and justification along the upper path are done as in the D-algorithm. We therefore restrict our attention to the consequences of a D_0 on gate 12. This results in D_1 and D_2 on the inputs of gate 12. Back-propagation then attempts to assign both $\overline{D}_1$ and $\overline{D}_2$ to the output of gate 9. Again, the conflict is resolved by assigning a fixed binary value to the output of gate 9. If a 1 is assigned, then one of the inputs must be chosen and set to 0.

When an input is chosen and set to a controlling value, for instance a 0 on the input to the AND or NAND gate, then in the general case it is desirable to choose that input which is easiest to control. However, in the present case an additional criterion may exist. If one or the other of the two inputs to gate 12 is found to have been tested by some previous test vector, then the flexible signal D_1 or D_2 corresponding to the untested input can be favored. When D_1 and D_2 converge at the output of gate 9, if it is found that the upper input to gate 12 has already been tested, then D_1 can be purged by assigning a 0 to the upper input of gate 9.

When a conflict occurs, its resolution usually requires that segments of D_i chains be deleted. In AALG this is accomplished by two subroutines called DROPIT and DRBACK[16]. DROPIT is called to purge a chain segment when the end closest to the PIs is known. It works forward toward the gate under test. It must examine fan-outs as it progresses, so that if two converging paths both have flexible signals, then both chain segments must be deleted. When a flexible signal is deleted, it may be replaced by a fixed binary signal. This signal, when assigned to the input of a gate, may be a controlling value for that gate and thus implies a logic value on the output. In that case, the output must then be further traced to the input of the gate(s) in its fan-out to determine whether this output value is a controlling value at the input of the gate in its fan-out.

In our example, when we assigned D_0 to the output of gate 12, we encountered a conflict at gate 9 and therefore assigned a 1 to its output. This forced us to assign a 0 on one of the inputs. As a result, we had to purge the D_2 chain from PI 5 to the input of gate 12. Since PI 5 has a 0 assigned, the output of gate 11 will be a 1; therefore the flexible signal D_2 initially assigned at the output of gate 11 must be purged and the path must be traced back another level. At gate 12 the enabling signal 1 is assigned to the lower input and the flexible signal D_1 is assigned to the upper input. Therefore DROPIT can stop at that point.

If D_j is controlling the output and one or more D_i are controlling the inputs, then it may be desirable to propagate the D_j toward inputs and purge the D_i signals. In that case the end of the chain farthest from the PIs is known and DRBACK is used to purge the chain. Working in the direction toward PIs, it may have to purge a considerable number of flexible signals since the signals were originally replicated when working toward the inputs.

The subroutines DROPIT and DRBACK are not always invoked independently of one another. When DROPIT is purging flexible signals and replacing them with fixed binary signals, it may be necessary to invoke DRBACK to purge other chain segments. This is seen in the upper branch of the circuit in Figure 2.22. PI 2 was assigned a 0 because of a conflict. Therefore DROPIT, working forward from PI 2, purges D_1 and replaces it with a 0. The 0 on the lower input to gate 7 blocks the gate and therefore DRBACK must pick up the chain segment on the upper input and delete it back to input 1 and replace it with x. Then DROPIT can regain control and proceed forward. The 0 on the input of gate 7 implies a 0 on the output and hence a 0 on the input to gate 13. Since a 0 on an OR gate is not a controlling value, the forward purge can stop, leaving gate 13 with $(0, D_1)$ on its inputs.

To help identify and purge unwanted chain segments, flexible signals are never implied forward in the direction of primary outputs during back-propagation. As an example, in Figure 2.22, when back-propagating from gate 7 toward PIs, any assignment to PI 2 will necessarily imply the inverse signal on the output of gate 6. However, if the flexible signal is assigned, then at some later point DROPIT may go unnecessarily along signal paths, deleting flexible signals and replacing them with controlling logic values where it may be unnecessary.

In measurements of performance, it has been found that AALG creates an input pattern with flexible signals in about the same time that the D-algorithm generates a single pattern. Overall time comparison for typical circuits shows that it frequently processes a circuit in about 30% of the time required by the D-algorithm. AALG is especially efficient, for reasons explained earlier, when working on circuits that have gates with large numbers of inputs, as is sometimes the case with programmable logic arrays (PLAs). The efficiency of AALG can be enhanced by selecting first primary outputs, and then gates with large numbers of inputs. Gates for which the output has not yet been tested are chosen next since they usually indicate regions where fault processing has not yet occurred. Finally, scattered faults are processed. On these faults AALG will occasionally default to the conventional D-algorithm.

2.11 SUMMARY

Numerous methods have been devised to develop test patterns for combinational logic. The methods range from topological to algebraic and they date from the early 1960s to the present. Some are effective and widely used; others are primarily of academic interest. They all have one thing in common: their objective is to create a set of inputs which cause the output response of a circuit to be dependent on the presence or absence of some hypothesized fault. Secondary objectives, which we did not address explicitly in this chapter, but which we will address in more detail in a later chapter, include:

Thoroughness (comprehensiveness)

Ease of implementation

Fault resolution (ability to identify *which* fault occurred)

Efficient operation

The algebraic techniques are quite thorough, complete, and accurate. They demonstrate the disparate ways in which to approach and solve a problem. But, in the final analysis, the methods that have gained acceptance have done so on the basis of some rather pragmatic considerations. Many companies involved in the manufacture of electronic products maintain a data base that contains a great deal of data describing their designs. This data is usually topological in nature because it is used, among other things, to specify the placement of components on higher level assemblies, such as chips on printed circuit boards, and it is used by programs which route wire to connect these components on the higher level assembly. The same data is available for test pattern generation. For each physical device employed in a design, a field or record exists to describe its logic function either directly or by association with another record in a library of elements where its function is more fully documented. The data in these data bases is more amenable to path-tracing methods than to symbol manipulation methods.

Among the path-tracing methods, the sensitized path approach was first to appear. Although R. D. Eldred advocated the modeling of stuck-at faults and creation of specific tests to detect these faults, the first suggestion for the use of the sensitized path is attributed, by D. B. Armstrong[14], to an unidentified attendee at a conference at the University of Michigan in 1961. Path sensitizing programs had already been well developed by C. B. Steiglitz and others[2] when the D-algorithm was introduced in 1966. The D-algorithm has, however, become the preferred method based on number of users[5]. Three distinct differences between the D-algorithm and the sensitized approach which make the D-algorithm more attractive to use are that it:

Uses a truth table rather than basic logic function

Provides a formal calculus

Pursues all fan-out paths

The truth table permits use of virtually any structure as a primitive. In conjunction with the D-calculus, the truth table permits a test to be created for just about any conceivable fault in a structure. The ability to pursue a fault along all fan-out paths insures that the D-algorithm will find a test for a fault if a test exists. The other path-tracing methods have their advocates. The critical path method, as employed in LASAR, has enjoyed considerable commercial success, and PODEM seems to be every bit as effective as the D-algorithm in general while executing more rapidly on circuits which contain much reconvergent fan-out. A refinement of PODEM, called FAN[17], speeds up the PODEM algorithm by performing special processing of fan-out points. The algebraic methods, relying on symbol manipulation, are not as readily implementable within the framework of a design automation system since

PROBLEMS

the design automation data bases are generally topological, describing physical entities and their interconnections rather than Boolean behavior.

PROBLEMS

2.1 Given a logic circuit with 64 inputs, if a tester applies tests at the rate of one test every 100 nanoseconds, how long will it take to apply 2^{64} patterns?

2.2 A four-input AND gate is exercised with the test pattern set: $(1,0,0,0)$, $(0,1,0,0)$, $(0,0,1,0)$, $(0,0,0,1)$, $(1,1,1,1)$. All input and output pins on the gate have switched. How many single stuck-at faults have been tested?

2.3 In the circuit of Figure 2.1, create a test pattern to detect a short across transistor Q_2.

2.4 Develop a sensitized path for a S-A-1 fault on input X_2 of the circuit in Figure 2.4.

2.5 Develop a sensitized path for a S-A-0 fault on the output of inverter J in the circuit of Figure 2.5.

2.6 Given that $a = (1, x, x, 0, 0)$ and $b = (1, x, x, x, 0)$, does a contain b or vice versa?

2.7 How many vertices are represented by the vector $(1, 0, D, x, 0, x, \overline{D}, x)$?

2.8 Given the following cubes; $a = (1, 0, x)$, $b = (x, 0, 0)$, $c = (1, 1, 1)$, $d = (x, x, 1)$, $e = (x, x, 0)$.

  (a) determine which cubes contain others
  (b) perform all pairwise intersections, using the table in Sec. 2.6.1.

2.9 Create singular covers for the inverter and the two-input OR, NOR, NAND, and exclusive-OR circuits. Create PDCFs for each of the stuck-at faults on inputs and outputs and create propagation D-cubes for these circuits.

2.10 For the AND-OR-Invert of Figure 2.9, create a Karnaugh map for the fault-free circuit and the circuit with (a) input 3 S-A-1 and (b) the upper AND gate output S-A-0. Using the Karnaugh maps, create tests for these faults.

2.11 Using the D-algorithm, devise a test to distinguish between the following two circuits:

$$F = abc + b\overline{c}d + \overline{a}\overline{b}d + \overline{a}cd$$
$$F^* = (\overline{a} + b)(c + d)$$

2.12 Using the D-algorithm, create a test for the output of gate F of Figure 2.5 S-A-0. Identify the complete sensitized path and all faults tested along the path.

2.13 Find a function for which 2^{2n-1} distinct propagation D-cubes exist.

2.14 Assign a 1 to the output of the circuit of Figure 2.15 and use the D-algorithm to create a sensitized path reaching to the inputs, through gate 7. Contrast the calculations with those made by the critical path method.

2.15 Reconverging fan-out may cause two or more D and/or $\overline{D}$ signals to converge at a primitive. Explain how you would determine what value from the set $(0, 1, D, \overline{D})$ propagates to the output of the primitive if the propagation table does not contain cubes with multiple D or $\overline{D}$ signals.

2.16 Create a NAND equivalent model of the circuit of Figure 2.5. Reduce it wherever possible by deleting consecutive pairs of one-input NANDs. Then use the critical path method (LASAR) to create a sensitized path back to I_1 and I_2.

2.17 Use PODEM to create a test for an S-A-1 fault on the output of gate 7 in Figure 2.15.

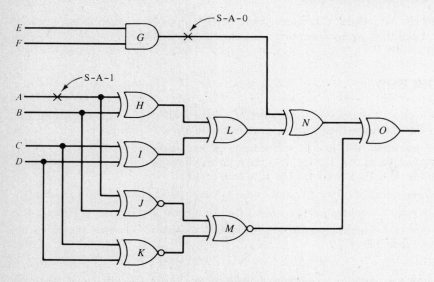

2.18 For the parity tree shown, use PODEM and the D-algorithm to find a test for each of the two faults indicated. Apply the D-algorithm in each of two ways: (a) propagate all the way to the output and then enter the justification phase once to justify all assignments, (b) justify the PDCF completely before doing any propagation, then during propagation justify the assignments on each gate before propagating the test forward to the next gate.

2.19 We saw, in the section on Boolean differences, that the expression $\overline{x}_2 \cdot x_4 \cdot (x_1 + x_3)$ will detect the shorted inverter. What is the equation that will detect the output of the inverter S-A-0? S-A-1?

2.20 We started an example to illustrate the chain rule for Boolean differences. We stopped when we obtained the expression $F = \overline{u + v}$. Complete the example; that is, obtain the expression which defines all tests for input 3 S-A-0 and S-A-1.

2.21 Show that Shannon's expansion

$$F(x_1, \ldots, x_i, \ldots, x_n) = x_i \cdot F(x_1, \ldots, 1, \ldots, x_n) + \overline{x}_i \cdot F(x_1, \ldots, 0, \ldots, x_n)$$

is equivalent to

$$F(x_1, \ldots, x_i, \ldots, x_n) = x_i \cdot F(x_1, \ldots, 1, \ldots, x_n) \oplus \overline{x}_i \cdot F(x_1, \ldots, 0, \ldots, x_n)$$

Then show that

$$D_i(F) = F(x_1, \ldots, 1, \ldots, x_n) \oplus F(x_1, \ldots, 0, \ldots, x_n)$$

Use the last result to compute a test for an S-A-1 fault on primary input 4 of the circuit in Figure 2.18.

2.22 Complete the details of the proof of property 4 of the Boolean differences. Prove the remaining nine properties, using property 4 wherever applicable.

2.23 Prove the Boolean difference chain rule.

2.24 Prove the independence theorem; that is, a necessary and sufficient condition that a function F be independent of x_i is that $D_i(F) = 0$.

2.25 Derive Poage's equation for $\overline{G^*}$ from the equation for G^*.

2.26 Prove: $((A \cdot 1_n + 1_1) \cdot (B \cdot 2_n + 2_1)) \cdot 3_n + 3_1$
$= [(A \cdot 1_n + 1_1) \cdot 3_n + 3_1] \cdot [(B \cdot 2_n + 2_1) \cdot 3_n + 3_1]$

2.27 Using the ENF, find a minimum test set for the circuit in Figure 2.22. Prove that your test set is minimal.

2.28 Use AALG to create tests for gate M in Figure 2.5. Assign D_0 to the output of gate M, assign D_1 and D_2 to its inputs, then back-propagate to the inputs. Then assign D_0 to each of K and L, in turn, and back-propagate D_i from the inputs, being sure to first propagate D_0 to the output.

REFERENCES

1. Eldred, R. D., "Test Routines Based on Symbolic Logic Statements," *J. ACM*, vol. 6, no. 1, Jan. 1959, pp. 33–36.
2. Case, P. W., et al., "Design Automation in IBM," *IBM J. Res. Dev.*, vol. 25, no. 5, Sept. 1981, pp. 631–646.
3. Schneider, P. R., "On the Necessity to Examine D-Chains in Diagnostic Test Generation—An Example," *IBM J. Res. Dev.*, vol. 10, no. 1, Jan. 1967, p. 114.
4. Roth, J. P., "Diagnosis of Automata Failures: A Calculus and a Method," *IBM J. Res. Dev.*, vol. 10, no. 4, July 1966, pp. 278–291.
5. Breuer, M. A., et al., "A Survey of the State of the Art of Design Automation," *Computer*, vol. 14, no. 10, Oct. 1981, pp. 58–75.
6. Thomas, J. J., "Automated Diagnostic Test Programs for Digital Networks," *Comput. Des.* Aug. 1971, pp. 63–67.
7. Sheffer, H. M., "A Set of Five Independent Postulates for Boolean Algebras," *Trans. Am. Math. Soc.*, vol 14, 1913, pp. 481–488.
8. Hayes, J. P., "A NAND Model for Fault Diagnosis in Combinational Logic Networks," *IEEE Trans. Comput.*, vol. C-20, no. 12, Dec. 1971, pp. 1496–1506.
9. Goel, P., "An Implicit Enumeration Algorithm to Generate Tests for Combinational Logic Circuits," *IEEE Trans. Comput.*, vol. C-30, no. 3, March 1981, pp. 215–222.
10. Lawler, E. W., and D. E. Wood, "Branch-and-Bound Methods—A Survey, "*Oper. Res.*, vol. 14, 1966, pp. 669–719.
11. Sellers, F. F., et al., "Analyzing Errors with the Boolean Difference," *IEEE Trans. Comput.* vol. C-17, no. 7, July 1968, pp. 676–683.
12. Akers, S. B., "On a Theory of Boolean Functions, "*J. SIAM*, vol. 7, Dec. 1959.
13. Poage, J. F., "Derivation of Optimal Tests to Detect Faults in Combinational Circuits," *Mathematical Theory of Automata*, Polytechnic Press, Brooklyn, N.Y., pp. 483–528.
14. Armstrong, D. B., "On Finding a Nearly Minimal Set of Fault Detection Tests for Combinational Logic Nets," *IEEE Trans. Electron Comput.*, vol. EC-15, no. 1, Feb. 1966, pp. 66–73.
15. Benmehrez, C., and J. F. McDonald, "Measured Performance of a Programmed Implementation of the Subscripted D-Algorithm," *Proc. 20th Design Automation Conf.*, 1983, pp. 308–315.
16. McDonald, J. F., and C. Benmehrez, "Test Set Reduction Using the Subscripted D-Algorithm," *Proc. 1983 Int. Test Conf.*, Oct. 1983, pp. 115–121.
17. Fujiwara, H. and T. Shimono, "On the Acceleration of Test Generation Algorithms," *IEEE Trans. Comput.*, vol. C-32, no. 12, Dec. 1983, pp. 1137–1144.

Chapter 3
Sequential Logic Test

3.1 INTRODUCTION

In the previous chapter we discussed methods for sensitizing paths in combinational logic from hypothesized faults to observable outputs. We now attempt to create tests for sequential circuits where the outputs are a function not just of present inputs but of past inputs as well. The objective will be the same: we wish to create a sensitized path from the point of a fault to an observable output. However, there are new factors which must now be taken into consideration. A path must now be sensitized not only through logic operators, but also through an entirely new dimension—time. The time dimension may be discrete, as in synchronous logic, or it may be continuous, as in asynchronous logic.

    We ignored the time factor when creating tests for combinational logic. We implicitly assumed that the output would stabilize before being measured with test equipment and we assumed that each test pattern was totally independent of its predecessors. As we shall see, we can no longer ignore the effects of time, for it can greatly influence the results of test pattern generation and can complicate, by orders of magnitude, the problem of creating tests. Assumptions about circuit behavior must be carefully analyzed to determine the circumstances under which they prevail.

3.2 DEFINITIONS

A sequential circuit can be conveniently represented by the Huffman model[1] depicted in Figure 3.1. The circuit consists of a combinational part and feedback

3.2 DEFINITIONS

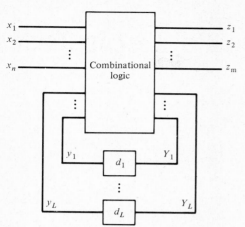

Figure 3.1 Huffman model.

lines $y_1, \ldots, y_L$, which pass through delay elements $d_1, \ldots, d_L$ and then act as additional inputs to the circuit. The set of values $\{y_1, y_2, \ldots, y_L\}$ at any time constitute the present state of the machine. Because there is a finite number of possible states, the circuit is called a *finite state machine*. The outputs z_i are a function

$$z_i = z_i(x_1, \ldots, x_n, y_1, \ldots, y_L)$$

of the values on the inputs and the present state. The delay elements $d_1, \ldots, d_L$ may be delay inherent in the logic devices, they may be devices specially designed to delay a signal by some known fixed amount, they may be flip-flops controlled by one or more clock signals, or there may be elements from each of these types. If the devices are all controlled by clock signals, then the circuit is *synchronous*, that is, its actions are synchronized by some external signal(s). If the delays are inherent in the devices, and not otherwise controllable by signals external to the circuit, the circuit is classified as *asynchronous*. A circuit which has both clocked and unclocked delays may be placed in either category, the distinction often depending on the exact purpose of the asynchronous signals. A circuit in which memory devices can be asynchronously set or reset, but which is otherwise completely controlled by clock signals, is usually classified as synchronous. Sequential circuits are also sometimes referred to as *cyclic*, a reference to the presence of feedback or closed loops, as distinguished from combinational circuits, which are termed *acyclic*.

A frequently used memory element is the cross-coupled latch, implemented by using either NOR gates or NAND gates, as depicted in Figure 3.2. These latches are widely used both by themselves and as building blocks in other memory devices. The value on the output Y at time t_{n+1} is determined by values on the Set and Reset input lines and by the present state of the latch. Given a present state y and values on its Set and Reset inputs, the next state can be determined from a *state table*. The value within the state table, at the intersection of a row corre-

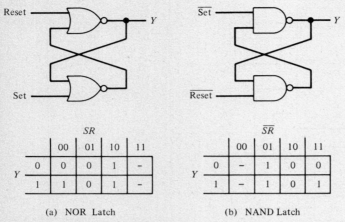

(a) NOR Latch (b) NAND Latch

Figure 3.2 Cross-coupled latches.

sponding to the present state and a column corresponding to the applied input value(s), specifies the next state to which the machine will make a transition. Entries containing dashes denote indeterminate states. For the NOR latch the column corresponding to (Set, Reset) = (1, 1) contains dashes. It would be illogical to set and reset the latch simultaneously and, if the combination (1, 1) were applied, followed by the combination (0, 0), the results would be unpredictable. A similar consideration holds if the sequence $\{(0, 0), (1, 1)\}$ were applied to the inputs of the NAND latch.

A latch may be preceded by gates which permit its signal changes to be controlled by another signal, called Enable or sometimes called Load. This is illustrated in Figure 3.3(a) and (b). In Figure 3.3(b) there is a single data input, called Data, whose value is inverted in one of its two paths so that the latch never sees an illegal input combination.

Clock-controlled flip-flops, also referred to as bistables or stored state variables in this text, are used quite extensively in digital circuits. Unlike the latch, where data changes on the input can pass directly through to the output when the Enable is active, the clocked flip-flop will not permit data changes on the input to pass through after the clock has gone active. The three basic building blocks of

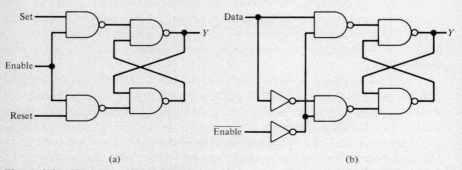

(a) (b)

Figure 3.3 Clock-controlled latches.

3.2 DEFINITIONS

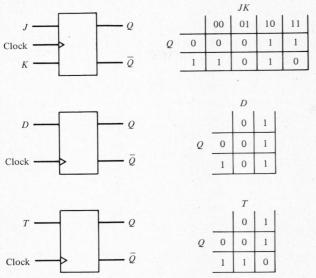

Figure 3.4 The standard flip-flops.

sequential circuits are the JK, D (delay), and T (toggle) flip-flops. The JK flip-flop behaves like the cross-coupled NOR latch but permits the input combination $(1,1)$. The delay flip-flop simply delays a signal for one clock period, and the toggle flip-flop switches state whenever the input is a 1. These, along with their state tables, are illustrated in Figure 3.4.

A well-known theorem in sequential machine theory states that any of these circuits can be configured to emulate any of the others. As an example, if the input signal to a D flip-flop is exclusive ORed with the output of the D flip-flop, then the resulting circuit behaves like the T flip-flop.

The flip-flops illustrated above can be implemented as *level-sensitive* flip-flops or as *edge-triggered* flip-flops. The level-sensitive flip-flop responds to a high or low level of the clock. The edge-triggered flip-flop responds to a rising or falling clock edge. We illustrate each of these two types. The flip-flop in Figure 3.5 is a

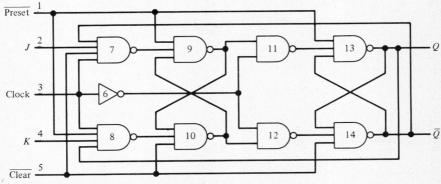

Figure 3.5 Level-sensitive JK flip-flop.

level-sensitive JK flip-flop implemented in a master/slave configuration. When Clock is high, data can enter the first stage or master. When Clock goes low the data in the first stage is latched and the second stage or slave latch is transparent, so the data in the first stage is now transferred to the outputs.

The edge-triggered D flip-flop, illustrated in Figure 3.6, is somewhat more complex in its operation[2]. It has $\overline{\text{Preset}}$ and $\overline{\text{Clear}}$ lines with which the output Q can be forced to either a 1 or 0 independent of the values on the Data and Clock lines. When the $\overline{\text{Preset}}$ and $\overline{\text{Clear}}$ are at 1, and Clock is low, then the complement of the value at the Data input appears at the output of N_4. Also, under these conditions, the output of N_1 has the same value as the Data input. Therefore, the input to N_2 matches the value on the Data line and the value on the input to N_3 is the complement of the value on the Data line.

When Clock goes high, the values at the inputs to N_2 and N_3 appear, inverted, at their outputs. They are then inverted once again as they go through N_5 and N_6 so that the output of N_5 matches the value on the Data line. There is an important point to note about this configuration: if Data is low when Clock goes high, then the output of N_3 goes low and prevents further changes in Data from propagating through N_4. If Data is high, then when Clock goes high, the high value at the output of N_1 causes a 0 to appear at the output of N_2, which blocks changes at the Data input from propagating through N_1 and N_3.

We see that the behavior of the circuit is sensitive to the rising edge of the clock input. Data cannot get through N_2 and N_3 when Clock is low, and shortly after Clock goes high the data is latched so that the flip-flop is immune to further changes on the Data input. However, changes on the Data input *during* the positive

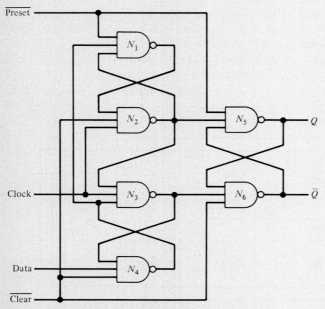

Figure 3.6 Edge-triggered delay flip-flop.

3.3 TEST PROBLEMS CAUSED BY SEQUENTIAL LOGIC

edge transition of Clock can cause unpredictable results. Therefore, these flip-flops are usually specified by their manufacturers with two key parameters: setup and hold time. *Setup* refers to the time interval during which a signal must be stable at an input terminal prior to an active transition occurring at another input terminal. *Hold* refers to the time interval during which a signal must be stable at an input terminal following an active transition at another input terminal. In the present case, these parameters measure the time duration during which the Data input must be stable relative to the Clock input.

3.3 TEST PROBLEMS CAUSED BY SEQUENTIAL LOGIC

Two new factors which complicate the task of creating tests for sequential logic include the presence of memory and circuit delay. In sequential circuits signals must not only be logically correct, they must occur at the correct time relative to other signals. The problem is further complicated by the fact that aberrant behavior can occur even in sequential circuits in which all of the components are fault-free and behaving completely in compliance with their manufacturer's specifications. We first consider the problems caused by the presence of memory; then we examine the effects of circuit delay time on the test pattern generation problem.

3.3.1 The Effects of Memory

In the first chapter we pointed out that, for combinational circuits, it was possible (but not necessarily reasonable) to create a complete test for logic faults by applying all possible binary combinations to the inputs of a circuit. This, as we shall see, is not true for circuits with memory. Not only may they require more than 2^n tests, they are also sensitive to the *order* in which stimuli are applied.

Test Vector Ordering Consider the cross-coupled NAND latch of Figure 3.2. To assess the effects of memory, we analyze four faults, these being the four NAND inputs S-A-1 (we number from top to bottom in the diagram). We apply all four possible binary combinations to the inputs in ascending order, that is, in the sequence $(\overline{S}, \overline{R}) = \{(0, 0), (0, 1), (1, 0), (1, 1)\}$. We get the following response for the good machine (GM) and the machine corresponding to each of the four input S-A-1 faults.

Input		Output				
$\overline{\text{Set}}$	$\overline{\text{Reset}}$	GM	1	2	3	4
0	0	1	0	1	1	1
0	1	1	0	1	1	1
1	0	0	0	0	0	1
1	1	0	0	0	1	1

In this table we find that fault number 2 responds to the sequence of input vectors with a set of outputs that exactly matches the good machine response. Quite clearly, this sequence of inputs will not distinguish between the good machine and one with input 2 stuck-at-1.

We next apply the sequence in exactly the opposite order. We get:

Input		Output				
Set	Reset	GM	1	2	3	4
1	1	?	?	0	1	?
1	0	0	0	0	0	?
0	1	1	0	1	1	1
0	0	1	0	1	1	1

The Indeterminate Value When the four input combinations were applied in reverse order, question marks were placed in some table positions. Exactly what significance do we attach to them? To answer this question, we must take note of a situation which did not exist when we were dealing only with combinational logic; the cross-coupled NAND latch has *memory*. In a typical application, such as an interrupt handler, the device that services interrupt requests will initially reset the latch and cause the output to go to the 0 state. A device requesting service puts a negative-going pulse on the $\overline{\text{Set}}$ line when it requires service. After the signal is removed, the latch remembers that a negative signal had appeared on the $\overline{\text{Set}}$ line; a logic 1 is latched on the output and persists until the interrupt handler applies a negative pulse to the $\overline{\text{Reset}}$ line.

Because of the memory, neither the $\overline{\text{Set}}$ line nor the $\overline{\text{Reset}}$ line need be held low any longer than necessary to effectively latch the circuit. However, when power is first applied to the circuit, it is not certain what value is contained in the latch. How do we perform calculations on a circuit when we do not know what value is contained in a device with memory? To solve this problem, we introduce the value x. It is used to represent a signal value on a net whenever that signal value is indeterminate. We initially assign the value x to all nets in the circuit. In order to calculate the results of operations which involve the x value in conjunction with the binary values 0 and 1, we introduce the following tables:

AND	0	1	x
0	0	0	0
1	0	1	x
x	0	x	x

OR	1	0	x
1	1	1	1
0	1	0	x
x	1	x	x

A brief glance at these tables reveals their derivation. For example, with a 0 on one input of a two-input AND, the output must be a 0 regardless of what value exists on the other input. However, if one input is at logic 1 and the other input is indeterminate, then the resulting output is indeterminate. We conclude, then,

3.3 TEST PROBLEMS CAUSED BY SEQUENTIAL LOGIC 75

that the question mark in the response table for the NAND latch, which precipitated this discussion of indeterminate values, represents an output that could be either a 1 or a 0; that is, it is indeterminate. We therefore initially assign an x to each of the nets in the circuit. Then, when we apply the input pattern (1, 1) to the $\overline{\text{Set}}$ and $\overline{\text{Reset}}$ lines, each of the NAND gates has a logic 1 and an x on its inputs, and the three-value operation table for the AND gate tells us that the result remains indeterminate (the inverse of an indeterminate value is indeterminate).

Upon examining the response tables for the two sequences of test vectors, we find that when the vectors are applied in ascending order, three of the four faults will be detected by virtue of the fact that machines with these faults differ from the good machine in their response to one or more of the input patterns. When the patterns are applied in descending order, only one of the four faults will be detected, namely fault 1. Machines with faults 2 and 3 agree with the good machine response in every instance where the good machine has a known response. On the first pattern the good machine response is indeterminate and the machine with fault 2 responds with a 0. The machine with fault 3 responds with a 1. Since we do not know what value to expect from the good machine, we cannot distinguish between the good machine and any machine with a stuck-at-1 fault on input 2 or 3.

Faulted machine 4 presents an additional complication. Its response is indeterminate for both the first and second patterns. However, because the good machine has a known response to pattern 2, we do know what to look for in the good machine, namely the value 0. Therefore, if a NAND latch is being tested with these stimuli, and it is faulted with input 4 S-A-1, it might come up initially with a 0 on its output, in which case the fault is not detected, or it could come up with a 1, in which case the fault will be detected.

Oscillations Another complication presented by memory is the presence of oscillations. Suppose that we first apply the test vector (0, 0) to the cross-coupled NAND latch. Both NAND gates produce a logic 1 on their outputs. We then apply the test vector (1, 1) to the inputs. Now the NAND gates have 1s on both of their inputs—but not for very long. The NAND gates transform these inputs into 0s on the outputs. The 0s show up on the NAND inputs and cause the NAND outputs to go to 1s. The cycle is repetitive; the latch is oscillating. We do not know what value is on the NAND gate outputs; the latch may continue to oscillate until some different stimulus is applied on the inputs or the oscillations may eventually subside. If the oscillations do subside, there is no way to predict, from a logic description of the circuit, the final state into which the latch settles. Therefore, the NAND outputs are set to the indeterminate x.

Probable Tested Faults When we analyzed the effectiveness of the sequence of binary combinations applied in descending order, we could not claim with certainty that the S-A-1 fault on input 4 of the NAND latch would be detected. In the circuit of Figure 3.3(b) one of the data inputs is inverted and, barring a fault in the circuit, the latch will not oscillate. However, when attempting to create a test for the circuit, we encounter another problem. If the $\overline{\text{Enable}}$ input is S-A-1, then the NAND gates connected to the inverter driven by the $\overline{\text{Enable}}$ input are permanently in the 1 state;

hence it is impossible to initialize the latch to a known state. We establish indeterminate states on the latch nodes prior to the start of test pattern generation and the states remain indeterminate for the faulted circuit. If we apply stimuli to the latch and look for the correct response, the faulted circuit may just happen to be in the same state as the unfaulted circuit.

The problem just described is inherent in any finite state machine (FSM). The FSM is characterized by a set of states $Q = \{q_1, q_2, \ldots, q_s\}$, a set of input stimuli $I = \{i_1, i_2, \ldots, i_n\}$, another set $Y = \{y_1, y_2, \ldots, y_m\}$ of output responses, and a pair of mappings

$$M: Q \times I \rightarrow Q$$
$$Z: Q \times I \rightarrow Y$$

which define the next state transition and the output behavior in response to any particular input stimulus. These mappings presume knowledge of the current state of the FSM at the time the stimulus is applied. When the initial stimulus is applied, that state is unknown unless some independent means such as a reset exists for driving the FSM into a known state.

In general, if there is no independent means for initializing an FSM, and if the Clock or Enable input is faulty, then we cannot apply just a single stimulus to the FSM and detect the presence of that fault. One approach used in industry is to mark a fault as *potentially detected* if the fault-free circuit has a known value and the faulted circuit has an indeterminate value. If a fault is potentially detected some arbitrary number of times, say six times, then there is a high probability that, if the fault actually occurs, it will be detected when the test is applied to the device.

The Initialization Problem Consider the circuit of Figure 3.7. We begin with the D flip-flop in an unknown state. We then apply the input combination $A = B = E = 0$ and clock the flip-flop. The $\overline{Q}$ output is now at 1. Therefore, we can

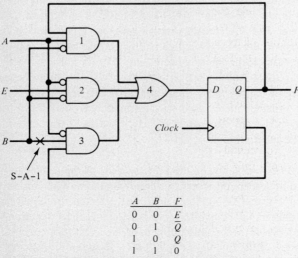

Figure 3.7 Initialization problem.

A	B	F
0	0	E
0	1	$\overline{Q}$
1	0	Q
1	1	0

3.3 TEST PROBLEMS CAUSED BY SEQUENTIAL LOGIC

clock the flip-flop a second time and we have a test for the middle input of gate 3 S-A-1. If it is S-A-1 we expect to get $F = 1$, and if it is fault-free we expect to get $F = 0$.

Unfortunately, our test has a serious flaw. If the middle input to gate 3 is S-A-1, the state of the flip-flop is indeterminate at the end of the first clock cycle. After the second clock pulse the value at F may agree with the good machine response despite the presence of the fault. The fallacy lies in assuming correct circuit behavior when setting up the flip-flop for the test. We depended upon correct behavior of the very net that we are attempting to test when setting up a test to detect a fault on that net.

To correctly establish a test we must initially assume an indeterminate value coming from the flip-flop and somehow drive the flip-flop into the 0 state ($\overline{Q} = 1$), without depending on either the input to gate 3 that is tied to the flip-flop or the faulted input. Then we can use the flip-flop value on a following test pattern, in conjunction with the primary inputs, to test for the S-A-1 fault. In this instance, we first set $A = B = 0, E = 1$. Then we clock a 1 from gate 2 into the flip-flop. This produces a zero at the $\overline{Q}$ output of the flip-flop, which can then be used with the assignment $A = B = E = 0$ to clock a 0 into the flip-flop. Now, with $\overline{Q} = 1$ and $A = B = E = 0$, another clock causes $\overline{D}$ to appear at the Q output of the flip-flop.

Notice that we used the B input. However, we used it to set up gate 2. If input B were faulted in such a way as to affect both gates 2 and 3, it could not have been used to set up the test.

3.3.2 Timing Considerations

Up to this point we have assumed that erroneous response on circuit outputs was the result of *logic* faults. These faults generally result from opens, shorts, or incorrect fabrication, for instance, an incorrect connection or a wrong component. Unfortunately, this assumption, while convenient, is an oversimplification. An error may indeed be a result of one or more logic faults, but it may also be the case that an error occurs and none of the above situations exists.

Defects exist which can prevent an element from performing in compliance with its specifications. Those faults which affect the performance of a circuit are referred to as *parametric* faults, in contrast to the logic faults that we have considered up to this point. Parametric faults can affect voltage and current levels and they can affect gain and switching speed of a circuit. Parametric faults in components can result from improper fabrication or from degradation as a consequence of a normal aging process. Environmental conditions such as temperature extremes, humidity, or mechanical vibrations may accelerate the degradation process. Design oversights can produce symptoms similar to parametric faults. Design problems include failure to take into account wire lengths, loading of devices, inadequate decoupling, and failure to consider worst-case conditions such as maximum or minimum voltages or temperatures over which a device may be required to operate. It is possible that none of these factors may cause an error in a particular design in a well-controlled environment, and yet all of these factors could, when a circuit is operating under adverse conditions, affect either the ability of the circuit to drive other circuits connected to it or the relative timing between signal paths.

Intermittent errors are particularly insidious because of their rather elusive nature, appearing only under particular combinations of circumstances. For example, a logic board may be designed for nominal signal delay for each component used in the circuit and then some fixed percentage, less than the worst case, added as a safety margin. Statistically, the delays should seldom accumulate so as to exceed a critical threshold. However, as with any statistical expectation, there will occasionally be a circuit which does exceed the maximum permissible value. Worse still, it may work well at nominal voltages and/or temperatures and fail only when voltages and/or temperatures stray from their nominal value. A new board substituted for the original board may be closer to tolerance and work well under the degraded voltage and/or temperature conditions. The original board may then, when checked at a depot on a board tester under ideal operating conditions, test satisfactorily.

Consider the effects of timing variance on the delay flip-flop of Figure 3.6. Correct operation of the flip-flop requires that the designer observe minimal setup and hold times. If the propagation delay along a signal path to the Data input of the flip-flop is greater than estimated by the designer, or if parametric faults exist, then the setup time requirement relative to the clock may not be satisfied and the clock attempts to latch the signal while it is still changing. Problems can also occur if a signal arrives too soon. The hold time requirement will be violated if a new signal value arrives at the data input before the intended value is latched up in the flip-flop. This can happen if a register directly feeds another register without any intervening logic.

The fact that logic or parametric faults can cause erroneous operation in a circuit is easy to understand, but the problems of digital logic test are further compounded by the fact that errors can occur in the operation of a device when it is behaving exactly as designed. Elements used in the fabrication of digital logic circuits contain delay. Ironically, although technologists are constantly trying to create faster circuits and reduce delay, sequential logic circuits could not function without delay; the circuits rely both on correct logical operation of the components in the circuit and on correct relative timing of signals passing through the circuit. However, this delay must be taken into account when designing and testing circuits. Suppose the inverter connected to the Data input in the circuit of Figure 3.3(b) has a delay of n nanoseconds. If the Data input makes a 0-to-1 transition and the Enable makes a transition from 0 to 1 approximately n nanoseconds later, the two cross-coupled NAND gates see an input of (0, 0) for about n nanoseconds followed by an input of (1, 1). This produces unpredictable results, as we have seen before. The problem is caused by the delay in the inverter. A solution to this problem is to put a buffer in the noninverting signal path so that the Data and $\overline{\text{Data}}$ signals reach the NANDs at the same time.

In each of the two situations just cited, the delay flip-flop and the latch, a race exists. A *race* is a condition wherein two or more signals are changing simultaneously in a circuit. The race may be caused by two or more simultaneous input signal changes, or it may be the result of a single signal change which

3.3 TEST PROBLEMS CAUSED BY SEQUENTIAL LOGIC

traverses two or more signal paths upon arriving at a fan-out point. Note that any time we have a latch or flip-flop we have a race condition, since these devices will always have at least one element whose signal both goes outside the device and feeds back to an input of the latch or flip-flop. Races may or may not affect the behavior of a circuit. A *critical race* exists if the behavior of a circuit depends on the outcome of the race. Such races can produce unanticipated and unwanted results.

3.3.3 Hazards

Unanticipated results can also occur as a consequence of logic circuit conditions which we have heretofore ignored, namely hazards. A *hazard* is the possibility of the occurrence on the output of a logic circuit of a momentary value opposite to that which is expected. Hazards can exist in combinational or sequential circuits, and they can be the result of the way in which a circuit is designed or they can be an inherent property of a function. When discussing tests for combinational logic circuits we ignored them because they generally posed no problem for testing in a strictly combinational logic environment. In sequential logic it is possible for unwanted and unexpected pulses to occur in combinational logic circuits which can lead to erroneous state transitions. Consider the circuit of Figure 3.8. If $A = B = R = 1$ and S changes from 1 to 0, then by virtue of the delay associated with the inverter, both AND gates, and subsequently the OR gate, will have a 0 output for a period corresponding to the delay of the inverter. After that period, the output of the OR gate goes back to 1, but the pulse may persist long enough to set the latch. That pulse, sometimes referred to as a *glitch*, can be avoided. By using a third AND gate to add the product term $A \cdot B$ to the sum, hence creating the function Set $= A \cdot \overline{S} + S \cdot B + A \cdot B$, the glitch is avoided.

The hazard just illustrated is called a static hazard. A *static* hazard exists if the initial and final values on a net are the same but at some intermediate time the net may assume the opposite value. If the initial and final values are 0(1), then the hazard is sometimes called a 0-hazard (1-hazard). A *dynamic* hazard exists if the initial and final values on a net are different and if, after achieving the final value, the net may assume the initial state one or more times. In other words, there is a dynamic hazard if it is possible to have $2n + 1$ transitions on a net for some integer n greater than 0. Note that the definition of a hazard only states that spurious transitions *may* occur; because of the variability of propagation delays, they may or may not actually occur.

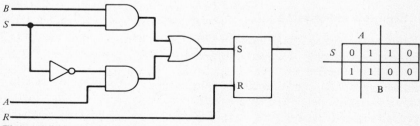

Figure 3.8 Circuit with hazard.

Hazards are also categorized as logic or function hazards. Given a function f, a p-variable *logic* hazard exists for a p-variable input change U to V if

1. $f(U) = f(V)$.
2. All 2^p values specified for f in the subcube defined by the p inputs that change are the same.
3. During the input change U to V a spurious hazard pulse may be present on the output.

The hazard illustrated in Figure 3.8 is a logic hazard. In the subcube specified by $A, S, B, R = (1, x, 1, 1)$ both values of f are 1. Eichelberger[3] proved that logic hazards can be eliminated by including all prime implicants in the implementation of a circuit. A *function* hazard exists for the function f and the input change U to V if and only if

1. $f(U) = f(V)$.
2. There exist both 1s and 0s specified for f within the 2^p cells of the subcube defined by the p inputs that changed.

Function hazards cannot be designed out of a circuit. Consider again the circuit of Figure 3.8. There is a function hazard when going from $A, S, B, R = (1, 0, 0, 1)$ to $A, S, B, R = (0, 1, 1, 1)$ because the input transition may go through the points $A, S, B, R = (0, 0, 0, 1)$ and $A, S, B, R = (0, 0, 1, 1)$ and the function f has value 0 at both points. The intermediate values assumed during operation will depend both on circuit delays and the order in which inputs change.

3.4 SEQUENTIAL TEST METHODS

We will now look at some methods that have been developed for the purpose of creating tests for sequential logic. The methods described here, though not a complete survey, are representative of the methods described in the literature and range from quite simple to very elaborate. To simplify the task, we will confine our attention in this chapter to errors caused by logic faults. Errors caused by hazards and timing-dependent errors caused by parametric faults will be discussed in subsequent chapters.

3.4.1 Seshu's Heuristics

A collection of heuristics was described in 1965 by Sundaram Seshu[4]. These involved use of trial patterns or sequences of patterns which were simulated in order to evaluate their effectiveness. The simulator was a computer program provided with a description of the components used in the circuit and their interconnections. With this description the simulation program could determine how each element in the circuit responds to an input pattern. For example, given an AND gate with two inputs, if both inputs are at a logic 1 level, it is a simple matter to deduce that

3.4 SEQUENTIAL TEST METHODS

the AND gate will respond with a logic 1 on its output. Likewise, if either or both inputs are at 0, the output will be at 0. If one input is at 1, and the other input has a logic 0 applied but is modeled to have a stuck-at-1 fault, then the AND gate again responds with a logic 1. The resulting value on the output of the AND gate is then used to evaluate the value on any other gates which have an input connected to the output of the AND gate. By proceeding in this fashion from primary inputs toward primary outputs, the response at the outputs to any logic combination applied at the inputs can be determined. In addition, by injecting faults into this computer model, we can determine the response of the circuit to the input pattern when one or more faults are present in the circuit. By comparing the response of the fault-free model to the response of the circuit model with a fault present, it can be determined whether the fault is detectable by the given input pattern. This is an oversimplification of the simulation process; we will describe simulation techniques in much greater detail in the next chapter, but for the moment we confine our attention to the methods used to create the test patterns.

Seshu identified four heuristics for creating test patterns. The test patterns created were actually trial test patterns whose effectiveness was evaluated with the simulator. If the simulator indicated that a given pattern was ineffective, then the pattern was rejected and another trial pattern was selected and evaluated. The four heuristics employed were:

> Best next or return to good
> Wander
> Combinational
> Reset

We briefly describe each of these.

Best Next or Return to Good   The best next or return to good begins by selecting an initial test pattern, perhaps one which puts the circuit into a reset state. Then, given a $(j-1)$st pattern, the jth pattern is determined by simulating all next patterns, where a *next pattern* is defined as any pattern which differs from the present pattern in exactly one bit position. The next pattern which gives best results is retained. Any other pattern which gives good results is saved in a push-down stack. If, at the jth step, no trial pattern gives satisfactory results, then the heuristic selects some other $(j-1)$st pattern from the stack and tries to generate the jth vector from it. If all vectors in the stack are discarded, the heuristic is terminated. A pattern on the stack which gave good results when placed on the stack may no longer be effective when simulating a sequential circuit because of the feedback lines. When the pattern is taken from the stack the values on the feedback lines may be totally different from the values that existed when the pattern was placed on the stack. Therefore, it is necessary to reevaluate the pattern to determine whether it is still effective.

Wander   The wander heuristic is similar to the best next in that the $(j-1)$st vector is used to generate the jth by generating all possible next vectors. However, rather than maintain a stack of good patterns, if none of the trial vectors is acceptable the heuristic "wanders" randomly. If there is no obvious choice for the next pattern, it selects a next pattern at random. After each step in the wander mode, all next patterns are simulated. If there is no best next pattern, again wander at random and try all next patterns. After some fixed number of wander steps, if no satisfactory next pattern is found, the heuristic is terminated.

Combinational The combinational heuristic ignores feedback lines and attempts to generate tests as though the circuit were strictly combinational logic by using the path sensitization technique (Seshu's heuristics predate the D-algorithm). The pattern thus developed is then evaluated against the real circuit to determine whether it is effective.

Reset The reset heuristic requires maintaining a list of reset lines. This strategy toggles some subset of the reset lines and follows each such toggle by a fixed number of next steps, using one of the preceding methods, to see if any useful information is obtained.

The heuristics were applied to some rather small circuits, the circuit limits being 300 gates and no more than 48 each of inputs, outputs, and feedback loops. In addition, the program could handle no more than 1000 faults. The best next or return to good was reported to be the most effective. The combinational heuristic was effective primarily on circuits with very few feedback loops. The system had provisions for human interaction. The test engineer could manually enter test patterns, which were then fault-simulated and appended to the automatically generated patterns. The heuristics were all implemented under control of a single program which could invoke any of them and could later call back any of the heuristics that had previously been terminated.

3.4.2 The Iterative Test Generator

The heuristics of Seshu are easy to implement but not effective for highly sequential circuits. We next examine the iterative test generator (ITG)[5], which can perhaps be thought of as an extension of Seshu's combinational heuristic. Whereas Seshu treats a mildly sequential circuit as combinational by ignoring feedback lines, the iterative test generator transforms a sequential circuit into an iterative array by identifying and cutting feedback lines in the computer model of the circuit. At the point where these cuts are made, pseudo-inputs SI and pseudo-outputs SO are introduced so that the circuit appears combinational in nature. The new circuit C contains the pseudo-inputs and pseudo-outputs as well as the original primary inputs and primary outputs. This circuit is replicated p times and the pseudo-outputs of the ith copy are identified with the corresponding pseudo-inputs of the $(i + 1)$st copy.

The D-algorithm is applied to the circuit C_p consisting of the p copies. A fault is selected in the jth copy and the D-algorithm tries to generate a test for the fault. If the D-algorithm assigns a logic value to a pseudo-input during justification,

3.4 SEQUENTIAL TEST METHODS

that assignment must be justified in the $(j-1)$st copy. However, the D-algorithm is restricted from assigning values to the pseudo-inputs of the first copy. These pseudo-inputs must be assigned the x state. The objective is to satisfy all requirements on feedback lines without assuming known values on any feedback lines at the start of the test for a given fault. From the jth copy, the test generator tries to propagate a D or $\overline{D}$ forward until, in some copy C_m, $m \leq p$, the D or $\overline{D}$ reaches a primary output. If a pseudo-output but not a primary output is reached, propagation must continue until either a D or $\overline{D}$ reaches a primary output or the last copy C_p is reached, in which case the test pattern generator gives up.

The first step in the processing of a circuit is to "cut" the feedback lines in the circuit model. To assist in this process, weights are assigned to all nets, subject to the rule that a net cannot be assigned a weight until all its predecessors have been assigned weights, where a *predecessor* to net n is a net connected to an input of the logic element which drives net n. The weights are assigned according to the following procedure:

1. Define for each net an intrinsic weight IW equal to its fan-out minus 1.
2. Assign to each primary input a weight $W = IW$.
3. If weights have been assigned to all predecessors of a net, then assign to the net a weight equal to the sum of the weights of its predecessors plus its intrinsic weight.
4. Continue until weights have been assigned to all lines that can be weighted.

If all nets are weighted the procedure is done. If there are nets not yet weighted, then loops exist. The weighting process cannot be completed until the loops are cut, but in order to cut the loops we must first identify them and then define a procedure that can identify points in the loops at which to make the cuts.

We define, for a set of nets S, a subset S_1 of nets of S to be a *strongly connected component* (SCC) of S if

> For each pair of nets l, m in S_1 there is a directed path connecting l to m
>
> S_1 is a maximal set

To find an SCC, select an unweighted net n and create from it two sets $B(n)$ and $F(n)$. The set $B(n)$ is formed as follows:

a. Set $B(n)$ initially equal to $\{n\} \cup \{$all unweighted predecessors of $n\}$.
b. Select $m \in B(n)$ for some m not yet processed.
c. Add to $B(n)$ the unweighted predecessors of m not already contained in $B(n)$.
d. If $B(n)$ contains any unprocessed elements, return to step b.

The set $F(n)$ is formed similarly, except that it is initially the union of n and its unweighted successors, where the *successors* of net m are nets connected to the outputs of gates driven by net m. When selecting an element m from $F(n)$ for processing, its unweighted and previously unprocessed successors are added to $F(n)$. The intersection of $B(n)$ and $F(n)$ defines an SCC.

Continue forming SCCs until all unweighted nets are contained in an SCC. At least one SCC must exist for which all predecessors, that is, inputs which originate from outside the loop, are weighted (why?). Once we have identified such an SCC, we make a cut and assign weights to all nets that can be assigned weights, then make another cut if necessary and assign weights, until all nets in S_1 have been weighted. The successor following the cut is assigned a weight that is one greater than the maximum weight so far assigned. Any other nets that can be assigned weights are assigned according to step 3 above. When the SCC has been completely processed, select another SCC (if any remain), using the same criteria, and continue until all SCCs have been processed.

The selection of a point in an SCC A at which to make a cut requires assignment of a period to each net in A. The *period* for a net k is the length of the shortest cycle containing k. Let B represent a subset of nets of minimum period within A. If B is identical to A, then select a net g in A which feeds a net outside A and make a cut on the net connecting g with the rest of A.

If B is a proper subset of A, then consider the set U of nets in $A - B$ that have some predecessors weighted. Let $U_1 \subseteq U$ be the set of nearest successors of B in U. Then U_1 is the set of candidate nets, one of whose predecessors will be cut. Select an element in U_1 fed by a weighted net of minimal weight. Since the weights assigned to nets indicate relative ease or difficulty of controlling the nets, gates with input nets that have low weights will be the easiest to control; hence a cut on a net feeding such a gate should cause the least difficulty in controlling the circuit.

EXAMPLE ■

We will use the JK flip-flop shown in Figure 3.5 to illustrate the cut process. We start, according to step 1, by assigning an intrinsic weight to each net. (We identify each net number with the number of the gate or primary input which drives it.)

1	2	3	4	5	6	7	8	9	10	11	12	13	14
2	0	2	0	2	1	0	0	1	1	0	0	1	1

Next, assign weights:

1	2	3	4	5	6	7	8	9	10	11	12	13	14
2	0	2	0	2	3								

From step 2 we determine that net 6 must be assigned a weight of 3. At this point no other net can be assigned. The unweighted successors of the weighted nets consist of the set

$$A = \{7, 8, 9, 10, 11, 12, 13, 14\}$$

A net is chosen and its SCC is determined. If we arbitrarily choose line 7, we find that its SCC is the entire set A. Since the SCC is the only loop in

3.4 SEQUENTIAL TEST METHODS

the circuit, all predecessors of the SCC are weighted, so processing of the SCC can commence.

We compute the periods of the nets in the SCC and find that nets 9, 10, 13, and 14 have period 2. Therefore, $B = \{9, 10, 13, 14\}$. In the set $A - B = \{7, 8, 11, 12\}$ all nets have at least one weighted predecessor, so $U = A - B$. It also turns out that $U_1 = U$ in this case. We select a net in U_1 that has a predecessor of minimal weight, say net 7. We therefore cut net 14 between gate 14 and gate 7.

The maximum weight assigned up to this point was 3. Therefore, we assign a weight of 4 to line 7. We cannot assign weights to any additional lines because loops still exist. The SCC is

$$A = \{8, 9, 10, 11, 12, 13, 14\}$$

We repeat the process and this time we make a cut from gate 13 to gate 8. We then assign a weight of 5 to line 8. This leaves two SCCs, $C = \{9, 10\}$ and $D = \{13, 14\}$. We must choose C because D has unweighted predecessors. We make a cut from 9 to 10. We assign a weight of 6 to line 10 and a weight of $2 + 4 + 6 + 1 = 13$ to line 9. We can now assign weights to lines 11 and 12. Line 11 is assigned a weight of $13 + 3 + 0 = 16$ and line 12 is assigned a weight of 9. Finally, a cut is made from 13 to 14. Line 14 is given a weight of 17 and 13 is given a weight of 36. ∎∎

We illustrate the ITG by means of the circuit in Figure 3.9. The original circuit had one feedback line from the output of J to the input of H which was cut and replaced by a pseudo-input SI and a pseudo-output SO. We will designate the logic gates and primary inputs with letters and will append a number to the gates to indicate which copy of the replicated circuit is being referenced during discussion.

We start by assigning an S-A-1 fault to the output of gate E. We want to get a $\overline{D}$ on that net; therefore, starting with replica 2, we assign $A_2 = 1$. The output of E fans out to gates F and G and here the ITG, as described by Putzolu and Roth[5], reverts to the sensitized path method and chooses a single propagation path based on weights assigned during the cut process. The weights influence the path selection process, with the objective being to attempt to propagate through the easiest apparent path. In this instance, the path through gate F_2 is initially selected. This requires a 0 from D_2, which in turn requires a 1 on input B_2. Propagation through K_2 requires a 1 from J_2 and hence 0s on input C_2 and gate H_2. The 0 on H_2 requires that the pseudo-input SI_2 be a 1. Because of the presence of a non-x value on a pseudo-input, we must back up to a previous time image.

To get a 1 on the pseudo-output of J_1 requires 0s on both of its inputs. A 0 from H_1 necessitates a 1 on one of its inputs. We avoid the pseudo-input and try to assign $G_1 = 1$. To get $G_1 = 1$ we need $E_1 = 0$, but E_1 is S-A-1. We cannot now, in this copy, assume that the output of E_1 is fault-free. Since it is assumed S-A-1, we could assign a $\overline{D}$, but that places a D and an x on H_1, a combination for which there is no entry in the D-algorithm intersection tables. The other alternative is to assign a 1 to the pseudo-input, but then we are no better off because

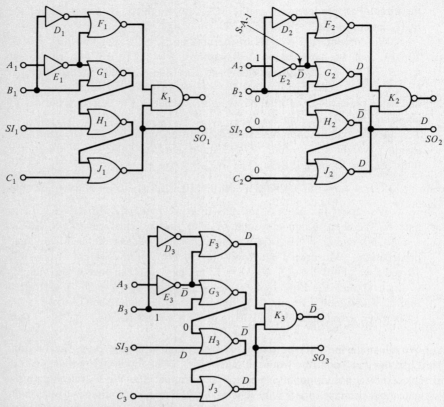

Figure 3.9 Iterated pseudo-combinational circuit.

we encounter the identical situation in the next previous time image. In practice, a programmed implementation may actually try to justify through the pseudo-input and go into a potential infinite loop. An implementation must therefore impose a maximum number of previous time images. If all assignments are not justified by the time it reaches the maximum time image, it must either give up on that fault or determine whether an alternative path exists through which to propagate the fault. In the present case, we can try to propagate through G_2.

To propagate through G_2 requires $B_2 = 0$. Then, propagation through H_2 requires a 0 on the pseudo-input and propagation through J_2 requires $C_2 = 0$. Now, however, by implication we find that $F_2 = 0$ so we cannot propagate through K_2. Therefore, we propagate through the pseudo-output SO_2. The 0 on SI_2 is justified by means of a 0 on J_1. That is justified by putting a 1 on primary input C_1.

We now have a D on the pseudo-input of time image 3. We assign $G_3 = 0$ and $C_3 = 0$; this places a D on the output of J_3. We set $B_3 = 1$ to justify the 0 from G_3 and we then try to propagate the D on J_3 through K_3 by assigning $F_3 = 1$. This requires $D_3 = E_3 = 0$. We again find ourselves trying to set the faulted line to a 0. But this time we set it to $\overline{D}$, which causes D to appear on the output of F_3 and hence both inputs to K_3 are D. Therefore, the output of K_3 is $\overline{D}$. Our final sequence of inputs is:

3.4 SEQUENTIAL TEST METHODS

	T_1	T_2	T_3
A	x	1	1
B	x	0	1
C	1	0	0

On the first time image T_1 we notice that two of the inputs have x values. We assign values to these inputs according to the following rule: if the jth coordinate of the ith pattern is an x, then set it equal to the value of the jth coordinate on the first pattern number greater than i for which the jth coordinate has a non-x value. If no pattern greater than i has a value in the jth coordinate position, assign the most recent preceding value. If the jth coordinate is never assigned, then set it to the dominant value; that is, if the input feeds an AND gate set it to 0 and if it feeds an OR gate set it to 1. The objective is to minimize the number of input changes required for the test and hence minimize or eliminate races.

The reader may have noted that after the xs were changed in time image T_1, the cross-coupled NOR latch received the input combination (1, 1). According to its state table that is an illegal input combination. Automatic test pattern generators occasionally assign illegal or illogical combinations when processing sequential circuits. It is one of the reasons why test patterns generated for sequential circuits are normally verified through simulation.

3.4.3 The Nine-Value Iterative Test Generator

In creating a test using the ITG it sometimes happens that we impose more restrictions than are absolutely necessary. Consider again the circuit of Figure 3.9. We started by propagating a test through gate F. That would not work, so we propagated through G. If we look again at the problem and examine the immediate effects of propagating a test through gate F, we notice that the faulted machine, because it produces a 0 on the upper input when $A = B = 1$, will produce a 1 on the output of K regardless of what value occurs on the lower input of K.

The D that was propagated to K implies that the fault-free machine will have a 1 on the upper input to gate K, and therefore the output of K, for the unfaulted machine, depends on the value at the lower input to K. Since we want a sensitized signal on the output of K, we require the fault-free machine to produce a 0 on the output; therefore we want a 1 on its lower input.

A logic 1 can be produced on the lower input to K by forcing J to produce a 1. This requires both of its inputs to be 0, and that in turn requires the output of H to be 0. Backing up one more step in the logic, we find that H is a 0 if either the pseudo-input or G is a 1. Gate G cannot be a 1 because primary input B is 1. Therefore, we must get a 1 from the pseudo-input. That is the point where we previously failed. The presence of the fault made it impossible to initialize the cross-coupled latch. Nevertheless, we will try again. However, this time we are going to ignore the existence of the fault in the previous copy since we are only concerned with justifying a signal for the good machine.

We create a previous time image and attempt to justify a 1 on its pseudo-output. That can be obtained with $C = 0$ and $G = 1$, which requires $B = E = 1$, and implies that $A = 0$. Therefore, a successful test is $I_1 = (1, 0, 0)$ and $I_2 = (1, 1, 0)$.

In order to distinguish between assignments required for faulted and unfaulted machines, a nine-value algebra is used[6]. The definition of the nine values is shown in Table 3.1. The dashes correspond to unspecified values. The final column shows the corresponding values for the D-algorithm. It is readily seen that the D-algorithm symbols are a subset of the nine-value ITG symbols. Table 3.2 defines the AND, OR, and Invert operations on these signals.

To illustrate the use of the tables, we employ the same circuit but start by assigning S_0 to the output of E (see Figure 3.10). The signal is propagated to the upper input of K, where, due to signal inversions, it becomes S_1. To propagate an S_1 through the NAND, we check the table for the AND gate. With an S_1 on one of its inputs, a sensitized signal S_1 can be obtained at the output of the AND by placing either an S_1, a G_1, or a 1 on the other input. The inversion then causes the output of the NAND to become S_0. The signal G_1 is the least restrictive of the signals that can be placed on the other input since it imposes no requirements on the input for the faulted machine.

During propagation we require a signal on the other input of F that will not block the sensitized signal. From the table for the OR, we confirm that propagation through F is successful with G_0 on the other input. That imples a G_1 on the input of gate D. Since the input to D is a primary input, the signal is converted to 1. To justify G_1 from J, we require G_0 from each of its inputs. Therefore, we need a G_0 from gate H, which implies that we need a 1 at an input to H. The output of G is 0 so we must get G_1 from the pseudo-input. We create a previous time image and require a G_1 from J. We then need G_0 from primary input C and also from H. That implies a G_1 from one of the inputs to H, which implies G_0 on both inputs to G. A G_0 from inverter E is obtained by creating G_1 on its input.

When justifying assignments, different values may be required on different paths emanating from a gate with fan-out. These may or may not conflict, depending on the values required along the two paths. If one path requires G_1 and the other requires S_1, then both requirements can be satisfied with signal S_1. If one path requires G_1 and the other requires S_0, then there is a conflict because G_1 requires the unfaulted machine to produce a logic 1 at the net and S_0 requires the unfaulted machine to produce a logic 0.

3.4.4 The Critical Path

The critical path method described in the previous chapter has sequential as well as combinational circuit processing capability. We will describe its operation on a sequential circuit by means of an example, using the JK flip-flop of Figure 3.5. Recall that the critical path begins by assigning a value to an output and then working its way to the input pins, creating a critical path along the way. Therefore, we start by assigning a 0 to the output of gate 13. This puts critical 1s on the inputs to gate 13, any one of which failing to the opposite state will cause an erroneous output.

We then select gate 11. To get a 1 from gate 11 we assign a 0 to gate 6. To make it critical we assign a 1 to gate 9. The assignment of a 0 to gate 6 forces assignment of 1s to input 3 and gate 12. Gate 14 is selected next. Since gate 13

3.4 SEQUENTIAL TEST METHODS

TABLE 3.1 SYMBOLS FOR NINE-VALUE ITG

Good	Faulted	ITG symbol	D symbol
0	0	0	0
0	x	G_0	—
0	1	S_0	$\overline{D}$
x	0	F_0	—
x	1	F_1	—
x	x	U	x
1	0	S_1	D
1	x	G_1	—
1	1	1	1

is a 0 and gate 12 is a 1, we can create a critical 0 by assigning a 1 to input 5. The presence of a 0 on gate 13 also implies a 1 on the output of gate 8, hence gate 10 has a 0 on its output. To insure that gate 9 has a 1, we assign a 0 to gate 7. That in turn requires input 2 be assigned a 1.

TABLE 3.2 LOGIC OPERATIONS ON NINE VALUES

	0	G_0	S_0	F_0	U	F_1	S_1	G_1	1
0	0	0	0	0	0	0	0	0	0
G_0	0	G_0	G_0	0	G_0	G_0	0	G_0	G_0
S_0	0	G_0	S_0	0	G_0	S_0	0	G_0	S_0
F_0	0	0	0	F_0	F_0	F_0	F_0	F_0	F_0
U	0	G_0	G_0	F_0	U	U	F_0	U	U
F_1	0	G_0	S_0	F_0	U	F_1	F_0	U	F_1
S_1	0	0	0	F_0	F_0	F_0	S_1	S_1	S_1
G_1	0	G_0	G_0	F_0	U	U	S_1	G_1	G_1
1	0	G_0	S_0	F_0	U	F_1	S_1	G_1	1

AND

	0	G_0	S_0	F_0	U	F_1	S_1	G_1	1
0	0	G_0	S_0	F_0	U	F_1	S_1	G_1	1
G_0	G_0	G_0	S_0	U	U	F_1	G_1	G_1	1
S_0	S_0	S_0	S_0	F_1	F_1	F_1	1	1	1
F_0	F_0	U	F_1	F_0	U	F_1	S_1	G_1	1
U	U	U	F_1	U	U	F_1	G_1	G_1	1
F_1	F_1	F_1	F_1	F_1	F_1	F_1	1	1	1
S_1	S_1	G_1	1	S_1	G_1	1	S_1	G_1	1
G_1	G_1	G_1	1	G_1	G_1	1	G_1	G_1	1
1	1	1	1	1	1	1	1	1	1

OR

X	0	G_0	S_0	F_0	U	F_1	S_1	G_1	1
Y	1	G_1	S_1	F_1	U	F_0	S_0	G_0	0

Invert

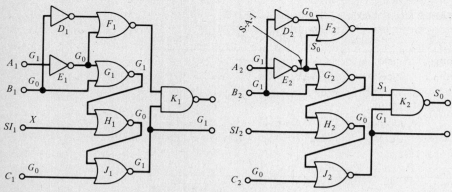

Figure 3.10 Testing with the nine-value ITG.

Notice that the loop consisting of {13, 14} has 1s on all predecessor inputs, while the loop {9, 10} is forced to its state by the 0 on gate 7. Since the inputs to loop {13, 14} cannot force it to its state, the loop must be initialized to its state by some previous pattern. Therefore, the loop {13, 14} becomes the initial objective of a preceding pattern. An assignment of 0 to input 5 and a 1 to inputs 1 and 3 forces the latch to the correct state.

One additional operation is performed here. The Clear input to gate 14 is made critical by reversing the values on the loop {13, 14} in a previous third time image. The Preset is set to 0 and the Clear is set to 1. The complete input sequence then becomes:

	T_1	T_2	T_3
1	0	1	1
2	x	x	1
3	1	1	1
4	x	x	x
5	1	0	1

The pattern at time T_1 resets the latch {13, 14}. The pattern at time T_2 sets the latch, hence the 0 on input 5 at time T_2 is critical. Then, at time T_3, there is a critical path from input 3, through gates 6, 11, and 13. A failure on that path will cause the latch {13, 14} to switch to the opposite state.

3.4.5 The Sequential Path Sensitizer

We next look at a system called the sequential path sensitizer (SPS)[7]. Like the iterative test generator, it creates tests by sensitizing paths through space and time. However, it does not model memory elements such as delay and JK flip-flops as interconnections of gate-level elements; it actually does just the reverse. Not only are functional memory elements retained in their functional form, but all combi-

3.4 SEQUENTIAL TEST METHODS

national logic between the memory device and the outputs of other memory devices or primary inputs is gathered up and combined with the destination flip-flop to create a "super flip-flop." In addition, all combinational logic from a primary output to primary input pins and memory elements is treated as a "super output block." D-cubes are defined for these super flip-flops. The basic memory element state transition properties are used to derive the state transition properties of the super flip-flops.

In another departure from convention, SPS does not explicitly model faults. Rather, it creates sensitized paths from outputs back to inputs via sequences of input vectors[8]. It then propagates 0 and 1 signals along the path. If an erroneous response occurs, then the fault lies along the path or on some attendant path used to set up the critical path. Path intersection is used to isolate the cause of the erroneous response.

The following assumptions are made concerning the nature of the circuit to be tested:

1. Memory elements are binary latches which are permitted to have unclocked Set and Reset lines.
2. All memory elements are explicitly identified.
3. State transitions do not depend on critical races in the combinational logic.
4. Memory elements along a path can be initialized.
5. State changes can be propagated along a path by using controllable input sequences which do not permit state changes to enter simultaneously via other routes,
6. Logic structures which require adaptive diagnosing experiments are precluded.

A word about adaptive diagnosing experiments—these are test sequences for which the inputs at time $n + 1$ are computed dynamically based on observations of the machine response to previous inputs. In fact, all of the test methods described in this chapter are based on this same assumption. We will have more to say about adaptive sequences in a following section.

We begin by considering the behavior of a JK flip-flop with output F and inputs J, K, R, S, and C, where the flip-flop is negative edge-triggered, that is, the 1-to-0 transition of the clock signal C latches the data on the J and K inputs, and the S and R inputs set or reset the flip-flop, respectively, when high. The JK flip-flop is capable of four distinct operations identified as Set, Reset, Toggle, and At-Rest, denoted by the symbols σ, ρ, τ, and α. Equations which express these actions are:

Set: $$\sigma = S \cdot \overline{R}(\overline{J \cdot \overline{K} \cdot C/\overline{C}}) + J \cdot \overline{K} \cdot \overline{S} \cdot \overline{R} \cdot C/\overline{C} \tag{3.1}$$

Reset: $$\rho = \overline{S} \cdot R(\overline{\overline{J} \cdot K \cdot C/\overline{C}}) + \overline{J} \cdot K \cdot \overline{S} \cdot \overline{R} \cdot C/\overline{C} \tag{3.2}$$

Toggle: $$\tau = J \cdot K \cdot \overline{S} \cdot \overline{R} \cdot C/\overline{C} \tag{3.3}$$

At-Rest: $$\alpha = \overline{J} \cdot \overline{K} \cdot \overline{S} \cdot \overline{R} + \overline{S} \cdot \overline{R} \cdot \overline{C/\overline{C}} \tag{3.4}$$

In these equations, $C/\overline{C}$ denotes a true-to-false transition of the clock and $\overline{C/\overline{C}}$ denotes absence of the true-to-false clock pulse. A complete set of state transitions can be expressed in terms of the preceding four equations. These yield:

$$F(i+1)/1 = \sigma + \tau \overline{F}(i) + \alpha F(i) \qquad (3.5)$$

$$F(i+1)/0 = \rho + \tau F(i) + \alpha \overline{F}(i) \qquad (3.6)$$

where $F(i)/1$ indicates that F is true at time i and $F(i)/0$ indicates that F is false at time i. Equation (3.5) states that we obtain a true output at time $i+1$ if we perform a set, or if we toggle the flip-flop when it was originally in a false state, or if it is true and we leave it at rest. Equation (3.6) is interpreted similarly.

From these equations primitive D-cubes can be derived which are then used to define local transition conditions for the super flip-flops. They constitute a covering set of cubes for the σ, ρ, τ, and α state control equations. Some of the D-cubes are listed in the following table:

F	S	R	J	K	C	Initial state F	Equation/term
D	D	0	x	x	0	0	3.1/1
D	D	0	0	0	x	0	3.1/1
D	D	0	x	x	1	0	3.1/1
D	0	0	D	0	1/0	0	3.1/2
D	0	0	1	0	$D/0$	0	3.1/2
D	0	0	1	1	$D/0$	0	3.3
$\overline{D}$	0	D	x	x	0	1	3.2/1
$\overline{D}$	0	D	0	0	x	1	3.2/1
$\overline{D}$	0	D	x	x	1	1	3.2/1
$\overline{D}$	0	0	0	$\overline{D}$	1/0	1	3.2/2
$\overline{D}$	0	0	0	1	$D/0$	1	3.2/2
$\overline{D}$	0	0	1	1	$D/0$	1	3.3

Corresponding to the D-cubes listed in the table is a set of inhibit D-cubes, which can be obtained by complementing all of the D and $\overline{D}$ terms. The final column in the table indicates the derivation of the D-cube. For example, the first D-cube was derived from the first term of Eq. 3.1. The interpretation of each entry is similar to the interpretation for the D-cubes of the D-algorithm. The first D-cube states that with the Clock and the Reset at 0, and the flip-flop output F at 0, the output F is sensitive to a D on the Set input. The coordinates within each cube are grouped in terms of output variables, internal variables, and controllable input variables. The cubes for a given output condition are arranged in hierarchical order, corresponding inversely to the number of non-x state memory variable coordinates in the cube required to facilitate generation of initializing sequences.

In all, four distinct activities are defined for the sequential path sensitizer; these are:

1. Identify the super flip-flops and super output block. Determine D-cubes for each of these super logic blocks.
2. Trace super logic block D-cubes to define sequential D-chains which define sequential machine propagation paths.

3.4 SEQUENTIAL TEST METHODS

3. Determine an exercise sequence for each sequential logic D-chain.
4. Determine an initialization sequence for each sequential logic D-chain.

In the first step, after defining the super logic blocks as described earlier and developing the D-cubes for the basic memory elements, this information is used to develop the D-cubes for the super logic blocks by extending the basic memory element D-cubes through the preceding combinational logic.

In the second step, beginning with a super logic block D-cube which generates an observable machine output, proceed as in the D-algorithm to chain D-cubes back to inputs. During this justification phase other super flip-flops may be reached which are inputs to the one being processed. The super flip-flops are chained as in the D-algorithm but by means of an extended set of symbols in order to permit computation of machine-level state transitions. The symbols used and their intersection rules are given in Table 3.3. An explanation of the symbols follows the table.

In Table 3.3 some symbols are identified as input symbols and some are identified as output symbols. The output symbols identify possible states of super flip-flops which correspond to possible states of the latch or JK flip-flop from which the super flip-flop was derived. Therefore, the outputs of these super flip-flops are expressed in terms of true and false final states, toggles, and at-rest conditions.

When using Table 3.3 to intersect an input value with an output value, the result provided by the table is a flip-flop output value which is compatible with input requirements on the element(s) driven by that flip-flop. For example, if element inputs connected to a net require a logic 1 in the present time frame, then

TABLE 3.3   INTERSECTION TABLE

	$\bar{D}$, 0	D, 1	x	$D/0$, $1/0$	$\bar{D}/1$, $0/1$	d	$\bar{d}$	T	$\bar{T}$	t	$\bar{\imath}$	A	$\bar{A}$
$\bar{D}$, 0	0	*	0	*	0/1	*	$\bar{d}$	*	$\bar{T}$	$\bar{A}$	$\bar{T}$	*	$\bar{A}$
D, 1	*	1	1	1/0	*	d	*	T	*	T	A	A	*
x	0	1	x	1/0	0/1	d	$\bar{d}$	T	$\bar{T}$	t	$\bar{\imath}$	A	$\bar{A}$
$D/0$, $1/0$	*	1/0	0/1	1/0	*	*	*	T	*	T	*	*	*
$\bar{D}/1$, $0/1$	0/1	*	0/1	*	0/1	*	*	*	$\bar{T}$	*	$\bar{T}$	*	*
d	*	d	d	*	*	d	*	*	*	*	$\bar{\imath}$	A	*
$\bar{d}$	$\bar{d}$	*	$\bar{d}$	*	*	*	$\bar{d}$	T	*	t	*	*	$\bar{A}$
T	*	T	T	T	*	*	T	—	—	—	—	—	—
$\bar{T}$	$\bar{T}$	*	$\bar{T}$	*	$\bar{T}$	$\bar{T}$	*	—	—	—	—	—	—
t	$\bar{A}$	T	t	T	*	*	t	—	—	—	—	—	—
$\bar{\imath}$	$\bar{T}$	A	$\bar{\imath}$	*	$\bar{T}$	$\bar{\imath}$	*	—	—	—	—	—	—
A	*	A	A	*	*	A	*	—	—	—	—	—	—
$\bar{A}$	$\bar{A}$	*	$\bar{A}$	*	*	*	$\bar{A}$	—	—	—	—	—	—

Inputs:
1 = true state
0 = false state
x = don't care
1/0 = true-to-false transition
0/1 = false-to-true transition
$\bar{D}, D, D/0, \bar{D}/1$ = D-states
$d, \bar{d}$ = asynchronous D-inputs

Outputs:
$\bar{\imath}$ = true final state
t = false final state
$\bar{T}$ = 0/1 toggle
T = 1/0 toggle
A = true at rest
$\bar{A}$ = false at rest
* = prohibited state

that value can be justified by a flip-flop which is true at rest, A, or one which is presently true but which will toggle to false on the next time frame, either t or T. The symbols t and T have identical meaning during the exercising sequence; they differ slightly during the initializing sequence, as will be explained later. The dashes indicate impossible conditions and the asterisks correspond to conflicting choices, as in the original D-algorithm.

When intersecting D-cubes, some rules must be followed. These include the following:

1. No memory variable output may be left with a $1/0$, $0/1$, $\overline{D}/1$, or $D/0$ state.
2. There must be no d or $\overline{d}$ terms left on the memory variable coordinates of a resultant cube.
3. Cubes that are asynchronously coupled via unclocked inputs must be intersected in the same time frame.

If a toggle state occurs, additional cubes must be combined with the original cube in order to completely define that step of the sequence. Cubes that are coupled by means of a d or $\overline{d}$ or by means of unclocked inputs must be combined via intersection.

We will use the circuit in Figure 3.11 to illustrate the techniques employed. Cubes are chained from the output back toward inputs, and these are used to create an initializing and exercising sequence for the propagation path.

We begin by identifying the super flip-flops and the super output block. The super output consists of a single AND gate, which we label as block Z. There are two JK flip-flops and a Set-Reset (S-R) latch. The JK flip-flops behave according to Eqs. (3.1)–(3.6). The S-R latch is at rest when both inputs are low. It is set (output high) or reset (output low) when the corresponding input is high. The S-R

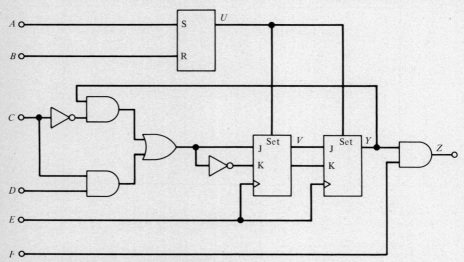

Figure 3.11 Circuit for sequential path sensitization.

3.4 SEQUENTIAL TEST METHODS

latch and flip-flop Y have no combinational logic preceding them. The JK flip-flop labeled V is preceded by an OR gate, two inverters, and two AND gates. These gates and flip-flop V are bundled together and processed as a single super flip-flop.

The second step is to create D-cubes for the four super flip-flops U, V, Y, and Z. These cubes are contained in Table 3.4 and are assigned names to facilitate the description that follows. The name consists of the letter U, V, Y, or Z that was originally assigned to the super flip-flop, complemented if necessary, followed by one of the symbols σ, ρ, τ, or α to indicate whether the action is a Set, Reset, Toggle or At-Rest. If there is more than one such entry, they are numbered.

Having created D-cubes for the super output block and the super flip-flops, we next identify sequential paths from the outputs to the inputs in order to construct an exercising sequence. If we select the first cube in the list, $Z\sigma_1$, corresponding to a true state on the output Z, we see that it requires a d on flip-flop Y, which we must now justify.

We justify the d by going across the top of Table 3.3 until we reach the column labeled d. Going down that column, we see that there appear to be six possible choices. However, only three of the entries in that column, $\bar{t}$, $\bar{T}$ and A, can be obtained from the output of a super flip-flop. Going across those rows to the left, we see that the $\bar{t}$, $\bar{T}$, and A can be created by intersection with $\bar{t}$, $\bar{T}$, and A. We then go to the set of D-cubes for Y in Table 3.4 and search for one that produces a $\bar{t}$, $\bar{T}$, or A without causing a conflict. For purposes of illustration we select $Y\sigma_2$. It requires a D from input V and a 0 from input U.

TABLE 3.4 SUPER FLIP-FLOP CUBES

Z	U	V	Y	A	B	C	D	E	F	Cube name
$\bar{t}$	x	x	d	x	x	x	x	x	1	$Z\sigma_1$
$\bar{t}$	x	x	1	x	x	x	x	x	d	$Z\sigma_2$
t	x	x	$\bar{d}$	x	x	x	x	x	1	$\bar{Z}\rho_1$
t	x	x	1	x	x	x	x	x	$\bar{d}$	$\bar{Z}\rho_2$
x	d	x	$\bar{t}$	x	x	x	x	x	x	$Y\sigma_1$
x	0	D	$\bar{t}$	x	x	x	x	1/0	x	$Y\sigma_2$
x	0	$\bar{D}$	t	x	x	x	x	1/0	x	$\bar{Y}\rho$
x	0	x	A	x	x	x	x	0	x	$Y\alpha$
x	0	x	$\bar{A}$	x	x	x	x	0	x	$\bar{Y}\alpha$
x	d	$\bar{t}$	x	x	x	x	x	x	x	$V\sigma_1$
x	0	$\bar{t}$	x	x	x	1	D	1/0	x	$V\sigma_2$
x	0	$\bar{t}$	x	x	x	1	1	$D/0$	x	$V\sigma_3$
x	0	$\bar{t}$	D	x	x	0	x	1/0	x	$V\sigma_4$
x	0	$\bar{t}$	1	x	x	$\bar{D}$	0	1/0	x	$V\sigma_5$
x	0	A	x	x	x	x	x	0	x	$V\alpha$
x	0	t	x	x	x	1	$\bar{D}$	1/0	x	$\bar{V}\rho_1$
x	0	t	$\bar{D}$	x	x	0	x	1/0	x	$\bar{V}\rho_2$
x	0	t	0	x	x	$\bar{D}$	1	1/0	x	$\bar{V}\rho_3$
x	0	$\bar{A}$	x	x	x	x	x	0	x	$\bar{V}\alpha$
x	$\bar{t}$	x	x	d	0	x	x	x	x	$U\sigma$
x	A	x	x	0	0	x	x	x	x	$U\alpha$
x	t	x	x	0	d	x	x	x	x	$\bar{U}\rho$
x	$\bar{A}$	x	x	0	0	x	x	x	x	$\bar{U}\alpha$

We go back to Table 3.3 to justify the D. From the column with header D, we see that a D occurs at the input to Y if V is true while at rest, A, or if it is presently true but toggles false, T, on the next time frame. Since there are no cubes in Table 3.4 with a T on the output of V, we check entries from Table 3.3 with A and find, by going across to the left, that they result from intersection with either an A or $\bar{\imath}$ on the output of V. From the D-cubes for V in Table 3.4 we select $V\sigma_4$. Finally, in similar fashion, we justify the 0 on U by means of cube $\bar{U}\alpha$.

We have now identified four cubes which extend a sensitized path back from the output Z to primary inputs and other elements. Before continuing, we point out that the sensitized path extends through both logic and time, since the cubes impose switching conditions as well as logic values. As a result, intersections are more complex and require attention to more detail than is the case with the D-algorithm. Some cubes must be intersected in the same time frame, and others, linked by synchronous switching conditions, are used to satisfy conditions required in the preceding time frame.

Consider the first D-cube selected, $Z\sigma_1$. It creates a $\bar{\imath}$ on the output of Z by assigning a 1 and a d to the inputs of the AND gate. The 1 is easily satisfied by assigning a 1 to the primary input F. The d, which is an asynchronous D, must be justified in the present time frame. This is done by intersecting $Z\sigma_1$ with the second cube previously selected, $Y\sigma_2$. Performing the intersection according to the rules in Table 3.3, we get:

$\bar{\imath}$	x	x	d	x	x	x	x	x	1	$Z\sigma_1$
x	0	D	$\bar{\imath}$	x	x	x	x	1/0	x	$Y\sigma_2$
$\bar{\imath}$	0	D	$\bar{\imath}$	x	x	x	x	1/0	1	

The resultant cube has a 0 on the Set input to flip-flop Y. The fourth cube previously selected, $\bar{U}\alpha$, which was selected to justify the 0 on the Set input, is asynchronously coupled to Y via the unclocked Set input. Therefore, according to our intersection rules, it must be intersected with the previous result.

$\bar{\imath}$	0	D	$\bar{\imath}$	x	x	x	x	1/0	1	
x	$\bar{A}$	x	x	0	0	x	x	x	x	$\bar{U}\alpha$
$\bar{\imath}$	$\bar{A}$	D	$\bar{\imath}$	0	0	x	x	1/0	1	

The remaining cube, $V\sigma_4$, was selected to justify a D on the input to Y. Since that input is synchronized to the clock, the cube $V\sigma_4$ becomes part of the preceding time frame.

Values on Z, U, V, and Y for this resultant cube are interpreted by using the legends at the bottom of Table 3.3. Super blocks Z, U, and Y have both a final value and a switching action specified. During an exercising sequence the $\bar{\imath}$ denotes a transition on the outputs of Z and Y from a present state of 0 to a final state of 1. The $\bar{A}$ on U denotes a super flip-flop that is false at rest; that is, its final value is false and, furthermore, its value did not change. Therefore, the Set input to Y is inactive. Super flip-flop V has a D, which is an input value; therefore no final

3.4 SEQUENTIAL TEST METHODS

value is specified for that super flip-flop. The interpretation, then, of the resultant cube is that there is an output of 1, 0, x, 1 at time $n + 1$ from the four super blocks. At time n the circuit requires the values $0, 0, 1, 0$ on the outputs of the super blocks and the values $A, B, C, D, E, F = (0, 0, x, x, 1/0, 1)$ on the primary inputs. Note that the clock value is specified as $1/0$ and is regarded as a single stimulus, although in fact it requires two time images.

We must now justify the values $(Z, U, V, Y) = (0, 0, 1, 0)$ required on the super blocks at time n. We start with our original third cube, $V\sigma_4$, which was selected to justify the D at the input to V. It puts a $\bar{t}$ on the output of V and requires a 0 on the input driven by U. Its combinational logic inputs require a 0 on input C and a D on the input from super flip-flop Y. The $\bar{t}$ represents a true final state on V and therefore satisfies the requirement imposed by the previously created pattern. However, we still need 0s on the other super flip-flops. We must justify these values without conflicting with values of the cube $V\sigma_4$.

There is already an apparent conflict. The cube specifies a D on Y and we have a requirement from the previously created cube for a 0 on Y. However, the D is an input to the super flip-flop at time $n - 1$ as specified by the cube $V\sigma_4$. The 0 is an output requirement at time n and the cube $V\sigma_4$ specifies that the flip-flop V is to perform a toggle. The apparent problem is caused by the fact that a loop exists.

We attempt to justify the 0 required on U. The cube $\overline{U}\rho$ will justify the 0. We then select $\overline{Z}\rho_1$ to get a 0 on Z and we select $\overline{Y}\rho$ to get a 0 on Y. The intersection of these cubes yields:

t	x	x	$\bar{d}$	x	x	x	x	x	1	$\overline{Z}\rho_1$
x	0	D	t	x	x	x	x	$1/0$	x	$\overline{Y}\sigma_2$
x	t	x	x	0	d	x	x	x	x	$\overline{U}\rho$
x	0	$\bar{t}$	D	x	x	0	x	$1/0$	x	$V\sigma_4$
t	$\overline{A}$	$\overline{T}$	T	0	d	0	x	$1/0$	1	

All columns except column 4, corresponding to super flip-flop Y, follow directly from the intersection table. As mentioned, the fourth column requires a $\bar{d}$ output from Y and a D input. In addition, the cube $Y\sigma_2$ requires a $1/0$ toggle. Therefore, we intersect D and t to get T and then intersect T with $\bar{d}$ to again get a T. We have completed the exercising sequence. The values $t, \overline{A}, \overline{T}, T$ satisfy the requirements for $0, 0, 1, 0$ that we set out to obtain but they in turn impose initial conditions of $1, 0, 0, 1$. We therefore must create an initialization sequence by continuing to justify backward in time until we eventually reach a point in which all of the super blocks have x states. To satisfy the assignments $1, 0, 0, 1$, we intersect the following:

$\bar{t}$	x	x	d	x	x	x	x	x	1	$Z\sigma_1$
x	t	x	x	0	d	x	x	x	x	$\overline{U}\rho$
x	0	t	x	x	x	1	$\overline{D}$	$1/0$	x	$V\rho_1$
x	0	D	$\bar{t}$	x	x	x	x	$1/0$	x	$Y\sigma_2$
$\bar{t}$	$\overline{A}$	T	$\bar{t}$	0	d	1	$\overline{D}$	$1/0$	1	

During creation of the initialization sequence, we are aided by an additional observation. The $\bar{\imath}$, which implied a true final state and a false start state while building the exercising sequence, still implies a true final state but implies an x state while constructing the initializing sequence. Therefore, we see that the values $\bar{\imath}, \bar{A}, T, \bar{\imath}$ on the super blocks satisfy the $1, 0, 0, 1$ requirement and also imply a previous state of $x, 0, 1, x$ on the super block outputs. Hence, two of the super blocks can be ignored.

To get the previous state in which $U = 0$ and $V = 1$, we intersect:

x	$\bar{A}$	x	x	0	0	x	x	x	x	$\overline{U}\alpha$
x	0	$\bar{\imath}$	x	x	x	1	D	$1/0$	x	$V\sigma_2$
x	$\bar{A}$	$\bar{\imath}$	x	0	0	1	D	$1/0$	x	

Again, the $\bar{\imath}$ satisfies the requirement for $V = 1$ and specifies a previous don't-care state. Since we are constructing an initializing sequence at this point, rather than an exercising sequence, we ignore the D and treat it as a logic 1. We now require a 0 on the output of super flip-flop U. We use D-cube $\overline{U}\rho$, which puts a t on the output of the flip-flop, hence a 0 preceded by don't-care state. The inputs for that cube are 0 and d. The d is again treated as a 1 because this is the initializing sequence. We are now done; at this point we go back and reconstruct the entire sequence. We get:

n	Z	U	V	Y	A	B	C	D	E	F
1	x	x	x	x	0	1	x	x	x	x
2	x	0	x	x	0	0	1	1	$1/0$	x
3	x	0	1	x	0	1	1	0	$1/0$	1
4	1	0	0	1	0	1	0	x	$1/0$	1
5	0	0	1	0	0	0	x	x	$1/0$	1
6	1	0	x	1						

3.5 SEQUENTIAL LOGIC TEST COMPLEXITY

The automatic creation of tests for sequential logic has proved to be an extremely difficult problem to solve. No theoretical basis exists comparable to the theory supporting the D-algorithm. Recall that the D-algorithm can find a test for any fault in a combinational circuit for which a test exists, given only a list of the components used in the circuit and a list of their interconnections. As we shall eventually see, no such claim can be made for sequential circuits, given the same set of conditions.

Our analysis of sequential circuits begins with the sequential circuit of Figure 3.12. Although it is a sequential circuit, there are no loops, apart from those which may exist inside the flip-flops. In fact, as we pointed out at the beginning of this chapter, memory devices need not be flip-flops at all. The circuit could be implemented with delay lines or it could be implemented with single-input AND gates in sufficient number to obtain the required amount of delay. The circuit would not behave in the same way as a circuit with clocked flip-flops, since delay lines have

3.5 SEQUENTIAL LOGIC TEST COMPLEXITY

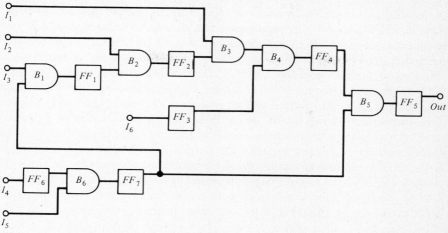

Figure 3.12 A loop-free sequential circuit.

fixed delays and the delay through a flip-flop depends on the relative timing between the clock and data signals. However, for some sequences of input signals, it might be difficult for an observer monitoring the outputs to determine whether the circuit is made up of delay lines or clocked flip-flops.

If the circuit is, in fact, made up of delay lines, then for testing purposes the circuit could be considered to be purely combinational logic. The signal on the output will initially fluctuate but will eventually stabilize and remain constant as long as the inputs are held constant. If a tester monitoring the output measures the response at a sufficiently late time relative to the total propagation time through the circuit, then the delay lines would have no more effect than wires having zero delay and could therefore be completely ignored.

Now supposing the delays are flip-flops, how much does the analysis change? Suppose we want to test for an S-A-1 fault on the input to gate B_4 coming from gate B_3. We can start by creating a test through the combinational logic in super flip-flop FF_4, where we use the concept of super flip-flop as it was used in the previous section to denote a flip-flop and all preceding combinational logic. A test for the input S-A-1 can be obtained by setting $I_1 = 0$, $FF_2 = x$, and $FF_3 = 1$. If we consider FF_4 to represent time image n, then we require a 1 on primary input I_6 in time image $n - 1$ in order to justify the 1 on FF_3 in time image n. In time image $n + 1$ we propagate through super flip-flop FF_5 by requiring $FF_7 = 1$. This can be justified by setting $I_5 = 1$ in time image n and $I_4 = 1$ in time image $n - 1$. The entire sequence then becomes:

Time	I_1	I_2	I_3	I_4	I_5	I_6	Out
$n - 1$	x	x	x	1	x	1	x
n	0	x	x	x	1	x	x
$n + 1$	x	x	x	x	x	x	x
$n + 2$							D

To summarize, we sensitized a fault in time image n, justified assignments backward in time to image $n-1$, and propagated forward in time to image $n+1$. The result appeared at the output in time image $n+2$. Of interest here is the fact that the test pattern could almost as easily have been generated by a combinational automatic test pattern generator (ATPG). The circuit has been redrawn in Figure 3.13; it has been levelized in time, and the time images are indicated. Because FF_7 fans out, it appears twice, as does its source FF_6. If we attempt to test the same fault in the redrawn circuit, we can ignore the flip-flops while computing the input stimuli and use the levelized circuit to determine time images at which the stimuli must occur. For test purposes, the complexity of this circuit is comparable to that of a combinational circuit. Since the number of test patterns for a combinational circuit with n inputs is upper-bounded by 2^n, the number of test patterns for this pseudo-combinational circuit is upper-bounded by $k \cdot 2^n$, where k is the depth of the circuit; that is, k is the maximum number of flip-flops in any path between an input and an output.

EXAMPLE ■

We create a test for the bottom input of B_4 S-A-1. The input stimuli are:

I_1	I_2	I_3	I_4	I_5	I_6
1	1	1	1/1	1/1	0

The double assignments for I_5 and I_6 represent the values at different times due to fan-out. If the destination flip-flops exist in different time images, we can permit what would normally be conflicting assignments. If the fan-out is to two or more destination flip-flops, all of which exist in the same time

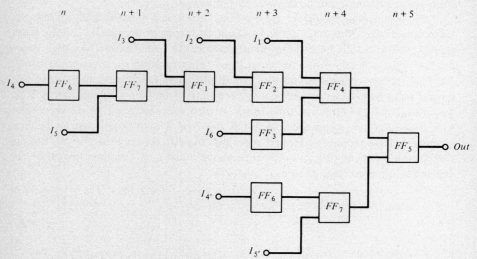

Figure 3.13 The loop-free circuit levelized.

3.5 SEQUENTIAL LOGIC TEST COMPLEXITY

image, then the assignments must not conflict. From the levelized circuit we see that the values must fall in the following time images:

Out	I_1	I_2	I_3	I_4	I_5	I_6	Time
x	x	x	x	1	x	x	n
x	x	x	x	x	1	x	$n+1$
x	x	x	1	x	x	x	$n+2$
x	x	1	x	1	x	0	$n+3$
x	1	x	x	x	1	x	$n+4$
x	x	x	x	x	x	x	$n+5$
D							$n+6$

The previously generated test sequence can be shifted three units forward in time and merged with the second test sequence to give:

Out	I_1	I_2	I_3	I_4	I_5	I_6	Time
x	x	x	x	1	x	x	n
x	x	x	x	x	1	x	$n+1$
x	x	x	1	1	x	1	$n+2$
x	0	1	x	1	1	0	$n+3$
x	1	x	x	x	1	x	$n+4$
D	x	x	x	x	x	x	$n+5$
D							$n+6$

We now make one alteration in the circuit. We eliminate input I_5 and add a connection from the output of B_5 to the input of B_6. With this one slight change the entire nature of the problem has changed. We have compounded by orders of magnitude the complexity of the problem we are trying to solve. In the original circuit the output was never dependent on inputs beyond six time frames. Furthermore, no flip-flop was ever dependent on a previous state generated in part by that same flip-flop. We have now changed all that. The four flip-flops FF_1, FF_2, FF_4, and FF_7 constitute a 16-state machine in which the present state may be dependent on inputs that occurred at any arbitrary time in the past. This can be better illustrated with the state transition graph of Figure 3.14. If we start in state S_1 the sequence 1011111... takes us to $S_2\{S_7, S_8, S_5, S_6\}*$, where the braces and asterisk denote an arbitrary number of repetitions of the four states in the braces. From the almost identical sequence 11011111..., we get the state sequence $S_2, S_3\{S_3, S_4, S_1, S_2\}*$. The corresponding output sequences are $0, 0\{0, 0, 0, 1\}*$ and $0, 1, 0\{1, 1, 0, 1\}*$, a significant difference in output response which will continue as long as the input consists of a string of 1s. In a machine with no feedback external to the flip-flops the output sequences will coincide within D time images, where D again represents the depth of the circuit.

How much effect does that feedback line have on the testability of the circuit? We will compute an upper bound on the number of test patterns required to test a state machine in which the present state is dependent on an input sequence of indeterminate length, that is, one in which present state of the memory cells is functionally dependent upon a previous state of those same memory cells.

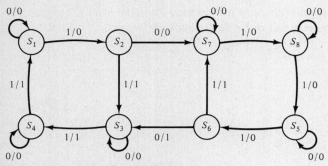

Figure 3.14 State transition graph.

Given a machine with n inputs and M states, $2^{m-1} < M \leq 2^m$, and its corresponding state table with M rows, one for each state, and 2^n columns, one for each input combination, there could be 2^n unique transitions out of each state. Hence, there could be as many as $M \cdot 2^n$, or approximately 2^{m+n}, transitions that must be verified. Given that we are presently in state S_i, and we want to verify a transition from state S_j to state S_k, it may require $M - 1$ transitions to get from S_i to S_j before we can even attempt to verify the transition $S_j \rightarrow S_k$. Thus, the number of test vectors required to test the state machine is upper-bounded by 2^{2m+n}, and that assumes we can observe the present state without requiring any further state transitions.

The argument was derived from a state table, but is there a physical realization requiring such a large number of tests? A realization can, in fact, be constructed directly from the state table. We implement the machine with m flip-flops. The outputs of these flip-flops are used to control m multiplexers, one for each flip-flop. Each multiplexer has M inputs, one for each row of the state table. Each multiplexer input is connected to the output of another multiplexer which has 2^n inputs, one corresponding to each column of the state table. The inputs to this previous bank of multiplexers are fixed at 1 and 0 and are binary m-tuples corresponding to the state assignments and the next states in the state table. In effecting state transitions, the multiplexers connected directly to the flip-flops select the row of the state table and the preceding set of multiplexers, under control of the input signal, select the column of the state table; thus the next state is selected by this configuration of multiplexers.

In this implementation there are $M \cdot 2^n$ m-tuples which must be verified, one for each entry in the state table. From the structure it can be seen that as many as $M - 1$ transitions of the state machine may be required to get the correct selection on the first bank of multiplexers. Consequently, the number of test patterns required to test this implementation is upper-bounded by 2^{2m+n}. This is not a practical way to design a state machine, but it is necessary to consider worst-case examples when establishing bounds. Of more significance, the implementation serves to illustrate the dramatic change in the nature of the problem caused by the presence of feedback lines.

3.5 SEQUENTIAL LOGIC TEST COMPLEXITY

EXAMPLE ■

Consider the machine specified by the following state table and flip-flop state assignments:

| | x | | | Q_1 | Q_0 |
	0	1			
S_0	S_0	S_2	S_0	0	0
S_1	S_3	S_2	S_1	0	1
S_2	S_1	S_0	S_2	1	0
S_3	S_2	S_3	S_3	1	1

We can implement this machine in the "canonical" form of Figure 3.15.

■ ■

3.5.1 Experiments with Sequential Machines

Early efforts[9] at testing state machines consisted of experiments aimed at determining the properties or behavior of a state machine from its state table. Such experiments consist of the application of sequences of inputs to the machine and observation of the output response. The input sequences are derived from analysis of the state table and may or may not also be conditional upon observation of the machine's response to previous inputs. Sequences in which the next input is selected using both the state table and the machine's response to previous inputs are called *adaptive* experiments. The selection of inputs may be independent of observations at the outputs. Those in which the entire input sequence is constructed from information contained in the state table, without observing machine response to previous inputs, are called *preset* experiments.

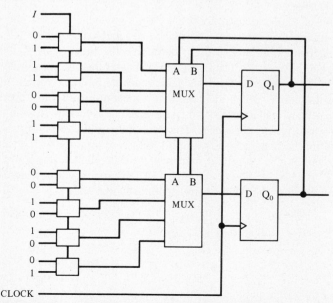

Figure 3.15 Canonical implementation of state table.

A sequence may be constructed for one of several purposes. It may be used to identify the initial or final state of a machine or it may be constructed for the purpose of forcing the machine into a particular state. Sequences which identify the initial state are called *distinguishing* sequences and those which identify the final state are called *homing* sequences. A sequence which is designed to force a machine into a unique final state independent of the initial state is called a *synchronizing* sequence (the definitions here are taken from Hennie[10]).

The creation of input sequences can be accomplished through the use of trees in which the nodes correspond to sets of states. The number of states in a particular set is termed its *ambiguity*. The root will usually correspond to maximum ambiguity, that is, the set of all states.

EXAMPLE ■

Consider the machine whose state transition properties are described by the state table of Figure 3.16. Can the initial state of this machine be determined by means of a preset experiment?

We wish to find an input sequence that can uniquely identify the initial state when we start with total ambiguity and can do no more than apply a precomputed set of stimuli and observe output response. From the state table we notice that if we apply a 0, the states A and D both produce an output of 1 and both go to state A. Quite clearly, if we create an input sequence starting with a 0, and the output response starts with a 1, it will never be possible to determine whether the machine started in state A or D. If the sequence begins with a 1, an output of 0 indicates a next state of B or E and an output of 1 indicates a next state of A, B, or C. Therefore, a logic 1 partitions the set of states into two subsets which can be distinguished by observing the output response of the machine.

Application of a second 1 further refines our knowledge because state B produces a 1 and state E produces a 0. Hence an input sequence of (1, 1) will, by working backward, enable us to determine the initial state if the output response begins with a 0. The 0 response indicates that the initial state was a C or E. If a second 0 follows, then the machine must have been in state E after the first input, which indicates that it must originally have been in state C. If the second response is a 1, then the machine is in state B, which indicates that it was originally in state E. But what if the initial response was a 1? Rather than repeat this analysis, we resort to the use of a tree which has a root with maximum ambiguity and branches corresponding to the inputs $I = 0$ and $I = 1$. We create subsets comprised of next states with set membership based on whether the output corresponding to that state is a 1 or 0.

	I	
	0	1
A	A/1	C/1
B	C/0	A/1
C	D/0	E/0
D	A/1	B/1
E	B/0	B/0

Figure 3.16 State table.

3.5 SEQUENTIAL LOGIC TEST COMPLEXITY

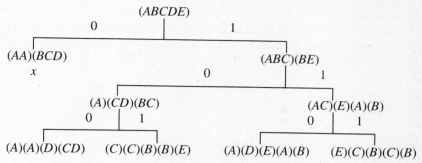

When we applied a 0 to the set with maximum ambiguity we immediately terminated the path because states A and D merged; that is, they produced the same output and went to the same next state, hence there is no reason to continue the path. When a 1 is applied, we obtain two subsets with no resulting state mergers in either subset. From this branch of the tree, if the second input is a 1, then a third input of either a 0 or 1 will take us to a leaf on the tree in which all sets are singletons. If the second input is a 0, then following that with a 1 will cause us to arrive at a leaf in which all sets are singletons. We conclude, therefore, that there are three preset distinguishing sequences of length three, namely $(1, 1, 0)$, $(1, 1, 1)$, and $(1, 0, 1)$. If we look at the result of applying the sequence $(1, 1, 0)$ to the machine in each of the five starting states, we see that we will get:

Start state	Output response	Final state
A	1 0 0	B
B	1 1 0	D
C	0 0 0	C
D	1 1 1	A
E	0 1 1	A

From the output response we can uniquely identify the start state. It must be noted that a state machine need not have a distinguishing sequence. In this example, if the machine responds with a 1 when an input of 1 is applied while in state E, then another merger would result and hence no distinguishing sequence would exist. Another terminating rule, although it did not happen in this example, is as follows: any leaf which is identical to a previously occurring leaf is terminated. There is obviously no new information to be gained by continuing along that path.

Because the distinguishing sequence identifies the initial state, it also identifies the final state, hence the distinguishing sequence is a homing sequence. However, the homing sequence is not necessarily a distinguishing sequence. Consider again the machine of Figure 3.16. We wish to find one or more input sequences that can uniquely identify the final state while observing only the output symbols. Therefore, we start again at the source node and apply a 0 or 1. However, we do not discard the path resulting from initial application of a 0 because we are now interested in final state rather than initial state, therefore state mergers do not cause loss of needed information.

EXAMPLE ■

We use the same state machine, but we only pursue the branch that was previously deleted, since the paths previously obtained are known to be homing sequences. We get:

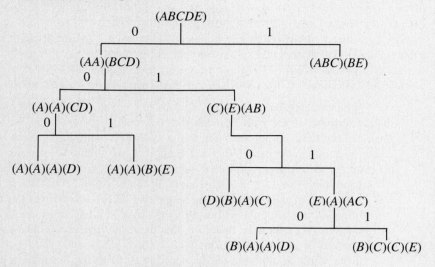

From this continuation of the original tree we see that we get several additional sequences of outputs that contain enough information to determine the final state. However, because of the mergers these sequences cannot identify the initial state and therefore cannot be classified as distinguishing sequences.
■ ■

The synchronizing sequence forces the machine into a known final state independent of the start state. We will again use the state machine of Figure 3.16 to illustrate the computation of the synchronizing sequence. As before, we start with the tree in which the root is the set with total ambiguity.

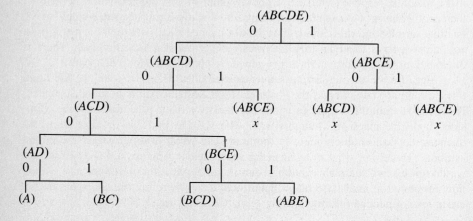

3.5 SEQUENTIAL LOGIC TEST COMPLEXITY

Starting with the total ambiguity set, we apply 0 and 1 and look at the set of all possible resulting states. With a 0 the set of successor states is (*ABCD*) and with a 1 the set of successor states is (*ABCE*). We then consider the set of all possible successor states that can result from these successor states. From the set of successor states (*ABCD*) and an input of 0 the set of successor states is the set (*ACD*). We continue until we either arrive at a singleton state or all leaves of the tree are terminated. A leaf will be terminated if it matches a previously occurring subset of states or if it properly contains another leaf that was previously terminated. In the example just given, we arrive at the state *A* upon application of the sequence $(0, 0, 0, 0)$. Other sequences exist; we leave it to the reader to find them.

We now use the same state machine to describe how we can create an adaptive homing sequence. Recall that the adaptive experiment makes use of whatever information can be deduced from observation of the output response. From the state table we know that if we apply a 0 and the machine responds with a 1, then it is in state *A* and we can stop. If a 0 response occurs, then it must be in *B*, *C*, or *D*. We can then choose to apply either a 0 or 1 as the second input. If we choose a 0, we find that with an output response of 1 the machine must again be in state *A* and with a response of 0 it must be in state *C* or *D*. Finally, with a third input we have enough information to uniquely identify the state of the machine. Adaptive experiments frequently permit faster convergence to a solution by virtue of their ability to use the additional information provided by the output response.

The distinguishing sequence permits identification of initial state by observation of output response. This is possible because the machine responds uniquely to the distinguishing sequence from each starting state. The existence of a distinguishing sequence can therefore permit a relatively straightforward construction of a *checking* sequence for a state machine. The checking sequence is intended to confirm that the state table correctly describes the behavior of the machine. It is required that the machine being evaluated not have more states than the state table that describes its behavior.

The checking sequence consists of three parts:

1. Put the machine into a known starting state by means of a homing or synchronizing sequence.
2. Apply a sequence that verifies the response of each state to the distinguishing sequence.
3. Apply a sequence that verifies state transitions not checked in step 2.

We illustrate by using the state machine of Figure 3.16. We first place the machine in state *A* by applying the synchronizing sequence. For the second step, we want to verify the response of the five states in the state table to the distinguishing sequence since that response will subsequently be used to verify state transitions. To do so we construct our sequence by appending the distinguishing sequence $(1, 1, 0)$ to the synchronizing sequence. If the machine is in state *A* it will respond to the distinguishing sequence with an output response of $(1, 1, 0)$. Furthermore, the machine will end up in state *B*. From there, state *B* can be verified by again applying the distinguishing sequence. This time the output response will be (1, 1, 0) and the machine will reach state D. A third repetition verifies state D and leaves the machine in state A, which has already been verified. Therefore, from state A a 1 is applied to put the machine into state C, where the distinguishing sequence is again applied to verify state C. Since the machine ends up in state C, a 1 is applied to cause a transition to state E. Then the distinguishing sequence is applied one more time to verify E. At that point the distinguishing sequence has been applied while the machine was in each of the five states. Assuming correct response by the machine to the distinguishing sequence when starting from each of the five states, the input sequence and resulting output sequence at this point are as follows:

```
             s.s    | d.s. | d.s. | d.s. | — | d.s. | — | d.s. |
input....0 0 0 0      1 1 0  1 1 0  1 1 0    1   1 1 0   1   1 1 0
output..........1 0 0 1 1 0  1 1 1    1   0 0 0   0   0 1 1
```

The synchronizing sequence is denoted by s.s. and the distinguishing sequence by d.s. The dashes (—) denote points in the sequence where inputs were inserted to effect transitions to states which had not yet been verified. The output values for the synchronizing sequence are unknown, hence they are omitted.

If the machine responds as indicated above, it must have at least five states because the sequence of inputs (1, 1, 0) occurred five times and produced five different output responses. Since we stipulated that it must not have more than five states, we assume that it has the same number of states as the state table. Now it is necessary to verify state transitions. We verified two transitions in step 2, namely, the transition from A to C and the transition from C to E; therefore eight state transitions remain to be verified. Since the distinguishing sequence applied when in state E leaves the machine in state A, we start by verifying the transition from A to A in response to an input of 0. We apply the 0 and follow that with the distinguishing sequence to verify that the machine made a transition back to state A. The response to the distinguishing sequence puts the machine in state B and so we arbitrarily select the transition from B to C by applying a 0. Again it is necessary to apply the distinguishing sequence after the 0 to verify that the machine reached state C from state B. The sequence now appears as follows:

```
            s.s    | d.s. | d.s. | d.s. |—| d.s. |—| d.s. |—| d.s. |—| d.s. |
input....0 0 0 0    1 1 0  1 1 0  1 1 0  1  1 1 0  1  1 1 0  0  1 1 0  0  1 1 0
output ..........1 0 0  1 1 0  1 1 1    1   0 0 0    0   0 1 1    1    1 0 0    0    0 0 0
```

We continue in this fashion until all state transitions have been confirmed. At this point six transitions have not yet been verified; we leave it as an exercise for the reader to complete the sequence.

3.5.2 A Theoretical Limit on Sequential Testability

We previously observed that the D-algorithm described by Roth[11] has been shown to be an algorithm in the strict sense. The algorithm requires no more than a structural description of the circuit, including the primitives that make up the circuit and their interconnections. In this section we show that such a claim cannot be made for sequential circuits under the same set of conditions.

3.5 SEQUENTIAL LOGIC TEST COMPLEXITY

For asynchronous circuits the pulse generator of Figure 3.17 serves as an example of such a circuit. If it comes up initially, after power is applied, in the 0 state, it will stay in the 0 state. If it comes up in the 1 state, it will reset to the 0 state by virtue of the logic 1 on the output being applied to the reset input (assumes an active high reset). Since it is known what state the circuit assumes when powered up, it can be tested for all testable faults. Simply apply power and check for the 0 state on the output. Then clock it and check the output for a positive-going pulse which returns to 0.

The simulator or ATPG, working with a structural model, initially sets all nets in the circuit to the x (don't-know) state. The x on the Q output prevents the circuit from ever going to a known state. If a simulator tries to clock in a 1, the x on the Reset line has the effect of telling it that the flip-flop's reset line may reset the output to a 0 or it may have no effect and permit the 1 to be clocked through. This inability to resolve the value on the output forces the simulator to set an x on the output. Thus, despite the fact that the circuit is testable, when only a gate or structural level description is available, the simulator or ATPG cannot get the circuit out of the unknown state.

For the class consisting solely of synchronous sequential machines, the delay flip-flop with the $\overline{Q}$ output connected to the data input, essentially an autonomous machine, is an example of a testable structure which cannot be tested by an ATPG. We know that there should be one transition on the output for every two transitions on the clock input. But again, when we initially set all nets to the indeterminate state, we preclude any possibility of predicting behavior of the circuit.

It is possible to define the self-resetting flip-flop as a primitive and specify its behavior as being normally at 0, with a pulse of some specified duration occurring at the output in response to a clock input. That, in fact, is frequently how the circuit is handled. The monostable, or single shot, is available from IC manufacturers as a single package and can be defined as a primitive.

If we model the self-resetting flip-flop as a primitive, and exclude the autonomous machine, can we show that synchronous sequential machines are testable under the same set of conditions as defined for the D-algorithm? To answer this question, we will start with the state transition graphs of Figure 3.18. One of them can be tested by a gate-level ATPG, using only structural information; the other cannot, even though both of them are testable.

The state tables for the machines of Figure 3.18(a) and (b) are shown in Figure 3.19(a) and (b), respectively. For machine A the synchronizing sequence $I = (0, 1, 0, 1, 0)$ will put the machine in state S_1. For machine B the synchronizing sequence $I = (0, 0)$ will put the machine in state S_3. The length and nature of the synchronizing sequence play a key role in determining whether the machine can be tested by the gate-level ATPG. Consider the machine shown in Figure 3.20,

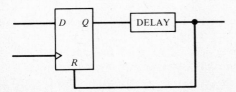

Figure 3.17 Self-resetting flip-flop.

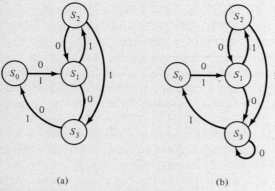

(a) (b)

Figure 3.18 State transition machines.

which is an implementation of the machine in Figure 3.18(a). Assign an initial value of (x,x) to the flip-flops labeled Q_1, Q_0. Because a synchronizing sequence of length 5 exists, we know that after the application of 5 bits we can force the machine into state S_1. However, upon application of any single stimulus, whether a 0 or 1, machine A has an ambiguity of at best 3 and possibly 4. Because the ambiguity is greater than 2, two bits are required to represent the complete set of successor states, hence simulation of any binary input value must leave both output bits, Q_1 and Q_0, being uncertain; that is, both Q_1 and Q_0 could possibly be in a 0 state or a 1 state, hence, both Q_1 and Q_0 remain in the x state.

In general, for any M-state machine, $2^{m-1} < M \leq 2^m$, implemented with m flip-flops, if a synchronizing sequence exists, the machine is testable. It is testable because the synchronizing sequence will drive it to a known state from which inputs can be applied that will reveal the presence of structural defects. In effect, the synchronizing sequence can be thought of as an extended reset; alternatively, the reset can be viewed as a synchronizing sequence of length 1. However, if no single input vector exists which can reduce ambiguity to 2^{m-1} or less, then all flip-flops are capable of assuming either binary state. Put another way, no flip-flop is capable of getting out of the indeterminate state.

Assuming that we have an input vector which can reduce ambiguity sufficiently to cause one flip-flop to assume a known value, after some additional number of inputs are applied, the ambiguity must again decrease if one or more additional flip-flops are to assume a known state. For an M-state machine implemented with m flip-flops, $2^{m-1} < M \leq 2^m$, the ambiguity must decrease to no more than 2^{m-2}. How many additional inputs can be applied before that level of ambiguity must be attained?

	0	1		0	1
S_0	S_1	S_1	S_0	S_1	S_1
S_1	S_3	S_2	S_1	S_3	S_2
S_2	S_1	S_3	S_2	S_1	S_3
S_3	S_0	S_0	S_3	S_3	S_0

Figure 3.19 State tables.

3.5 SEQUENTIAL LOGIC TEST COMPLEXITY

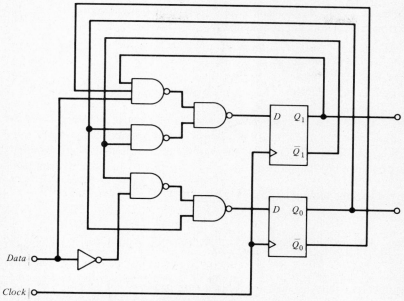

Figure 3.20 Implementation of the state machine.

Consider the situation after one input has been applied and exactly one flip-flop is in a known state. Ambiguity is then 2^{m-1}. From this ambiguity set it is possible to make a transition to a state set wherein ambiguity is further reduced, that is, additional flip-flops reach a known value, or the machine may revert back to a state in which all flip-flops are in an unknown state, or the machine may make a transition to another state set in which exactly one flip-flop is in a known state. (In practice, the set of successor states cannot contain more states than its predecessor set.) For a machine with m flip-flops, there are at most $2m$ transitions such that a single flip-flop can remain in a known state, 0 or 1. After $2m$ transitions, it can be concluded, if the ambiguity is not further resolved, that it will not be resolved, because the machine will at that time be repeating a state set which it previously visited.

Given that i flip-flops are in a known state, how many state sets exist with ambiguity 2^{m-i}? Or, to put it another way, how many distinct state sets with i flip-flops in a known state can the machine go through before ambiguity is further reduced or the machine repeats a previous state set? To compute this number, consider a single selection of i positions from an m-bit binary number. There are 2^i values that these i positions can assume. Furthermore, there are $\binom{m}{i}$ ways in which these i positions can be selected. Hence there are $2^i \cdot \binom{m}{i}$ state sets with ambiguity 2^{m-i}. Thus, the machine could conceivably transition through $2^i \cdot \binom{m}{i}$

state sets before either repeating a state set or reducing ambiguity. Hence, the synchronizing sequence is upper-bounded by

$$\sum_{i=1}^{m-1} 2^i \cdot \binom{m}{i} = 3^m - 2^m - 1$$

From the preceding, we have:

Theorem. Let M be a synchronous, sequential M-state machine, $2^{m-1} < M \leq 2^m$, implemented with m binary flip-flops. A necessary condition for M to be testable by a gate-level ATPG using only structural data is that a synchronizing sequence exists having the property that, with i flip-flops in a known state, the sequence reduces the ambiguity to 2^{m-i-1} within $2^i \cdot \binom{m}{i}$ input stimuli[12].

Corollary. The maximum length for a synchronizing sequence that satisfies the theorem is $3^m - 2^m - 1$.

The theorem tells us that a synchronizing sequence of length $\leq 3^m - 2^m - 1$ permits design of an ATPG-testable state machine. It does not tell us how to accomplish the design. In order to design the machine so that it is ATPG-testable, it is necessary that state assignments be made such that if ambiguity at a given point in the synchronizing sequence is 2^{m-1}, then the state assignments must be made so that the 2^i states in each state set with ambiguity equal to i all have the same values on the 2^{m-1} flip-flops with known values.

EXAMPLE ■

The following state machine has a synchronizing sequence of length 4.

	0	1
S_0	S_0	S_2
S_1	S_1	S_3
S_2	S_0	S_0
S_3	S_1	S_2

The synchronizing sequence is $I = (0, 1, 1, 0)$. The state sets which result from the synchronizing sequence are:

$$\{S_0, S_1\} \rightarrow \{S_2, S_3\} \rightarrow \{S_0, S_2\} \rightarrow \{S_0\}$$

If we assign flip-flop $Q_1 = 0$ for states S_0 and S_1, $Q_1 = 1$ for states S_2 and S_3, $Q_0 = 0$ for states S_0 and S_2, then simulation of the machine, as implemented in Figure 3.21, causes the machine to go into a completely specified state at the end of the synchronizing sequence. ■ ■

The importance of the proper state assignment is seen from the following assignments.

3.5 SEQUENTIAL LOGIC TEST COMPLEXITY

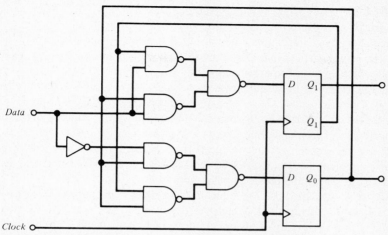

Figure 3.21 Machine with length 4 synchronizing sequence.

	Q_1	Q_0
S_0 —	0	1
S_1 —	1	0
S_2 —	1	1
S_3 —	0	0

From the synchronizing sequence we know that the value 0 puts us in either state S_0 or S_1. However, with this set of state assignments, Q_1 may come up as a 0 or 1; the same applies to Q_0. Hence, the synchronizing sequence is not a sufficient condition.

We showed the existence of a state machine with a synchronizing sequence that could not be tested by an ATPG when constrained to operate solely on structural information. It remains to show that there are infinitely many such machines. The following family has an infinite number of members, each of which has a synchronizing sequence but, when implemented with binary flip-flops, cannot be driven from the unknown to a known state because the ATPG, starting with all flip-flops at x, cannot get even a single flip-flop into a known state.

	I	
	0	1
S_0	S_1	S_0
S_1	S_2	S_0
.	.	.
.	.	.
.	.	.
S_{n-1}	S_n	S_{n-2}
S_n	S_0	S_{n-1}

3.6 SUMMARY

The presence of memory introduces a new level of complexity into the problem of automatic test pattern generation. A successful test now requires a sequence of inputs, applied in the correct order, to a circuit in which some or all of the memory elements may initially be in an unknown state. The types of faults that must be considered are now expanded. We must now be concerned not only with logic faults, but also with parametric faults, because proper behavior of a circuit depends on memory elements being updated with correct values, and that in turn frequently depends on control signals arriving in the correct order. We also saw that it is possible for a circuit to operate exactly as designed and still produce erroneous or unexpected results due to the presence of hazards in the circuit.

We looked at four methods for sequential test pattern generation, and briefly revisited the critical path method, which we examined in the previous chapter. Seshu's heuristics are primarily of historical interest, although the concept of using multiple methods, usually a random method followed by a deterministic approach, is still in use. The iterative test generator permits application of the D-algorithm to sequential logic and may well be the most commonly used test method for sequential logic. A version of the iterative test generator called the general logic analyzer simulator (GLAS), developed within IBM by R. Rasmussen, was used by this author in 1968. The nine-value ITG can minimize computations for developing a test where a circuit has fan-out. The sequential path sensitizer (SPS) extends the D-calculus to create sensitized paths through logic functions in time. Its functional latch and flip-flop models can be quite effective for sequencing of clock and data lines. Because SPS creates a sensitized path from an input, through latches and flip-flops, extending all the way to the outputs, and then attempts to propagate both a 0 and a 1 along the path, it can be effective at detecting faults in latches and flip-flops, such as the stuck-at clock line, which normally cause the model to remain in an indeterminate state.

There are other methods for sequential test pattern generation which were not covered here, some of which incorporate ideas from two or more existing techniques, such as using macro-level flip-flops within the framework of the iterative test generator. Another one which we did not discuss extends the Boolean algebra technique to sequential logic[13]. The SALT (sequential automated logic test)[14] programs detected loops in logic circuits and created state tables, where possible, for latches made up of the loops. Hence the subcircuit consisting of the loop was, in a sense, equivalent to a macro-level flip-flop. Although we have omitted some systems, the methods discussed here are representative and the reader who understands them should have less difficulty understanding other methods described in the literature.

Despite the numerous attempts to create automatic test pattern generators capable of testing sequential logic, the problem has remained intractable. While some sequential circuits are reasonably simple to test, others are quite difficult to test and some simply cannot be tested by gate-level ATPGs. A major cause of the difficulty in testing sequential circuits is the presence of feedback loops which cause the present machine state to become dependent on inputs extending an indefinite period into the past. By using the concept of synchronizing sequences, it can be shown that entire classes of testable sequential circuits exist which cannot be tested within the same set of ground rules specified by the D-algorithm. These require either that functional information be provided to the ATPG or that circuit modifications be made to permit putting the circuit into a known starting state. The implication of this is that designers must make an effort to understand testability problems and design circuits which can be testable with the tools available. We will have more to say about design for testability in a later chapter.

PROBLEMS

3.1 For the cross-coupled NAND latch, find a separate test for each of the four input S-A-1 faults. Merge these tests to find the minimal length test that can detect all four faults.

3.2 Identify a fault and a corresponding input combination in the circuit of Figure 3.2 (b) that can cause the circuit to oscillate.

3.3 For the following Karnaugh map,
(a) Identify a 1-hazard.
(b) Identify all transitions for which 1-hazards can be avoided.
(c) Find a dynamic hazard.

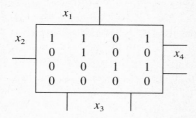

3.4 Identify a fault in the circuit of Figure 3.3(b) which will change it from a synchronous to an asynchronous circuit.

3.5 Using the D flip-flop of Figure 3.6,
(a) Create a pseudo-combinational circuit model by making cuts in the model.
(b) Assign weights to the various nodes.
(c) Create a test for the Data input S-A-1.
(d) Create a test for an S-A-1 fault on the $\overline{\text{Clear}}$ input to N_4.

3.6 Create an intersection table for the nine-value ITG. Show all possible intersections of each of the nine values with all of the others. Indicate unresolvable conflicts with a dash.

3.7 Apply the nine-value ITG to the S-A-0 fault on the output of gate 6 in Figure 2.7. Propagate the test forward along a single path.

3.8 Apply the nine-value ITG to the indicated fault in Figure 3.7. Use the circuit in Figure 3.3(b) in place of the D flip-flop and cut all loops according to the procedure described in Sec. 3.4.2.

3.9 Use the critical path method to create a test for an S-A-1 fault on the Data input of the D flip-flop of Figure 3.6.

3.10 Create a set of D-cubes for a D flip-flop primitive. Then create super flip-flops for the circuit of Figure 3.12 and use them to create a test for the bottom input of gate B_6 S-A-1.

3.11 Substitute the D flip-flop for the JK flip-flop in the circuit of Figure 3.11. Assume the existence of a set input. Duplicate the process for the path exercised in the test, using the circuit just created.

3.12 Find a synchronizing sequence for the following machine:

	0	1
S_0	S_0	S_4
S_1	S_1	S_5
S_2	S_2	S_6
S_3	S_3	S_7
S_4	S_0	S_2
S_5	S_1	S_3
S_6	S_0	S_0
S_7	S_0	S_1

3.13 The machine of figure (a) below has the synchronizing sequence 101. If the machine of figure (a) starts in state C, and the machine of figure (b) starts in state A, then the input sequence 101 causes identical responses from the two machines. Assuming the application of the sequence 101 to the two machines under the conditions just stated, find a sequence which exercises each state transition in the machine of figure (a) at least once, without verification, and which causes an identical output response from machine (b), that is, show that step 2 of the checking sequence is necessary.

	0	1
A	C/0	B/0
B	A/0	B/1
C	B/1	C/0

(a)

	0	1
A	A/0	B/0
B	C/1	C/0
C	B/0	A/1

(b)

3.14 Define an algorithm for finding a preset distinguishing sequence.

3.15 Complete the checking sequence for the example that was started in Sec. 3.5.1

REFERENCES

1. Huffman, D. A., "The Synthesis of Sequential Circuits," *J. Franklin Inst.*, vol. 257, 1954, pp. 161–190 and 275–303.
2. *The TTL Data Book*, 2nd ed., Texas Instruments Inc., Dallas, Tex., pg. 6–48.
3. Eichelberger, E. B., "Hazard Detection in Combinational and Sequential Switching Circuits," *IBM J. Res. Dev.*, vol. 9, no. 2, March 1965, pp. 90–99.
4. Seshu, S., "On an Improved Diagnosis Program," *IEEE Trans. Electron Comput.*, vol. EC-14, no. 2, Feb. 1965, pp. 76–79.
5. Putzolu, G., and J. P. Roth, "A Heuristic Algorithm for the Testing of Asynchronous Circuits," *IEEE Trans. Comput.*, vol. C-20, no. 6, June 1971, pp. 639–647.

REFERENCES

6. Muth, P., "A Nine-Valued Circuit Model for Test Generation," *IEEE Trans. Comput.*, vol. C-25, no. 6, June 1976, pp. 630–636.
7. Kriz, T. A., "A Path Sensitizing Algorithm for Diagnosis of Binary Sequential Logic," *Proc. Symposium on Switching and Automata Theory*, 1970, pp. 250–259.
8. Kriz, T. A., "Machine Identification Concepts of Path Sensitizing Fault Diagnosis," *Proc. 10th Symposium on Switching and Automata Theory*, Waterloo, Canada, Oct. 1969, pp. 174–181.
9. Moore, E. F., "Gedanken-Experiments on Sequential Machines," *Automata Studies*, Princeton University Press, Princeton, N.J., 1956, pp. 129–153.
10. Hennie, F. C., *Finite-State Models for Logical Machines*, Wiley, New York, 1968.
11. Roth, J. P., "Diagnosis of Automata Failures: A Calculus and a Method," *IBM J. Res. Dev.*, vol. 10, no. 4 July 1966, pp. 278–291.
12. Miczo, A., "The Sequential ATPG: A Theoretical Limit," *IEEE Int. Test Conf.*, 1983, pp. 143–147.
13. Hsiao, M. Y., and D. K. Chia, "Boolean Difference for Fault Detection in Asynchronous Sequential Machines," *IEEE Trans. Comput.* vol. C-20, Nov. 1971, pp. 1356–1361.
14. Case, P. W., et al., "Design Automation in IBM," *IBM J. Res. Dev.*, vol. 25, no. 5, Sept. 1981, pp. 631–646.

Chapter 4
Simulation

4.1 INTRODUCTION

Simulation is an imitative process. It is used to study relationships between parameters which interact in a system. In some cases simulation may point out errors in a design which can cause the design to be unworkable. In other cases it permits optimization of a design for maximum performance or economy of operation or construction. In still other instances, the system may be so complex that simulation is the only way in which variables affecting the design, and their interaction with each other, can be controlled and studied.

In order to imitate the behavior of a product or system, simulation employs models. A model is an imperfect replica. It must be accurate enough to imitate the behavior of the variables of interest in the process or system being studied, but not so complex as to obscure details of the variables and their relationships or so intricate that it can only be created through use of exorbitant amounts of resources.

In this chapter we will be concerned with the simulation of logic circuits. Our primary interest is with simulation as it is related to fault detection and diagnosis, but it has a growing role in design verification, where logic designers have traditionally built prototypes of systems being designed. The prototype is a physical mockup of the system being designed in which connections are made by wire wrap or other means that are easily altered to correct design errors. It is used to evaluate the logical correctness and the timing characteristics of a design. The prototype is attractive because it can run at design speed and can be evaluated under actual operating conditions. Because it runs at operating speed, it can be run with virtually unlimited amounts of stimuli. Various types of test equipment can be hooked up to the design to evaluate its performance and determine relative timing margins and voltage levels. If the system configuration includes operational software and diagnostic tests, these can be developed and debugged on the prototype.

The prototype has its drawbacks. It may require many months and great expenditure of resources to build the prototype[1]. The prototype can usually accommodate only a single experiment at a time and it may require a considerable amount of time to set up the experiment. If the prototype goes down for any length of time because of failure or damage to a critical part, the entire design team may be idled. Furthermore, with increasing amounts of logic being incorporated into single integrated circuits (ICs), the accuracy of prototypes fabricated with discrete transistors and/or other integrated circuits diminishes.

The ever increasing number of ICs, the availability of programmable logic such as programmable logic arrays (PLAs) and gate arrays, and the accessibility of IC fabrication facilities to increasing numbers of logic designers have accelerated the use of simulators. There is an increasing need to be right the first time. Whereas a design error on a logic board using numerous small scale integration (SSI) parts could be corrected by adding or deleting wires and components, the IC with a design error is usually scrapped.

The increasing interest in and use of simulators has resulted in enhancements that improve their efficiency, effectiveness, and ease of use so that they are being applied to a wider range of applications. Among the advantages cited for simulation[2] is its ability to permit several designers to evaluate different parts of a design simultaneously. It accommodates the designer who wants to engage in "what if" speculations. Ideas can be quickly evaluated and either discarded or adopted, depending on their merit. The idea that would have resulted in physical damage to a prototype is easily corrected in the model.

Simulation permits evaluation of a part before it is available. The effect of the unavailable part can be studied within the simulation model long before the actual part becomes available from the manufacturer. Simulation also permits substitution of a range of components into a signal path to study the effects of different timing. Measurement of these effects is rather simple compared to similar measurements in the prototype. Other benefits of simulation include the ability to "see" inside ICs, so that the designer can examine the contents of registers and flip-flops during simulation. This can be useful when the wrong answer occurs and it is necessary to determine the cause. The simulator can also be designed to detect glitches caused by hazards. It is possible that a glitch may not affect a prototype but does affect some production units. The simulator can be designed to detect instances where two or more tri-state devices are simultaneously active. Finally, as we shall see, simulation can be used to evaluate the quality of test programs.

4.2 THE SIMULATION HIERARCHY

Digital systems can be described at several levels of abstraction, ranging from the behavioral to the geometric level, as seen in Figure 4.1[3]. Simulation capability exists at all of these levels. The *behavioral* model is the highest level of abstraction. At this level a system is described in terms of the algorithms that it performs. The development of a large system may begin with a behavioral simulation. The user, when simulating at this level, is interested in such things as determining the optimal

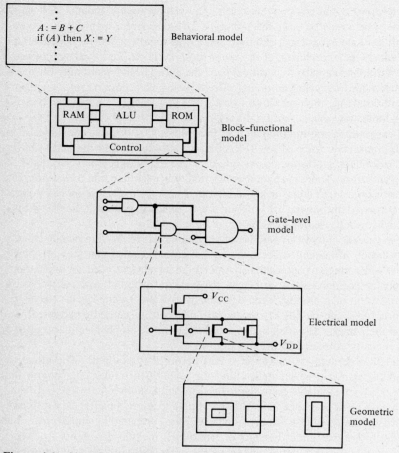

Figure 4.1 Simulation hierarchy. (© 1979 IEEE.)

instruction set mix by studying the effects of sequences of instructions on data flow. The flow of data through system elements may also be studied at this level to detect potential bottlenecks. For example, it serves no useful purpose to put a more powerful central processor unit (CPU) into a system in which the present CPU already outperforms the memory or input/output (I/O) unit. Simulation at the behavioral level can detect such mismatches. When the behavioral simulation is complete, the results of that simulation can serve as a specification for the system design.

Once the system has been specified, a *functional* model, also called a *register transfer level* model, can be used to describe the flow of data and control signals within and between functional units. The circuit description is made up of building blocks such as flip-flops, registers, multiplexers, counters, encoders, decoders, arithmetic logic units (ALU), and elements of similar level of complexity. The building blocks and their controlling signals must be interconnected so as to perform in a manner consistent with the preceding behavioral level simulation.

4.2 THE SIMULATION HIERARCHY

A *logic* model describes a system as an interconnection of switching elements or gates. At this level the designer's interest is in logical correctness of designs intended to implement functional building blocks and units. Other concerns at this level include performance or timing of the design. Closely related to the logic model is the *switch-level* model used to describe logic behavior of metal oxide semiconductor (MOS) circuits[4]. A switch-level network consists of nodes connected by transistors. Each node has value 0, 1, or x and each transistor is open, closed, or indeterminate. Logic processing is enhanced by additional capabilities required to resolve apparent conflicts caused by bidirectional pass transistors, differences in signal strengths of multiple elements driving a net, and other unique properties of MOS digital circuits.

A *circuit* level model is used on individual gate and functional level devices to verify their behavior. It describes a circuit in terms of such things as resistances, capacitances, and current sources. The simulation user is interested in knowing what kind of switching speeds, voltages, and noise margins to expect. Finally, the *geometric* level model describes a circuit in terms of physical shapes.

Simulation at a high level of abstraction requires less detailed processing, hence simulation speed is greater and more input stimuli can be evaluated in a given amount of CPU time. In most cases the loss of detail is known and accepted. However, in some cases the designer may be unaware of the loss of information whose absence may obscure details essential to proper understanding of the circuit's

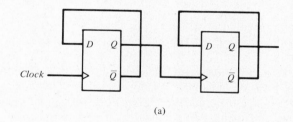

(a)

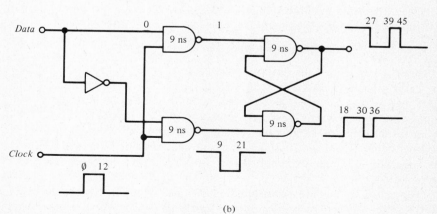

(b)

Figure 4.2 Frequency divider with spurious pulse.

behavior. The importance of the information may depend on whether the product being designed is synchronous or asynchronous. In a synchronous design, clocking of bistable devices is usually controlled in such a way as to make them immune to unexpected pulses caused by transient signals. In an asynchronous design, where the designer has the freedom to create clock pulses for flip-flops and latches, circuits are more likely to exhibit erratic behavior.

EXAMPLE ■

The circuit of Figure 4.2(a), when viewed as two functional level flip-flops connected as a frequency divider, appears to be well-behaved. But a pulse can be seen in Figure 4.2(b) which the designer may not have expected[5]. If the pulse contains enough energy, it will clock the following flip-flop more frequently than expected. ■ ■

4.3 THE FAULT SIMULATOR: AN OVERVIEW

The use of fault simulation is motivated by a desire to minimize the amount of defective product shipped to customers. From the equation for defect level given in Chap. 1,

$$DL = 1 - Y^{1-T}$$

it is obvious that the amount of defective product can be reduced by improving yield or by improving the test. To improve the test, we must be able to quantify it. But how?

The quality of a test is measured or quantified by means of fault simulation. The test stimuli that will eventually be used to evaluate the product are themselves first evaluated by applying them to circuit models which have been altered slightly to imitate the effects of faults. If the circuit output response, as determined by simulation, differs from the response of the circuit model without the fault, then the fault is detectable by those stimuli. After the process has been performed for sufficient numbers of modeled faults, an estimate

$$T = (\text{no. of faults detected})/(\text{no. of faults simulated})$$

is obtained which reflects the quality of the test stimuli.

Fault simulation uses a system model called a *structural* model. This expresses the fact that the model describes the system in terms of realizable physical components. The term can, however, refer to any level except behavioral, depending on whether the designer was creating a circuit using geometric shapes or functional building blocks. The fault simulator can be thought of as a structural level simulator in which some part of the structural model has been altered to represent a fault.

The gate-level model is most often used in performing fault simulation, but fault simulation can also be performed with functional or circuit level models. The stuck-at fault models, used with logic gates, make it quite easy to automatically

4.3 THE FAULT SIMULATOR: AN OVERVIEW

inject faults into the model by means of a computer program. The fault simulation serves several purposes besides evaluating the stimuli; it:

- Confirms detection of a fault for which an automatic test pattern generator (ATPG) claims that a successful test was created
- Computes fault coverage for a given test pattern
- Provides diagnostic capability
- Indicates areas of a circuit where fault coverage is inadequate

Confirm Detection Consider first the test for a specific fault. When creating the test, the ATPG often makes simplifying assumptions. By restricting its attention to logic behavior and ignoring element delay times, the ATPG may create test patterns which are susceptible to the effects of races and hazards. A simulator, taking into account element delays and using hazard and race detection techniques, may detect anomalous behavior caused by the pattern and conclude that the fault cannot be detected with certainty.

Compute Fault Coverage The ability to determine all faults detected by each pattern can reduce the number of iterations through an ATPG. If the simulator determines that other faults were detected by a pattern originally created to detect a particular fault, then there is no need to create test patterns to detect these other incidentally detected faults.

EXAMPLE ■

Consider the circuit of Figure 4.3. Suppose that an ATPG has created a pattern $A, B, C, D, E, F = (0, 1, 1, 1, 0, 0)$ to test for the output of gate H S-A-1. Simulating the fault-free circuit produces an output of 0. Simulating the same circuit with an S-A-1 on the output of H produces a 1 on the circuit output, hence the fault is detected. But when we simulate the effects of an S-A-1 on the upper input of gate G using the same pattern, we find that this fault also causes a circuit output of 1 and therefore is detected by the same pattern. Several other faults are detected by the pattern. We leave it as an exercise for the reader to find them. ■■

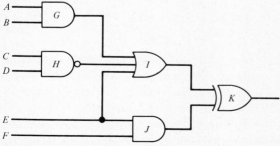

Figure 4.3 Circuit to be simulated.

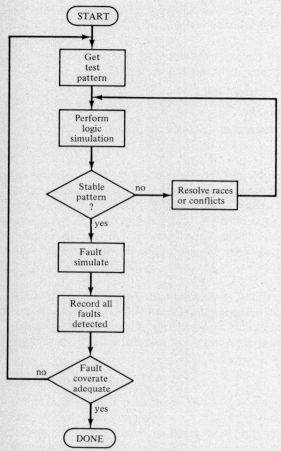

Figure 4.4 Test stimuli evaluation.

Diagnose faults Fault diagnosis consists of determining which fault actually caused an erroneous output response. If several faults are suspect, then repairing a circuit may consist of either replacing several devices or obtaining better fault resolution in order to reduce the number of suspected elements. As an example, if the previous input pattern is followed by $(0, 0, 1, 1, 0, 1)$ and the circuit output correctly responds with a 0, then there could not have been an S-A-1 fault on the output of gate H. By eliminating gate H as a candidate, we have reduced the list of components that must be replaced.

Diagnostic capability was perhaps more relevant several years ago, when more discrete parts were used. It is desirable, when repairing a board, to obtain as small a list of suspect parts as possible. However, diagnosis can still be useful in debugging mask sets for integrated circuits. When a set of anywhere from 7 to 12 masks are used to create an IC with 10,000 or more switching elements, and the mask set has a flaw which causes ICs to be manufactured incorrectly, fault diagnosis via fault simulation can sometimes be helpful in isolating the cause of the problem.

Identify Areas of Untesteds When a test engineer is writing test stimuli for a circuit, he may expend much effort in one area of the circuit but very little effort in another area. The fault simulator can provide a list of faults not yet detected by test stimuli and thus encourage the engineer to work in an area of the circuit where very few faults have been detected. Writing test patterns specifically for faults in these areas usually gives a quick boost to the fault coverage.

The overall process of test pattern generation, in conjunction with a fault simulator, is illustrated in Figure 4.4. The test patterns may be created by an ATPG or supplied by the logic designer or a diagnostic engineer. The ATPG is fault-oriented; it selects a fault from a list of fault candidates and attempts to create a test for the fault by using the methods described in previous chapters. Because stimuli created by the ATPG are susceptible to races and hazards, a logic simulation may precede fault simulation in order to screen the test stimuli. If application of the stimuli causes numerous critical races, it may be preferable to correct the stimuli before proceeding with fault simulation.

After each test vector has been fault-simulated, the faults which cause an output response different from the good machine output are checked off in the fault list and their response at primary outputs may be recorded in a data base for diagnostic purposes. The circuits used here for illustrative purposes usually have a single output, but real circuits have numerous outputs and several faults may be detected in a given pattern, with each fault possibly producing a different response at the set of primary outputs. By recording the output response to each fault, diagnostic capability can be significantly enhanced. After recording the results, if fault coverage is not adequate, the process is continued.

4.4 THE COMPILED SIMULATOR

Simulators for logic circuits, like high-level languages for computers, can be characterized as compiled or interpreted. The *compiled* simulator is created by converting a network description directly into a series of machine language instructions which reflect the functions and interconnections of the individual elements in the circuit. For each logic element there exists a series of one or more machine language instructions and a corresponding entry in a circuit value table which holds the current value of that element. The *interpretive* simulator, also called *table-driven*, operates on a circuit description contained in a set of tables, without first converting the network into a machine language image. We will first examine the compiled simulator.

The compiled simulator is constructed by using instructions from a host machine's repertoire of machine language instructions. Each element in the circuit is simulated by one or more of the instructions of the host computer and the results are stored in a table that contains an entry corresponding to each logic element being simulated. The instructions which simulate each device obtain their required input values from this table; therefore the logic must be levelized. This simply means that if element m is an input to element n, then the circuit elements are ordered such that element m is simulated first (see Sec. 3.4.2). Simulation values assigned to primary inputs for a simulation pass are stored in the table and then each element is simulated only after the gates connected to its inputs have been simulated.

EXAMPLE ■

We will create a simulator for the circuit of Figure 4.3 using assembler language instructions for the Z80 microprocessor.

```
LD      A,(IX)
AND     (IX+1)
LD      (IX+6),A
LD      A,(IX+2)
AND     (IX+3)
XOR     0FFH
LD      (IX+7),A
LD      A,(IX+4)
AND     (IX+5)
LD      (IX+8),A
LD      A,(IX+6)
OR      (IX+7)
OR      (IX+4)
LD      (IX+9),A
LD      A,(IX+9)
XOR     (IX+8)
LD      (IX+10),A
RET
```

The network is compiled into machine code by a preprocessor that receives a description of the circuit in terms of logic elements and interconnecting nets. The circuit value table, which we shall simply call TABLE, contains an entry for each primary input and logic gate in the circuit. A correspondence is established between logic gates and locations in TABLE by the preprocessor. In this example, locations 0 through 5 in TABLE are used to store the current values on primary inputs A through F. Locations 6 through 10 are used to store the current value on logic gates G through K. Prior to execution the base address of TABLE is loaded into index register IX. Then, during simulation, updated values are stored in TABLE locations 0 through 5 whenever inputs are due to change. Values for the logic gates are computed and stored in TABLE by the simulator. Displacements relative to the contents of IX are used to access various entries in TABLE. The notation $(IX + i)$ refers to the contents of byte number $i + 1$ of TABLE. In those instructions with two addresses, the first argument is the destination and the second argument is the source argument.

The first simulator instruction loads the value in TABLE corresponding to primary input A into the A register or accumulator. The second instruction performs an AND operation on the contents of the accumulator and the value in $(IX + 1)$ which contains the value on primary input B. Then, in the third instruction, the resulting value in the accumulator is stored in $(IX + 6)$, corresponding to gate G. The second set of calculations performs an AND operation on the values corresponding to inputs *C* and *D*. However, before storing the result in TABLE, the result is complemented by performing an exclusive-OR using a byte containing all 1s.

The remaining gates are processed in this fashion and then the simulator returns to the calling program. Note that in one instance, the exclusive-OR gate, the simulator stores a result and then immediately loads the same value into the accumulator. Since the simulator is called repetitively with many input vectors, it may be advantageous to optimize its performance. This can be done when levelizing the circuit. If a gate output is connected to another gate for which all other inputs have already been processed, then the destination gate satisfies the levelizing criteria and can be the next gate to be simulated. Therefore, the value in the accumulator can be used without being reloaded. It is usually still necessary to save the calculated result in TABLE if there is fan-out or if the control program must provide the ability to inspect intermediate simulation results on internal circuit nets after a simulation pass.

■■

The compiled simulator can also be implemented by using two tables or arrays: the READ array and the WRITE array. In this implementation it is not necessary to levelize a circuit. When simulating an input pattern, changes on primary inputs are stored in the READ array. Then each element is simulated as before, except that they may occur in random order. When an element is simulated, its inputs are obtained from the READ array and its result is stored in the WRITE array. After all elements have been simulated, the contents of the READ and WRITE arrays are compared. If they differ, the contents of the WRITE array are transferred to the READ array and the circuit is again simulated. [In practice, it is simpler to exchange names; the READ (WRITE) array in pass n becomes the WRITE (READ) array in pass $n + 1$.] Eventually, after a finite number of passes, contents of the two arrays must match if simulating a combinational circuit and the simulator can then process the next input vector. Although this obviates the need to levelize the circuit, it may take several passes before all changes on inputs reach the outputs.

4.4.1 Three-Valued Simulation

As we saw in the previous chapter, when processing sequential circuits it is necessary to be able to process a third value, the indeterminate state. In order to do this, it is necessary to use two binary values to represent the three simulation values. For this purpose we establish the following correspondence between the three simulation values and the 2-bit vectors:

$$0—0,0$$
$$1—1,1$$
$$x—0,1$$

The simulation program must be expanded accordingly, but first we must define the operations on these 2-bit vectors. It turns out that the processing is similar to processing of single-bit values in most cases. For example, to AND a pair of arguments, we AND individual bit positions. The OR operation is treated similarly. Operations that invert arguments, such as the inverter and the exclusive-OR, require special attention because a $(1, 0)$ is not the complement of an x. The inverter can be processed by complementing the individual bits and swapping them. The exclusive-OR of variables A and B is complicated by the fact that A and B could both be x. The computation may best be processed as $A \cdot \overline{B} + \overline{A} \cdot B$.

4.4.2 Simulating Sequential Circuits

Sequential logic requires additional processing before the compiled simulator can proceed. Consider the cross-coupled NAND latch of Figure 4.5. Before we can simulate gate 1 we need a value from gate 2. But simulation of gate 2 requires a value from gate 1. We could extract the latch in its entirety from the circuit and create a state table model for the circuit. Then, after simulation reached the point where all inputs to the latch were evaluated, the state table model could be used to determine the new values on the output of the latch. For a NAND latch the state table is not difficult to derive. For an asynchronous state machine comprised of many states the task of creating a state table is formidable. Another approach is to cut the feedback line, as was done in the iterative test generator. If a cut is made from gate 1 to gate 2, we obtain the circuit of Figure 4.5(b).

After all loops in the circuit have been cut, the network is compiled. The circuit is now a pseudo-combinational circuit and the pseudo-inputs are treated in the same way as primary inputs when levelizing and compiling the circuit.

Before the initial call to the simulator, the control program sets all pseudo-inputs to the x state. Then, during any single pass through the compiled simulator, each element is simulated once. It occasionally happens that the value on one or more pseudo-outputs is not the same as the value on the corresponding pseudo-inputs, in which case the values at the pseudo-outputs are transferred to the corresponding pseudo-inputs and simulation is again performed. If pseudo-outputs and pseudo-inputs still fail to match after some predetermined number of passes, it is concluded that the circuit is oscillating and the pseudo-inputs and pseudo-outputs that are oscillating are set to the x state. The control program then permits additional passes through the simulator, each time setting to x any additional pseudo-inputs

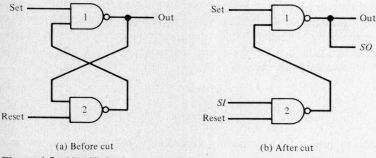

(a) Before cut   (b) After cut

Figure 4.5 NAND latch.

4.4 THE COMPILED SIMULATOR

which did not agree with their corresponding pseudo-outputs until, eventually, the circuit stabilizes with a number of pseudo-inputs in the x state.

The pseudo-inputs and pseudo-outputs are analogous to having READ and WRITE arrays, but only for feedback lines. In fact, if the entire circuit is simulated using READ and WRITE arrays, then not only is it not necessary to levelize the circuit, it is also not necessary to cut the loops. It is, however, still necessary to detect oscillations and inhibit them with the x state.

4.4.3 Hazard Detection

The compiled simulator simply performs logic simulation. It makes no use of inherent delays in circuit elements. Furthermore, the cutting of feedback lines presumes that delay is lumped into that particular point where the cut occurred. Consider again the NAND latch with the cut feedback line. If a transition occurs in which both $\overline{\text{Set}}$ and $\overline{\text{Reset}}$ lines change from 0 to 1, then the simulation result is totally dependent on where the cut occurred. With the cut illustrated in Figure 4.5(b), gate 2 will be simulated first and the latch will "stabilize" at $Q, \overline{Q} = (1, 0)$. If the cut were made from gate 2 to gate 1, then gate 1 would be simulated first and the latch would stabilize at $Q, \overline{Q} = (0, 1)$. This problem results from the assumption that the delays were lumped at one point. By moving the cut, in effect lumping the delay at another point in the circuit model, the simulator computed a different answer. In actual circuits, delay is distributed and the circuit could, in fact, oscillate if the input changes occurred sufficiently close together.

In Chap. 3 we pointed out that hazards can affect the behavior of a circuit. Hazards are a consequence of delay in circuit elements. The static hazard, which may cause a momentary change to the opposite value on signal lines which should remain unchanged, could be of sufficient duration to cause a NAND latch to change state. If the inputs are $\overline{\text{Set}} = \overline{\text{Reset}} = 1$ and the present state is $Q = 0$, then a momentary 1–0–1 glitch on the $\overline{\text{Set}}$ line could cause it to latch up in the $Q = 1$ state. But the compiled logic simulator will not detect the glitch if it is only simulating logic values. To address this problem, Eichelberger[6] proposed use of the ternary algebra, consisting of the symbols $(0, 1, x)$. The values were already in

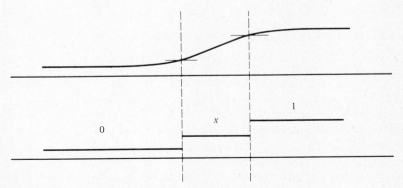

Figure 4.6 The transition region.

use to handle unknown values associated with feedback lines. However, Eichelberger advocated their use on inputs whenever a change occurred. In effect, the ternary algebra describes the transition region in switching devices. It permits an approximation to continuous signals, as illustrated in Figure 4.6, by representing the "in between" time when a signal is neither 0 nor 1. In fact, if a signal fans out from a source, that signal could simultaneously represent a 0 to one device and a 1 to another device due to differences in switching characteristics of the driven devices.

The ternary algebra tables for the AND gate and the OR gate are as follows:

	AND				OR		
	0	x	1		0	x	1
0	0	0	0		0	x	1
x	0	x	x		x	x	1
1	0	x	1		1	1	1

The following two lemmas follow directly from the ternary algebra tables.

Lemma 4.1. If one or more gate inputs are changed from 0 to x, or 1 to x, the gate output will either remain unchanged or change to x.

Lemma 4.2. If one or more gate inputs are changed from x to a known value, the gate output will either remain unchanged or change from x to a known value.

From the lemmas, the following theorems result:

Theorem 4.1. If one or more ternary inputs to a combinational logic network change from 1 to x or 0 to x, then the network output either remains unchanged or changes to x.

Theorem 4.2. If one or more ternary inputs to a combinational logic network change from x to 1 or x to 0, then the network output either remains unchanged or changes from x to 1 or x to 0.

Theorem 4.3. The output $f(a_1, \ldots, a_n)$ of a combinational logic network may change as a result of changing inputs $a_1, \ldots, a_p$ if and only if

$$f(x, \ldots, x, a_{p+1}, \ldots, a_n) = x$$

With these theorems Eichelberger then defines a pair of procedures for determining whether or not a circuit will be affected by static hazards, critical races, or essential hazards during a given input state change. Using the Huffman model, proceed as follows:

Procedure A: Determine all changing Y signals. Changing inputs are first changed to x. If any Y_i outputs change to x, change the corresponding y_i inputs and resimulate. Continue until no additional Y_i changes are detected.

4.4 THE COMPILED SIMULATOR

Procedure B: Determine which Y signals stabilize. Set changing inputs from x to their new binary state and simulate. If any Y_i changes from x to 1 or 0, then change the corresponding y_i and resimulate. Continue until no additional Y_i changes occur.

Theorem 4.4. If feedback line $Y_k = 1\,(0)$ after applying procedure A and procedure B to a sequential circuit for a given input-state change starting in a given internal state, then the Y_k feedback signal must stabilize at 1 (0) for this transition regardless of the values of the (finite) delays associated with the logic gates.

These theorems tell us that if we use the ternary algebra when simulating, and handle unstable feedback lines as prescribed in procedures A and B, then:

1. Hazards, races, and oscillations are automatically detected.
2. For a circuit with n feedback lines, at most only $2n$ simulation passes are required.

EXAMPLE ■

For the NAND latch of Figure 4.5(b), the original input $\overline{Set} = \overline{Reset} = 0$ causes a 1 on the pseudo-input SI. Now, with ternary simulation, the $\overline{Set}$ and $\overline{Reset}$ lines both go first from 0 to x, and then from x to 1. We first apply procedure A. Gate 2 is simulated first and the 1 and x values on the inputs cause an x on the output. This value is input to gate 1 and, together with the x on the other input, causes an x from gate 1. This x then appears on the pseudo-output. Since the value on SO differs from the value on SI, the value on SO is transferred to SI and the circuit is resimulated with the x values on the $\overline{Set}$, $\overline{Reset}$, and pseudo-input. The circuit is now stable with an x on SI and SO. Procedure B is now applied. The inputs are changed to 1 and the circuit is resimulated. Now, however, the x on the pseudo-input causes an x to occur on the output of gate 2, and that in turn causes an x on the output of gate 1 and, subsequently, on the pseudo-output SO. The circuit is "stable" in the unknown state. ■ ■

4.4.4 Parallel Fault Simulation

The simplest form of fault simulation that can be implemented in the compiled simulator is *serial* fault simulation. In this method a single fault is injected into the circuit model and simulated with the same stimuli that were applied to the fault-free model. When done, the fault is removed, another fault is injected and simulation again performed. This is done for all faults of interest. Then, as explained before, the testability figure T is computed. Fault injection can be achieved for a logic gate input by simply deleting the input. For example, when a net connected to the input of an AND gate is deleted from the list of inputs to that AND gate, the logic value present on that net has no effect on the AND gate, hence the AND gate behaves as though that input were stuck-at-1. Similarly, deleting an input to the OR gate causes it to behave as though it were stuck-at-0.

When the Z80 compiled simulator processed a circuit, it manipulated bytes of data. For ternary simulation, one bit from each of two bytes was used to represent a logic value. The *parallel* fault simulator takes advantage of the unused bits to simulate faulted machines in parallel with the good machine. It does this by letting each bit in the byte represent a different machine. The leftmost bit (bit 7) represents the fault-free machine. The other seven bits represent faulted machines corresponding to seven faults in the fault list.

In order to use the extra bits they must somehow be made to represent the values that exist on the faulted machines. This is accomplished by the process of "bugging the simulator." Faults must be injected into the simulator in such a way that individual faults affect only a single bit position. For example, suppose the OR gate in Figure 4.3 is to be modeled with an S-A-0 fault on its upper input. Let bit 7 represent the fault-free machine and bit 6 represent the faulted machine with the OR gate input S-A-0. Then, prior to simulation, the control program makes an alteration in the compiled simulator.

The instruction which loads the value from (IX + 6) into register A is replaced by a call to a subroutine. This subroutine loads the value from (IX + 6) into register A and then performs an AND operation on that value with the 8-bit mask 10111111. The subroutine then returns to the compiled simulator.

This method of bugging the model causes the OR gate to always receive a 0 on its upper input for the machine corresponding to bit position 6, regardless of what value is generated by the AND gate. Hence, if $A = B = 1$ and $C = D = E = F = 0$, the good machine, bit 7, will simulate the OR gate with inputs of $(1, 0, 0)$ and the machine corresponding to bit 6 will simulate the OR with inputs of $(0, 0, 0)$. Therefore, they will produce different values at the primary output.

In practice, the bugging operation will use all 7 bits of the byte. In the example previously used, bit 5 could represent the fault corresponding to the center input of the three-input OR gate S-A-0. Then, when the program does an OR operation from (IX + 7), it again calls a subroutine. However, this time it uses the mask 11011111. When bugging a gate output, the value is masked before being stored in TABLE. If modeling an S-A-1 fault on an input, the program performs an OR instruction using a mask containing 0s in all bit positions except the one corresponding to the faulted machine, where it uses a 1.

When a pass through the simulator has been completed, and no disagreements exist between pseudo-outputs and corresponding pseudo-inputs, the circuit value table entries corresponding to circuit observable outputs are checked by the control program. Values in any bit positions which differ from bit 7, the good machine output, indicate faulted machines in which the fault caused an output to differ from the good machine output. However, before claiming that the fault is detected by the input pattern, the differing values must be examined further. If the good machine response is x, and a faulted machine responds with a 0 or 1, we cannot claim detection of that fault.

In the Z80 program, parallel simulation can be performed on the good machine and seven faulted machines simultaneously. In general, the number of faults that can be simulated in parallel is a function of the machine architecture. A machine that can perform logic operations in a storage-to-storage mode may be

4.4 THE COMPILED SIMULATOR

able to process several hundred machines simultaneously. Regardless of machine architecture, a circuit of reasonable size will contain more faults than can be simulated in parallel. Therefore, several passes through the simulator will be required. On each pass a fault-free copy of the simulator is obtained and bugged. The number of passes is equal to the total number of faults to be simulated divided by the number of faults that can be simulated in a single pass. It is interesting to note that, although we adhere to the single-fault assumption, it is relatively easy to bug the simulator to permit multiple-fault simulation.

4.4.5 Performance Enhancements

The compiled simulator is efficient with memory, a factor which was once quite important but became less so as memory became cheaper and more readily available. Augmented with just a circuit value table and a small control program, the compiled simulator can simulate circuits of reasonable size. Unfortunately, it suffers from poor performance. As circuits grow in size, simulation time increases disproportionately faster. The simulator grows in proportion to the number of elements in the circuit. The larger circuit has more faults and therefore more fault simulation passes are necessary. Finally, more patterns are required because of the increased number of faults. As a result of these three factors, simulation time could grow in proportion to the third power of circuit size, although in practice the degradation in performance is not quite that severe.

    A number of techniques were developed to reduce simulation time. Some faults are equivalent in their behavior. For example, in the circuit of Figure 4.3, the output of AND gate G S-A-1 and the input of OR gate *I* S-A-1 will produce identical symptoms, regardless of the test pattern applied. Therefore, it is pointless to simulate both of these faults. It is more economical to simulate one of them and, if it is detected, the simulator identifies a class of faults as being detected, where the fault class in this instance consists of the two faults just cited. As we shall see in Chap. 5, fault classes can be established on the basis of fault equivalence and also on a concept called fault dominance.

    Simulation time can be reduced by using *stimulus bypass*[7]. This technique avoids execution of machine code for devices when they cannot possibly change. As an example, if the simulator contains machine code to simulate a gate-level model of a positive edge-triggered delay flip-flop with a reset input, then when the clock is low and the reset is inactive, the output of the flip-flop will not change. By placing a few instructions at the beginning of the flip-flop model, it is possible to check for activity on the clock and reset lines and branch past the code for the flip-flop if the clock and reset are inactive. This requires that the compiler be able to recognize the delay flip-flop as a functional unit and compile the machine code accordingly.

    Circuit partitioning is useful in reducing simulation time. If the sets of logic elements which drive two distinct outputs have very few gates in common, then it becomes more efficient to simulate them as separate circuits. The faults that occur in only one of the two circuits will not necessitate simulation of the other circuit.

A circuit partitioning can be performed by means of a backtrace from a primary output as follows:

1. Select a primary output.
2. Put gates that drive the primary output onto a stack.
3. Select an unmarked gate from the stack and mark it.
4. Put its unmarked driving gates onto the stack.
5. If there are any unmarked entries on the stack, go back to step 3.

The gates on the stack constitute a subcircuit, sometimes called a *cone*, which can be processed as a single entity. Where two outputs define nearly disjoint circuits of approximately the same size, the simulator for each circuit is about half its former size, and there are half as many faults, hence perhaps as few as half as many vectors for each circuit. Thus, total fault simulation time could decrease by half or more.

A practice called *fault dropping* was also developed to speed up fault simulation performance. The simulator drops faults from the fault list and no longer simulates them after they have been detected. Continued simulation of detected faults can be useful for diagnostic purposes, as we shall see later, but it requires additional simulation time.

4.5 THE EVENT-DRIVEN SIMULATOR

Stimulus bypass took advantage at a local level of the fact that quite frequently the flip-flop would not experience any internal state change as a result of activity on the inputs. The local optimization was made possible because the flip-flop could be modeled as an integral body of machine code for which the first few instructions checked key inputs to determine whether internal activity would occur. As it turns out, the amount of activity within a circuit at a given time is often minimal and can terminate abruptly, as when changing signals converge on flip-flops at which the clock lines are inactive.

Since the amount of activity in a time frame is minimal, why bother to simulate the entire circuit? Why not simulate only those elements which are involved in signal changes? This strategy, employed at a global level, requires a simulator that can operate on tables of information which include a list of the elements in the circuit and their interconnections. The interconnection list includes a list of the inputs and a list of the destination elements for each circuit device. The list is used to link elements together for the purposes of passing simulation results between elements and facilitating scheduling of elements for simulation.

When a signal change occurs on a primary input or the output of any circuit element, then an *event* is said to have occurred on the net driven by that primary input or element. When an event occurs on a net, then all elements driven by that net are simulated. If a signal change on the input to a device does not cause a change on the device output, then simulation is terminated along that signal path.

Event-driven simulation can be performed in either the nominal delay or the

4.5 THE EVENT-DRIVEN SIMULATOR

zero delay environment. The *nominal delay* simulator assigns delay values to logic elements based on manufacturers' recommendations or measurements with precision instruments. The *zero delay* simulator totally ignores delay time within a logic element and simply simulates the logic function performed by the element. Some simulators, in trying to strike a compromise between the two, perform a *unit delay* simulation in which each logic element is assigned a fixed delay, and since the elements are all assigned the same delay, the value 1 (unit delay) is as good as any other. The nominal delay simulator can give precise simulation results but at a cost in CPU time. The zero delay simulator will usually run faster but does not indicate when events occur. The unit delay simulator lies between the other two in range of performance. It records time units during simulation, so it does more computation than zero delay simulation, but the mechanism for scheduling events in future time is simpler than for time-based simulation. However, regarding all element delays as being equal can occasionally produce inaccurate results in timing sensitive circuits and provide a false sense of security.

4.5.1 Zero Delay Simulation

We first consider simulation in the zero delay environment, where we are not concerned with the delay of individual elements but only with the order in which elements change value. In this simulation method, it is not necessary to levelize the circuit. Before simulating the first input pattern, all nodes are initialized to x. Then, whenever an element assumes a new value, whether it be a primary input changing as a result of new stimuli being applied or an internal element changing as a result of simulation, any elements driven by that element are simulated.

The Event-Driven, Zero Delay Simulator A direct way to implement an event-driven, zero delay simulator, if employing the READ/WRITE arrays described earlier, is to associate a flag bit with each entry in the arrays. If a change occurs on an element, elements in its fan-out are identified and flagged for simulation in the next pass. When no elements are flagged at the end of a pass, the circuit is stable. Alternatively, elements that must be simulated in the next pass can be placed on a first-in, first-out (FIFO) stack, unless they are already on the stack. When the stack is empty at the end of a pass, the circuit is stable.

EXAMPLE ∎

We will illustrate with the circuit of Figure 4.7. At the first time interval, t_0, simulation is performed on all elements driven by inputs 1, 2, 3, and 5. Simulation of gates 6 and 7 causes their outputs to change from x to 0. Simulation of gate 8 results in a 1 on its output. These changes cause gate 9 to be scheduled for simulation and it produces a 1 on its output. At time t_1 input 1 changes from a 1 to 0. However, there is no change on the output of gate 6, so simulation for time t_1 is done. Input 2 changes at time t_2, causing gate 9 to be simulated. The output of gate 9 does not change. Gate 7 is simulated at time t_3, but again no output activity occurs. At time t_4 all gates must be simulated. ∎∎

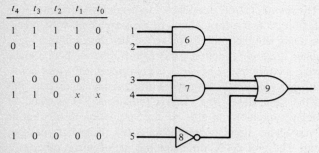

Figure 4.7 Zero delay simulation.

In this tiny example it is difficult to appreciate the value of event-driven simulation, but in a circuit of many hundreds or thousands of gates, a situation such as occurred in time t_1, can happen frequently and can provide substantial savings in computer time. The simulation at time t_1 was terminated almost immediately because a single input change occurred which had virtually no effect on the circuit.

Hazard Detection Using Multiple Values The Eichelberger three-valued hazard analysis is effective with event-driven simulation, and can be used without making cuts in the circuit model. Simply perform an intermediate x value simulation on all changing inputs and the circuit will stabilize. However, the three-valued simulation will not detect dynamic hazards. To detect dynamic hazards, Fantauzzi[8] proposed the nine-valued simulation. The nine values denote various combinations of stable and changing signals. The values are used in conjunction with operator tables for the basic logic operations. The symbols used and the operation table for the AND gate are as follows:

Symbol	Meaning	Complement
0	constant 0	1
1	constant 1	0
/	dynamic hazard 0–1	\
\	dynamic hazard 1–0	/
∧	0–1 transition, hazard-free	∨
∨	1–0 transition, hazard-free	∧
M	0–0 static hazard	W
W	1–1 static hazard	M
*	race condition	*

AND	0	1	M	W	*	∧	∨	/	\
0	0	0	0	0	0	0	0	0	0
1	0	1	M	W	*	∧	∨	/	\
M	0	M	M	M	M	M	M	M	M
W	0	W	M	W	*	/	\	/	\
*	0	*	M	*	*	*	*	*	*
∧	0	∧	M	/	*	∧	M	/	M
∨	0	∨	M	\	*	M	∨	M	\
/	0	/	M	/	*	/	M	/	M
\	0	\	M	\	*	M	\	M	\

4.5 THE EVENT-DRIVEN SIMULATOR

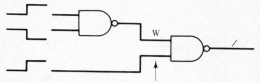

Figure 4.8 Creation of dynamic hazard.

From this table any pair of incoming signals to a two-input AND gate can be processed to determine whether the result will cause a static or dynamic hazard. For instance, we see that if one of the inputs is a constant 0, the output must always be a constant 0. With a static 0–0 hazard on one input, there will always be a static 0–0 hazard on the output unless another input to the AND gate has a constant 0. The circuit of Figure 4.8 illustrates the creation of a dynamic 0–1 hazard in a pair of NAND gates. The table for the AND gate is easily extendable to n, $n \geq 2$, since the AND operation is commutative and associative.

The table for the NAND latch is given below. In this table the columns correspond to values on the Reset input and the rows correspond to values on the Set input. The entries in the lower right quadrant of this table contain two values. The actual value assumed at the output depends on the previous state of the latch. If the Q output is presently true, then the first value is assumed. If false, then the second value is assumed.

		Reset							
		0	∧	M	/	*	1	∨	M	\
Set	0	1	∨	W	\	*	0	∧	M	/
	∧	1	*	W	*	*	0	∧	*	/
	M	1	∨	W	\	*	0	∧	M	/
	/	1	*	W	*	*	0	∧	*	/
	*	1	*	W	*	*	*	*	*	*
	1	1	1	1	1	*	1–0	1–∧	1–*	1–∧
	∨	1	∨	W	\	*	∨–0	W–∧	\–M	W–/
	W	1	*	W	*	*	∨–0	W–∧	*	W–/
	\	1	∨	W	\	*	∨–0	W–∧	\–M	W–/

4.5.2 Nominal Delay Simulation

Zero delay simulation with three or nine values can give simulation results that are correct because it can accurately predict hazards and races. However, it is also worst-case or pessimistic because it ignores the time dimension and collapses all computations into zero time. Consequently, it will see conflicts that do not actually exist. A designer may knowingly design an asynchronous state machine to receive two or more input changes during the same input pattern. In being aware of what he is doing, the designer will use the delay of the devices in the design and, if necessary, incorporate additional delay into the signal paths to insure that the signals arrive in the correct sequence. The simulator, in ignoring delay information, concludes that a race exists, so it puts the state machine into the indeterminate state rather than risk predicting the wrong state.

Delay Models There are several kinds of delays that exist in switching devices, and there are some that do not exist but are used, nonetheless, in simulation. The zero delay is obviously not a delay at all; the term simply denotes a simulation environment in which delay time is ignored. Unit delay is only slightly more real; it assumes that all elements have identical switching times.

Propagation delay is a measure of the time interval between the appearance of a change on an input to a device and the appearance of a corresponding change at an output. It is measured at specified reference points on the input and output waveforms. When the output waveform is changing from a low to a high level, the delay is referred to as *rise time*. A transition from high to low is called *fall time*. Rise and fall times frequently differ because of capacitance and storage effects of transistors used to implement switching circuits. The delay times for a given device are specified for a particular set of operating conditions. These conditions include specification of ambient temperature and load resistance and/or capacitance on the output of the device. When computing propagation time for that element in a circuit to be simulated, it may be very important, for the sake of accuracy, to adjust these numbers based on the actual environment in which the device is operating.

Manufacturers may specify individual rise and fall times for switching elements used in a given technology, but for specific ICs containing many switching gates the manufacturers will usually specify a nominal or average expected delay for signals relative to the input and output pins of a device. They may also specify maximum and/or minimum delays between the same points. The difference between the nominal and maximum or minimum delays is referred to as *ambiguity delay*. Another delay of logic gates that is sometimes of interest is *inertial delay*, which is the minimum amount of time during which a signal must persist at the input of a device in order to effect a change at the output.

Delay is also introduced through the interconnecting wires of a circuit. The length of time required to propagate a signal from one physical point to another through wire is called *media delay*; this time is approximately 1 nanosecond per foot of wire. As circuits continue to shrink and devices continue to switch at higher speeds, the media delay becomes a significantly larger percentage of the total elapsed time in a circuit and it is not unusual for media delay to account for 20 to 50% of the cycle time on a high-performance machine.

When testing a logic board on a tester, ambiguity delay can result from skew at the tester pins. Although the ATPG may specify that two or more signals change at the same time, the signal changes on the tester may actually occur many microseconds, or even milliseconds, apart due to architectural considerations within the tester. Since the ATPG is usually configured to be general purpose, rather than designed for a particular tester, the ATPG assumes that the signal changes occur simultaneously and it may be necessary to use the simulator to determine whether this skew or ambiguity delay represents a problem.

The Scheduler Nominal delay simulation makes use of the inherent delay in logic elements. However, because of variability in the delay of individual elements, they cannot simply be placed into an FIFO queue as they are encountered. The element

4.5 THE EVENT-DRIVEN SIMULATOR

presently being simulated may have an output change that occurs sooner than some elements previously scheduled and later than others. Hence, it must be scheduled so as to be processed at the right time relative to other changes. The simplest way to do this is through the use of a linear linked list. In this structure an *event notice* is used to describe an activity that must be performed and the time at which it must be performed. The notices are arranged in the order in which they are to be performed. Included in each event is a link which points to the next event notice in the list. When an event is to be scheduled it is first necessary to find its proper chronological position in the linked list. Then the link in the preceding event is altered so that it points to the newly inserted event and the link that was in the preceding event is inserted into this newly inserted event so that it now points to the next event.

To insert an event in this linked list requires searching an average of half the elements in the linked list and then modifying two pointers. As the number of events grows, due to increased system size or increased activity, the average search time grows. To reduce this time, the scheduling mechanism shown in Figure 4.9 is used[9]. It is a combination of a vertical time mapping table, also called a delta-t loop or "timing wheel"[10], and a number of horizontal lists. The vertical list represents integral time slots at which various events occur. If an event is to occur

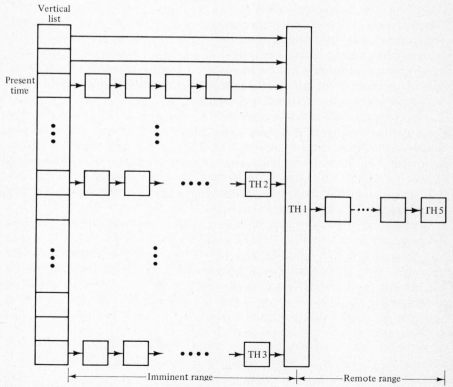

Figure 4.9 The converging lists scheduler.

at time i, then either it is the first event, in which case a pointer is inserted at slot i to identify the event that is to be processed, or other events may already have been scheduled, in which case the present event is appended to the end of the list. Note that the event may be the result of a gate simulation, in which case the event is to be processed at a future time, or the event may be a print request or some similar request for service.

A further refinement, called nonintegral event timing[11], defines the slots in the vertical list as intervals. If an event occurs within the time interval represented by that slot, then it must be inserted into its correct position in the horizontal linked list. Therefore, the search through a linked list must again be performed. However, the search is through a much smaller list. Performance is enhanced by making the vertical list as large as is practical, although not so big that a large average number of slots go unused.

To handle events that must occur far in the future, imminent and remote ranges are used. These are implemented by means of thresholds shown in the converging lists scheduler of Figure 4.9. All but two of the wheel slots link directly to threshold TH1. The remaining two slots link first to TH2 and TH3, and then to TH1. From TH1, the linked list terminates on TH5, which represents infinity. The thresholds are control notices, which are events that can be scheduled like elements and represent requests for service such as printout of simulation results. When inserting an item into a horizontal list, if TH1 is encountered, then the item is inserted between TH1 and the item previously linked to TH1. If time of occurrence of an event exceeds imminent time, then it is inserted into its appropriate slot in the remote list. During simulation, if TH2 or TH3 is encountered, then imminent time is increased, the new maximum imminent time is stored in control notice TH1, and items from the remote range are retrieved and inserted (converged) into their proper place in the imminent range.

In order to obtain correct simulation results when an event is simulated, it is necessary that any change on the inputs cause a simulation using the values that existed on the other inputs at the time when the changing signal arrived at the given input. Therefore, the input change is simulated immediately. But the output value is not altered until a future time corresponding to the delay of the element. This emulates the behavior of a logic element with finite, nonzero delay. A signal change appears at a gate input and at some future time, depending on element delay, the effects (if any) of that input change appear on the element output and then propagate to the inputs of gates in the fan-out of that gate.

If simulation does not result in a change on the output of an element, it is tempting to assume that nothing further need be done with that element. However, it is possible for a simulation to indicate no change, but a previously scheduled change occurred and has been recorded on the scheduler. For example, suppose a two-input AND gate with propagation delay of 10 nanoseconds has values (1, 0) on its inputs at time t when a positive pulse of duration 3 nanoseconds reaches the second input. The simulation result at time t is a 1, which differs from the 0 presently on the output, so the event is placed on the scheduler for processing at time $t + 10$. At $t + 3$, when simulating the change to 0, the simulator computes a

4.5 THE EVENT-DRIVEN SIMULATOR

0 on the output, which matches the present value. Therefore, the simulator may conclude erroneously that no scheduling is required.

One solution to this problem is to always put the event on the scheduler regardless of whether or not there is a change on the output. Then, when it is processed later, if its output value is equal to its present value, drop it from further processing. In the example just given, the AND gate is simulated at time t and placed on the scheduler. It is simulated at $t + 3$ and again placed on the scheduler. At $t + 10$ it is retrieved from the scheduler and its output is checked. The current value is 0 and the new value is 1, so the element output is updated in the descriptor cell and the result propagated forward. At $t + 13$ the process is repeated, this time with the present value equal to 1 and the computed value switching back to 0.

Another approach, which can save scheduling time, makes use of a schedule marker. It is used as follows:

> Simulate the input event, and then
> if there is an output change, schedule the change and increment the schedule marker;
> if there is no output change, and the schedule marker is 0, no activity is required;
> if there is no output change, and the schedule marker is >0, schedule the change and increment the marker.

When an output event is processed, decrement the schedule marker.

It occasionally happens that a change on the output of an element is followed almost immediately by another change such that the duration of the pulse is less than the inertial delay of the element. In that case, the user may want to retain the glitch and propagate it to succeeding logic to determine whether it could cause a problem. While the glitch should not occur if all elements have delay values exactly equal to their nominal values, delay values that vary slightly from nominal can cause the pulse to exceed the inertial delay of the element.

It may be the case that the glitches are in data paths where, even if they could occur, they are not likely to cause any problems and their presence clutters up the output. In that case it would be desirable to suppress their effects. Consider a two-input AND gate with t_p nanoseconds propagation delay and suppose its present input values are $(1, 0)$. If it has inertial delay of t_i nanoseconds and a pulse of duration $t_g, t_g < t_i$, appears on its lower input, then it is scheduled for a change at $t + t_p$ and again at time $t + t_p + t_g$. In that case, it would be desirable to delete the change at $t + t_p$ from the scheduler before it is processed since it would otherwise cause unwanted changes to be scheduled in successor elements.

If the time at which an element is placed on the scheduler is recorded, that information can be used to determine whether the duration of the output signal value exceeds the inertial delay. In the situation just described, the time $t + t_p$ is recorded. When the next output change occurs at $t + t_p + t_g$, it is compared with the previous change and, if the signal duration does not exceed the inertial delay, the recorded time of the previous change is used to search the appropriate linked list on the schedule for the event to be deleted. If a previous change occurred but its time was not recorded, it would be necessary to search all time slots on the scheduler between $t + t_p$ and $t + t_p + t_g$.

The Descriptor Cell During simulation and scheduling, information concerning each element in a circuit is contained in a descriptor cell. The cell contains permanent information, including pointers for each input and output, and descriptive information about the element it represents, such as its function and delay values. It also contains data that change during simulation, including the schedule marker and values on the inputs and outputs of the element. The basic cell is illustrated in Figure 4.10(a) for an element with one output. The first few entries point to devices which drive the inputs of the element represented by this descriptor cell. There is an entry for each input and a field for each entry which indicates its displacement from the descripter information. Since input changes are stored in the descriptor cell, the input number is used to access and update the correct bits in the descriptor cell during simulation. The last entry points to destination input(s) that are driven by this element.

For an element with two or more outputs, there is a corresponding number of output entries in the descriptor cell. A simple circuit and its descriptor cell model are illustrated in Figures 4.10(b) and (c), respectively. Each descriptor cell corre-

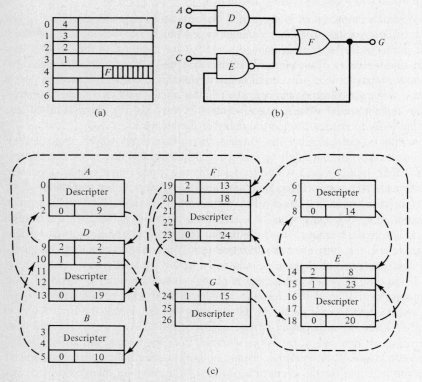

Figure 4.10 The descriptor cell.

4.5 THE EVENT-DRIVEN SIMULATOR

sponds to an element in the circuit being modeled and the nets which interconnect circuit elements are represented in the model by linked lists which thread their way through the descriptor cells. For example, input A goes to input 1 of gate D. Therefore, the output pointer of descriptor cell A points to location 9, which corresponds to the first entry of D. Gate F fans out to two places so the linked list extends through entries in three descriptor cells. When traversing the linked list to find the fan-out elements for a particular device, the traversal is halted when a word is encountered in which the high-order field is 0.

To illustrate the scheduling process using the scheduler and descriptor cells, suppose we want to schedule input A for a change at time t_i. To do so, we check A's schedule marker. If it is not busy, we take the output pointer from cell A, location 2, and attach it to the linked list at scheduler slot t_i (assumes an integral timing scheduler). If nothing is scheduled at time t_i, then schedule location t_i contained a pointer to one of the thresholds TH1, TH2, or TH3. The threshold pointer is placed in location 2, while schedule location t_i receives the value 9.

If other elements are already scheduled for time t_i, then this operation automatically links the descriptor cells. Suppose C had already been scheduled. Then the schedule contains the value 14 and location 8 contains the threshold pointer. To schedule a change on A, its output pointer is exchanged with the slot on the vertical list. Slot t_i on the vertical list then contains 9, location 2 contains the value 14, and location 8 contains a pointer to threshold TH1, TH2, or TH3. Therefore, at time t_i a change on the first input of D will be simulated, as will a change on E. When processing for time t_i is complete, all pointers are restored to their original values.

If an element is busy, as indicated by its schedule marker, and it must be scheduled a second time, it becomes necessary to obtain an unallocated memory cell for scheduling this second event. The address of the spare cell is placed in the schedule and the spare cell contains a pointer to the cell to be scheduled. If other events are scheduled in the time slot, then this spare cell must also contain a link to the additional events.

EXAMPLE ■

We use the circuit of Figure 4.11 to illustrate nominal delay simulation. The alphabetic characters inside the logic symbols represent the gate name and the numbers represent gate delay. All nets are initially set to the x state. At time t_0 input D changes from x to 0. At time t_2 the inputs are set to the values $(x, 1, 1, 1)$. At time t_4 input A changes from x to 0 and input C changes from 1 to 0. At t_8 input C changes back to 1.

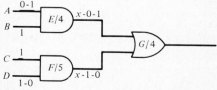

Figure 4.11 Circuit to illustrate timing.

	A	B	C	D	E	F	G	Comments
t_0:	x	x	x	0	x 0	x 1	x 0	Simulate F, schedule it for t_5
t_2:	x	1	1	1	x 0	x 2	x 0	Simulate E and F, schedule F for t_7
t_4:	0	1	0	1	x 1	x 3	x 0	Simulate E and F, schedule E for t_8, F for t_9
t_5:	0	1	0	1	x 1	0 2	x 0	$F \leftarrow 0$, simulate G, no change
t_7:	0	1	0	1	x 1	1 1	x 1	$F \leftarrow 1$, simulate G, schedule G for t_{11}
t_8:	0	1	1	1	0 0	1 2	x 1	$E \leftarrow 0$, simulate F and G G unchanged, schedule F for t_{13}
t_9:	0	1	1	1	0 0	0 1	x 2	$F \leftarrow 0$, simulate G, schedule G for t_{13}
t_{11}:	0	1	1	1	0 0	0 1	1 1	$G \leftarrow 1$
t_{13}:	0	1	1	1	0 0	1 0	0 1	$G \leftarrow 0$, $F \leftarrow 1$, schedule G for t_{17}
t_{17}:	0	1	1	1	0 0	0 0	1 0	$G \leftarrow 1$

In this table, the times at which activities take place are indicated, as well as the values on the inputs and the gates at those times. For each of the logic gates there are two values; the first is the logic value on the output of the gate and the second is the value of the schedule marker. The comments indicate what activity is occurring. For example, at time t_0 input D changes, so gate F is simulated; its output changes from x to 0, so it is scheduled for time t_5 and its schedule marker is incremented to 1. At time t_2, E and F are both simulated because of input changes. There is no change on the output of E and its schedule marker is at 0, so it is not scheduled. However, F does change from its present value so it is scheduled for update at time t_7 and its schedule marker is again incremented. The remaining entries are similarly interpreted. Note that at t_8, the output of F has the value 1, and it is simulated with (1, 1) on its inputs. Although the simulation result is a 1, F is put on the scheduler because its schedule marker is nonzero. ∎

Evaluation Techniques A number of techniques exist for performing element evaluation for the basic logic gates. For AND gates and OR gates, evaluation can be performed on inputs, two at a time, using AND and OR operations of the host computer's machine language instruction set. As we saw, it also works for ternary algebra. It is also possible to assign numerical values to the ternary values as follows:

$$0 — 1$$
$$x — 2$$
$$1 — 3$$

Then the AND of several inputs is the minimum value among all inputs and the OR is the maximum value among all inputs.

For binary values, it is possible to count 1s on AND gates and count 0s on OR gates. If an n-input AND gate has $n - i$ inputs at 1, for $i > 0$, then the output evaluates to 0. Whenever an input changes, the number of inputs at 1 is incremented or decremented. If the number of inputs at 1 reaches n, the output is assigned the value 1. A similar approach works for an OR gate except that it is necessary to count 0s.

Logic gates can also be evaluated using a truth table. This approach has the advantage that it is flexible enough that it will work with any combinational structure whose behavior can be described by a truth table. It is quite efficient when the input values are grouped together in the descriptor cell so that the processing program can pick up the input values as a group and use them to index directly into a table containing the output response corresponding to that input combination. Table lookup is not restricted to binary valued simulation; it can also be used for n-valued simulation, $n \geq 3$. It requires $\log_2(n)$ bits for each input and the table can become excessively large, but the simulation is quite rapid. For example, when evaluating logic gates such as the OR and the AND, three- or four-valued simulation would require two bits for each input. A six-input gate then requires a truth table, or lookup table, of 4096 two-bit entries. Only one table would be necessary because an AND (OR) gate with fewer than six entries could be computed with the same table by setting unused inputs to 1s (0s). Furthermore, since AND and OR are both associative operations, a gate with more than six inputs could be computed by using successive lookups.

The zoom table takes the truth table one step further. Rather than examine the function code to determine the gate type, truth tables for all the primitive functions are placed in contiguous memory. The function code is appended to the input values by placing the function code adjacent to the input values in the descriptor cell. Then, the catenated function/input value serves as an index into a much larger truth table to find the correct output value for a given function and set of inputs. The program implementation is more efficient because there are fewer decisions to be made; one simple access to the value table produces the value regardless of the function.

Race Detection in Nominal Delay Simulation The zero delay simulator resorted to multiple value simulation to detect transient pulses caused by hazards. These unwanted signals are caused by delay in physical elements and can be detected by the nominal delay simulator using just the logic values {0, 1} and individual element delay values—if the transients occur for nominal delay values. However, a hazard is only the possibility of a spurious signal and the transient may not occur at nominal delay values. But individual physical elements usually vary from nominal ratings, and some combination of real devices, each varying from its nominal value, may combine to cause a transient which would not have occurred if all elements possessed their nominal values. To further complicate matters, a transient may be innocuous or it may cause erroneous state transitions. In a circuit with many hundreds or thousands of elements, how do we decide what delay values to simulate? Do we simulate only nominal delays? Do we also simulate worst-case delays?

Consider again the cross-coupled NAND latch. Erroneous behavior can occur if unintended pulses arrive at either the Set or Reset input. If the latch is presently reset and a negative pulse of sufficient duration occurs on its Set input, it becomes set. This situation may occur only for delay values that are significantly beyond nominal values. Furthermore, in a circuit with many thousands of gates there may only be a few asynchronous latches that are susceptible to glitches.

Potential problems can be addressed by identifying nets connected to the inputs of a latch and grouping these nets together as a set. If an input to an asynchronous latch changes value during simulation, then other nets in the same set are checked for their most recent change. If another net has previously changed within some user-specified interval, a race exists. The race may be critical if some combination of delay variances causes the first input change to occur later than the second input change. To determine whether this is the case, trace the changing signals back to primary inputs or to a common origin. Increase the delay of all elements along the path whose signal change occurred first. Decrease the delay of the elements along the path to the latch input that changed last, then resimulate. If this causes a reversal in the order in which the two inputs change, then a critical race exists.

Subsequent action depends on the reason for the simulation. For design verification, an appropriate course of action is to provide a message to the user that primary input signals are changing too close together or that a signal change at a gate with fan-out causes a critical race. If patterns are being developed for test pattern generation, then a state transition which is dependent upon the order in which two or more inputs change indicates a problem because it may become impossible to obtain repeatable tests that are usable on automatic test equipment. One possible solution is to alter the input stimuli by delaying one or more of the input stimuli to a later time period. This is sometimes referred to as "deracing." If the race results from a signal change at a common fan-out point, then somewhere along one of the two paths it may be possible to identify a gate by means of which a signal change can be inhibited. This is illustrated in Figure 4.12. A signal change reaches both the Set and Reset inputs of a latch. One path goes through an OR gate, the other path goes through other combinational logic. The signal change through the OR gate may be inhibited by first setting a 1 on the other input.

Min-Max Timing The earliest and latest possible times at which a signal may appear at some point in a circuit can be determined through the use of min-max timing simulation. In this method, each element has a minimum and a maximum switching time associated with it. During simulation, the earliest and latest switch-

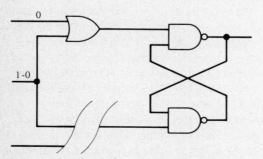

Figure 4.12 Inhibiting a propagation path.

4.5 THE EVENT-DRIVEN SIMULATOR

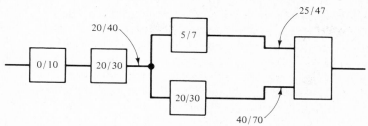

Figure 4.13 Min-max timing.

ing times of a changing signal are computed as the signal propagates through the circuit. As each element is simulated, its minimum and maximum values are added to the cumulative min and max, respectively. The time interval between the earliest and latest times at which a signal switches at a given point in the circuit is called the *ambiguity region*.

The circuit in Figure 4.13 illustrates the computation of minimum and maximum delay times. The first block contains the numbers 0 and 10. These could represent the range of uncertainty as to when a signal arrives at a board from a backplane or from a tester due to skew caused by wiring, fixtures, and so on. The next block represents logic with a timing range of 20 to 30 nanoseconds, after which the circuit fans out to two other blocks. The upper path has a cumulative delay ranging from 25 to 47 nanoseconds by the time it arrives at the last block, and the bottom path has a cumulative delay of 40 to 70 nanoseconds. If the rightmost block represents an AND gate and if the signal arriving at the upper input is a falling signal, and the signal arriving at the lower input is a rising signal, then the numbers indicate a time region from 40 to 47 when there is uncertainty about the output value, because the numbers imply that the signal at the lower input may rise as early as time 40 and the signal at the upper input may not fall until time 47.

A more careful analysis of the circuit reveals that there is a component 20/40 which is common to both signal paths. This is called *common ambiguity*. If we subtract the common ambiguity, we see that the upper path will arrive at the AND gate no later than 7 nanoseconds after it fans out from the common element. The signal on the lower path will not arrive until at least 13 nanoseconds after the upper input change arrived. If this common ambiguity is ignored, then a pulse is predicted on the output of the gate and propagated forward when it could not possibly occur in the actual circuit. This pulse could result in considerable unnecessary simulation activity in the logic forward of the point where the pulse occurred.

If the block on the right is an edge-triggered delay flip-flop in which the upper input is the Data input and the bottom input is the Clock input, then results of the common ambiguity may be more catastrophic. Because of the common ambiguity, it is impossible to determine whether the data arrived prior to the clock or after the clock. Hence, it is necessary to set the flip-flop to x. To get accurate results, the common ambiguity must be removed.

A common ambiguity region can be identified with the help of the *causative link*.[12] This is simply a pointer included in the descriptor cell which points back to the descriptor cell of the element that caused the change. If two inputs change on a primitive and there is overlap in their ambiguity regions, then the simulator traces back through the causative links to determine whether there is a common fan-out point which caused both signal changes. If a common source is found, then the ambiguity at that point is subtracted from the minimum and maximum change times of the two signals in question. If there is still overlap, then the block currently under consideration is set to x during the interval when the signals overlap if it is a logic gate or its state is set to x if it is a flip-flop.

4.5.3 Unit Delay Simulation

Unit delay simulation operates on the assumption that all elements possess identical delay time. A major advantage of unit delay simulation is the fact that it is easier to implement than is nominal delay simulation. The unit delay simulator does not require the timing wheel; a simple stack is sufficient for scheduling purposes. In fact, when every element has unit delay the READ/WRITE array implementation described earlier is sufficient since each pass through the simulator corresponds to advancement of signal changes through one level of logic. Primary inputs can be changed while other changes are still propagating to outputs. When copying the WRITE array into the READ array, if the entries that have changed during the simulation pass are flagged, then performance can be enhanced by simulating only the elements driven by elements whose output value has changed.

When creating test stimuli for a timing-sensitive circuit, the unit delay simulator can give a false sense of security. Timing for the actual circuit may not resemble the results predicted by the unit delay simulator so it may be necessary, when using the simulator to create a test file for automatic test equipment, to insert additional gates with unit delay into a circuit model in order to force the simulator to predict correct circuit response to a given set of input stimuli.

4.6 TIMING VERIFICATION

The zero delay simulator can detect hazards during simulation by using multiple values. However, because it collapses all computations into zero time, it cannot predict their relative time of occurrence. The nominal delay simulator, with its additional computational power, can determine relative time of occurrence of signal changes on circuit elements and can explicitly identify multiple signal changes which occur within a specified interval. In either case, multiple signal changes appearing at latch or flip-flop inputs must be detected so as to avoid ambiguous simulation results at the latch or flip-flop outputs. The need to observe signal changes on all latch or flip-flop inputs is characteristic of asynchronous logic.

As systems grow larger, and design, simulation, test, and maintenance grow more difficult, synchronous design techniques become more attractive. The use of master clock(s) to synchronize events reduces or eliminates the need for accurate timing simulation and makes it possible to simulate logical and functional behavior in a zero delay environment. If, in addition, the system is provided with a master

4.6 TIMING VERIFICATION

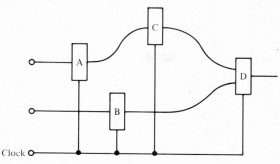

Figure 4.14 Synchronous circuit.

reset that forces all memory elements into a known initial state, then it is possible to dispense with the x value and restrict simulation to the Boolean values 0 and 1. The key feature of this design methodology is the fact that *all* registers and flip-flops are controlled by one or more clock signals that either are not gated with combinational logic or are gated only within the framework of a very closely controlled set of design rules. This operation is illustrated in Figure 4.14 for a circuit with a single clock. The elements labeled A, B, C, and D may be registers or single flip-flops. At no time in this circuit does any clock pulse result from operations performed in combinational logic. Since the clock signals are closely controlled, their behavior is predictable and not subject to circuit hazards or critical races. This obviates the need for race and hazard detection in simulation software.

Zero delay simulation requires some means for computing propagation delay along signal paths to insure that they are not too short or too long. With excessive delay, a signal will not reach its destination before the clock pulse. If the delay is too short, the hold time requirements for a flip-flop may be violated. Two methods have been developed for performing timing verification: path enumeration and block-oriented analysis[13].

4.6.1 Path Enumeration

The path enumeration techniques start at particular elements, either I/O pins or stored state variables, and trace through the logic until some termination point is reached, either an I/O pin or a stored state variable. The delays of the elements encountered along the paths are added to a cumulative total as the program traces the path. Rise and fall times may both be used to precisely calculate propagation time[14].

EXAMPLE ■

We use the circuit of Figure 4.15 to illustrate the method. To calculate the propagation time required for a signal originating at E to reach L, we start at L and work back toward the inputs. We assume a rising signal has reached L. Therefore we begin by using the rise time for gate K as our initial sum. We add to this the rise time for gate I and the rise time for gate J. We add the fall time for G because the 0-to-1 transition at the output of gate J requires

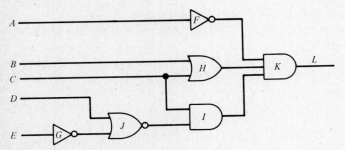

Figure 4.15 Path analysis circuit.

a 1-to-0 transition at its input. We next calculate the propagation time for a falling signal to reach gate L. We add the fall time for gate K, for gate I, and for gate J. We then add the rise time for gate G. The larger of the two sums becomes the propagation time from E to L. ■ ■

An important point in the rationale for timing verification is the fact that, at some point during operation of a circuit, the signal along the path being calculated will be the controlling signal for that output, in the sense of being a sensitized path. For example, if inputs A, B, C, and D in Figure 4.15 are assigned the values $(0, 0, 1, 0)$, then the output is totally dependent on the value assigned to input E. If it has value 0 (1), then output L has value 0 (1). When the path being analyzed is the controlling signal, we want to determine which signal originating at the input, 0 or 1, takes longer to propagate to the output. We then want to find, among all paths into a bistable, that path which has maximum propagation delay when it is the controlling signal. The implicit assumption that all other signals are set up to propagate the signal whose delay is being calculated permits us to ignore the logic function performed by the elements along that path. It is only necessary to know the rise and fall delays of each element and whether or not the element inverts the signal.

4.6.2 Block-Oriented Analysis

In this method the program starts at some assumed time with signals at primary inputs and bistables. Furthermore, required arrival times are assigned to destination elements. The elements, or blocks, which are driven by the primary inputs and bistables are processed to find the earliest and latest time at which a signal could propagate through them. Then elements driven by these elements are processed. In general, no element is processed until all elements driving its inputs are processed. This requires that the circuit be levelized.

The block-oriented method identifies the worst path leading up to each block and feeds this information forward. This is continued until a primary output or bistable is reached. Then the difference between the required arrival time and the propagation time is computed. This value is called *slack*. A negative slack indicates excessive propagation time.

4.6 TIMING VERIFICATION

After all paths have been propagated forward, computations are performed in the opposite direction. The propagation value at the element which drives the primary output or bistable is subtracted from the required arrival time to determine when the signals must arrive at the inputs to this block. The previously computed propagation numbers are subtracted to find the slack at the inputs to this block, and the process is continued until the source elements are reached.

EXAMPLE ■

We again use Figure 4.15. Suppose each of the elements has an identical rise and fall delay of 5 units. Also, suppose input changes occur at time 0 and the required arrival time is 18 units later. Gates F and H can both be processed to give a delay of 5 units on their outputs, but J cannot be processed until G is processed. After G is processed, the delay at the output of J is the greater of the values on D and G plus the delay of J. Since the delay at G is 5 units, the delay at J is 10 units. In similar fashion, the delay at I is 15 units and the delay at primary output L is 20 units, which results in a slack of -2 at the output.

The computations are now performed in reverse, starting with the required arrival time and using the previously calculated propagation times. The slack numbers on the inputs to K are $+8$, $+8$, and -2, derived by computing the required arrival time at the inputs to K, $18 - 5 = 13$, and subtracting from that the propagation delay at the outputs of F, H, and I. The required arrival time at the inputs to F, H, and I is $13 - 5 = 8$. The slack at the inputs to F and H is 8 and the slack numbers at the inputs to I are $+8$ and -2. Continuing, we find that the slack at E is -2 and a critical path with excessive propagation time has been identified. ■ ■

If looking for early arrival times, the computations use minimum arrival times. If separate rising and falling times are used, then pairs of numbers are maintained and inverting elements must be identified. The falling edge on the inverting element output is computed by taking the greater of the rising propagation delays at its input and adding the falling delay of the element.

The object of timing verification is to find signal paths having long (or short) delay times. If the propagation time along such paths is excessive, the path delay can be reduced by redesigning the logic, by selecting faster components, or by assigning different physical dimensions to elements within an IC. A consequence of redesigning circuits to switch faster is that they may then consume more power. This increased power consumption may be offset by finding other signal paths where the timing margin is greater than it needs to be and, if the technology permits, redesigning the devices to switch more slowly and thus conserve power[15].

A major benefit of timing verification is the fact that signal paths are not overlooked because test patterns were not written to exercise those paths; all paths are analyzed. However, some practical considerations must be taken into account. Path enumeration can generate large amounts of data. It may be necessary to reduce the amount of data provided to the user. Therefore, options must be provided which permit the user to specify printout only of paths which fall within some user-defined range, either above or below some threshold value. For engineering design changes, it is not necessary to recompute all paths; therefore the user should have an option to specify signal paths of interest. Other considerations include the ability to detect latches which are not specifically identified as latches, or to prohibit them in the design since a feedback path could cause the program to enter an infinite loop.

Clock circuits which enter a board or IC often fan out to many destination elements, requiring that the clock be distributed through repowering circuits in order to have sufficient drive capability. These repowering circuits must be analyzed to determine whether they shorten or lengthen the clock pulse width. Clock skew must be factored into the overall analysis somewhere, since the time required for the clock signal to reach numerous devices throughout a design, whether a chip or board, can vary significantly. The user may have to be careful to spot paths which are called problem paths but which require logic combinations that cannot occur in practice, as well as paths which exceed a clock period but which are known to require two or more clock cycles to complete their operation.

4.7 MULTIPLE-VALUED SIMULATION

We have already seen multiple-valued simulation used when necessary to represent indeterminate values or to detect hazards. It is also useful in applications where it is necessary to simulate circuit behavior that does not strictly conform to the rules of Boolean algebra.

4.7.1 Tri-State Logic

A tri-state circuit is one in which the output may go to a logic 1 or logic 0 state, or it may have no effect on the remainder of the circuit. In this third state, the output assumes a high impedance and in effect is disconnected from the circuit. This circuit is used when the outputs of two or more devices are connected together and alternately drive a common electrical point, frequently called a bus. A circuit employing two tri-state drivers is illustrated in Figure 4.16.

When the tri-state control line A is high, the AND gate controlled by A behaves as an ordinary two-input AND gate. When $A = 0$ the output assumes the high-impedance state, represented by the symbol Z. With the high-impedance capability, two or more of these outputs can be tied directly together. However, if this is done, one rule must be observed. Two tri-state control devices must not be active at the same time. In Figure 4.16, A and D must not be simultaneously high.

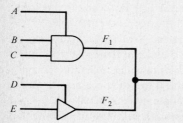

Figure 4.16 Circuit employing tri-state drivers.

4.7 MULTIPLE-VALUED SIMULATION

If they are both high, and if the output of one device is low and the output of the other is high, then there is a virtual short circuit from ground to V_{CC} and in a very short time one or both of the devices may overheat and be permanently damaged.

Circuit designs normally do not permit two or more tri-state outputs to be active simultaneously. However, design errors do occur and the designer may want to employ simulation in order to identify those errors that could lead to simultaneously active drivers. This imposes the requirement that a simulator be able to correctly predict the behavior of bus-oriented circuits in which two or more tri-state drivers are connected to the bus. Another potential for conflicting drive signals exists when creating tests for logic boards. Sometimes the control logic for the tri-state devices is located on another board. In the environment in which it is intended to operate, it will not be possible for any pair of tri-state controls to be simultaneously active. But when it is being tested, the board is operating in an artificial environment as a stand-alone entity. When this happens, an ATPG may assign signals to board edge pins which simultaneously activate two or more tri-state drivers. It is important that this situation be identified and corrected.

To resolve problems that may occur when the outputs of tri-state drivers are tied together, the LASAR simulator[16] augments a three-state $\{0, 1, x\}$ simulator with seven additional values to create a ten-state simulator. Consider again the circuit of Figure 4.16. For the various combinations of values of inputs D and E, the result on output F_2 is given by the following table:

D	E	F_2
1	0	$AC:0$
1	1	$AC:1$
1	x	$AC:x$
x	0	$AC?0$
x	1	$AC?1$
x	x	$AC?x$
0	—	Z

In this table the first three entries, $AC:0$, $AC:1$, and $AC:x$, correspond to situations where the tri-state device is actively driving the net connected to its output. The next three, $AC?0$, $AC?1$, and $AC?x$, depict the situation where the device may or may not be controlling the output, depending on the value assumed by the control line. The last entry corresponds to the situation in which the device is definitely known to be inactive and the output of the device assumes the high-impedance state regardless of the value on input E.

To evaluate the outcome when two or more tri-state devices are driving a common electrical point, the ten states are prioritized. The Z state is given lowest priority. When one device is in a high-impedance or Z state, any other device driving the same net with other than a Z signal will dominate. The next priority is the 1 state. Then the remaining eight states constitute the highest priority states. These states will dominate a Z state or the 1 state on another device connected to the same net. By assigning output values as a function of the control and data inputs, and by establishing a table to resolve the outcome when two or more drivers inject signals into the same net, it is possible to determine the output and to identify real or potential conflicts. For example, in Figure 4.16, if F_1 is at $AC:0$ and F_2 is at $AC:1$, then a conflict exists. Four types of bus conflicts are identified: bus state, potential bus state, bus activity, and potential bus activity.

A bus state conflict occurs when two or more signals attempt to drive a bus to opposite states. If the inputs are such that the outputs could be opposite but one or more of the control lines are indeterminate, then a potential state conflict exists. A bus activity conflict exists when two or more sources drive the bus to the same state, and potential bus activity exists when the inputs are such that the outputs are at the same state but one or more tri-state control signals are indeterminate.

EXAMPLE ■

Referring to Figure 4.16, the conflict messages are as follows for the indicated output signals on F_1 and F_2.

F_1	F_2	Conflict message	
0	Z	None	
$AC:0$	$AC:1$	State	(S)
$AC:0$	$AC?1$	Potential state	(PS)
0	$AC:0$	Activity	(A)
$AC:0$	$AC?0$	Potential Activity	(PA)

■ ■

In general, a state conflict occurs if outputs from two tri-state drivers differ. It is a potential state conflict if the possibility exists that the outputs from two or more tri-state drivers differ. An activity conflict occurs if two or more drivers are trying to drive the net to the same state. The conflict is potential if it cannot be verified that the devices are actually driving the net. The response to these conflicts is to inform the user that they exist. The user may decide that activity conflicts are not detrimental. In fact, the simulator resolves activity conflicts by assigning the state value created by the tri-state driver(s) to the net. In the above table, when F_1 is actively controlling the net to a 0 and F_2 may be controlling the net to a 0, the simulator sets the net to 0 and continues with simulation.

4.7.2 MOS Simulation

The MOS technology employs transmission gates. A transmission gate behaves as a switch that is closed when its control input is on and open when its control input is off. They can be used as unidirectional and bidirectional switches. In Figure 4.17 a transmission gate holds a charge on a MOS circuit.

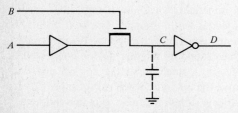

Figure 4.17 MOS circuit with transmission gate.

4.7 MULTIPLE-VALUED SIMULATION

A signal at input A appears at net C if the signal on input B permits it to pass through the transmission gate. When the signal at B is turned off a charge is held at net C for some period of time determined by the RC time constant of stray capacitance and element resistances. In the following table, inputs are applied to A and B in the time sequence indicated and the response is indicated at nets C and D.

Time	1	2	3	4	5
A	0	0	1	1	1
B	1	0	0	1	1
C	0	Z_0	Z_0	1	Z_1
D	1	1	1	0	0

The signals Z_0 and Z_1 represent the logic signal at a net where the transmission gate is off and are called floating logic states[17]. The signal Z_x is used to represent an indeterminate floating signal. Since these signals represent charge stored in the capacitance of the circuit, they are dominated by the normal logic states generated by MOS transistors. If a net is being driven by transmission gates, one of which has a floating logic value and the other has a normal logic value, then the net with the normal logic value provides an escape path for the trapped charge. When a floating logic state controls an inverter, as illustrated in Figure 4.17, not only is the signal inverted when going through the inverter, but it is converted from a floating logic state to a normal logic state.

In Figure 4.18, if the signal at net E is off, then the signal at nets E and F can be computed independently of each other. If it is on, then the signals at F and G are computed as though there were a direct connection from F to G. The transmission gate between them is bidirectional; a signal can pass in either direction. When E is on, the resulting signal on nets F and G can be computed from the following table:

	0	1	x	Z_0	Z_1	Z_x
0	0	x	x	0	0	0
1	x	1	x	1	1	1
x	x	x	x	x	x	x
Z_0	0	1	x	Z_0	Z_x	Z_x
Z_1	0	1	x	Z_x	Z_1	Z_x
Z_x	0	1	x	Z_x	Z_x	Z_x

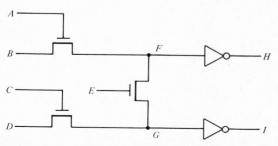

Figure 4.18 Bidirectional transmission gate.

Another factor that requires attention during MOS simulation is *relative* signal strength. A logic device may dominate another even when they are both actively driving a node[18]. Figure 4.19 depicts a random access memory (RAM) cell. In this RAM cell the drive of gate A exceeds that of gate B. This is necessary to permit the values in memory to be overwritten.

The outcome when two signals A and B drive a node and signal A dominates signal B is given in the following table:

		B		
A	0	1	x	Z
0	0	0	0	0
1	1	1	1	1
x	x	x	x	x
Z	0	1	x	Z

4.8 FAULT SIMULATION

We previously examined parallel fault simulation and noted that simulation time is sensitive to circuit size. The efficiency of parallel fault simulation can be improved by implementing it in a table-driven mode and by employing event-driven simulation. However, an input signal change can cause activity in a faulted circuit where none exists in the fault-free circuit, such as when AND gate inputs change from $(0,0)$ to $(1,0)$. An S-A-1 fault on the second input causes activity on the output of the faulted circuit even though the fault-free circuit has no activity. The larger the number of faults simulated in parallel, the greater the likelihood that fault activity occurs without corresponding activity in the fault-free circuit. Efforts to accelerate the parallel fault simulation algorithm are hampered by the fact that individual gates must be redundantly computed numerous times for each set of input signal changes due to the many fault simulation passes performed. This inherent weakness has led to the development of other fault simulation algorithms.

4.8.1 Testdetect

It is possible to determine whether or not an individual fault has been detected by examining signal paths from the fault to observable outputs. This method, called

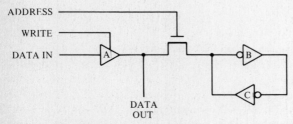

Figure 4.19 RAM cell.

4.8 FAULT SIMULATION

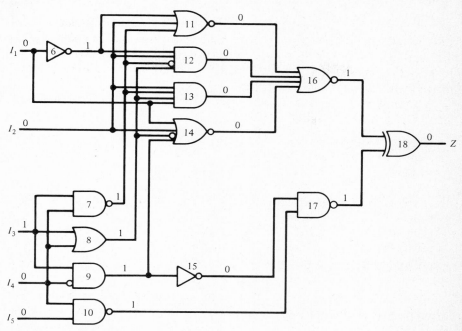

Figure 4.20 Testdetect example.

Testdetect, was developed by Paul Roth[19] and can be thought of as a converse of the *D*-algorithm. To illustrate, consider the circuit of Figure 2.5, reproduced here as Figure 4.20. We apply the inputs $I_1, I_2, I_3, I_4, I_5 = (0, 0, 1, 0, 0,)$. The signal value on each internal net is indicated on the diagram. Obviously, if the output is S-A-1, it is detectable. Some other detectable faults include S-A-0 faults on either input to gate 18 and S-A-1 faults on both the upper input to gate 17 and the output of gate 15. We cannot detect S-A-0 faults on any of the inputs to gate 16 since the good machine value is 0 for all inputs, but we can detect S-A-1 faults on the output of any gate driving gate 16.

We started at the output and noted any signal change on any net, caused by a stuck-at fault, which would ultimately cause the output to change value. Testdetect formalizes this approach. It selects a fault and determines whether a *D*-chain can be extended from this fault to an observable output. However, in this converse *D*-algorithm, all signal values are fixed. The objective is not to create a test but rather, having created a test, to determine what other faults are detectable by the input vector. Therefore, we only want to determine, for a given fault, if its effects propagate through a given gate and eventually to an output.

The method employs a *D-list* to maintain a list of gates in the *D*-frontier while progressing toward primary outputs. A gate is selected from the *D*-list and it is determined whether the fault will propagate through the gate. If not, then the *D*-chain has died on that path and if the *D*-list is empty the fault will not be detected by that test vector. If the fault does propagate through the gate, then the gate or gates in the fan-out from that gate are placed in the D-list. This continues until either

1. A primary output is encountered, or
2. The D-list becomes empty.

A third criterion for stopping exists:

Lemma. If at any stage in the computation for failure F, the D-frontier reduces to a single net L and there is no reconvergent fan-out beyond the D-frontier, then F is testable if and only if L is testable[20].

The rules for determining whether or not a fault propagates through an element are the same as those used in the D-algorithm. For an AND gate with a D or $\overline{D}$ on an input (or inputs), if the other inputs are all 1s, then the D or $\overline{D}$ will propagate to the output of the gate. In general, if the good machine signal causes a 1 (0) on the output of the gate and the fault causes a 0 (1), then the fault signal propagates to the output of the gate.

EXAMPLE ■

For the circuit of Figure 4.20, the bottom input of gate 18 has a 1. If it is S-A-0, then the values $(1, D)$ on its inputs cause a $\overline{D}$ on the output so the S-A-0 is detected. If the upper input of gate 17 is S-A-1, then $(\overline{D}, 1)$ produces a D on the output. By the lemma, the fault is detected. Likewise, faults on gate 15 are detected. However, the output of gate 9 must be analyzed to the output because there are two gates, 14 and 15, in its D-list.

We assign a D to the output of gate 9, corresponding to the output S-A-0, and place 14 and 15 in the D-list. We assume that the circuit has been leveled, and require that when there are two or more entries in the D-list, the lower numbered gate be selected first. Therefore, gate 14 is selected for processing. The inputs to gate 14 are $(0, 0, 1, D)$. Since the 1 on the third input is inverted (indicated by the bubble) the output of 14 is a $\overline{D}$. This causes gate 16 to be placed in the D-list. Gate 15 is processed next. The D on its input causes a $\overline{D}$ on its output, so gate 17 is placed in the D-list. Gate 16 is processed and a $\overline{D}$ appears on its output, causing gate 18 to be placed in the D-list. When gate 17 is processed, the D on its output is placed on the other input to gate 18. The D and $\overline{D}$ signals on the inputs to gate 18 cancel each other, hence the fault on the output of gate 9 is not detected by this test pattern. ■ ■

In Figure 4.21 we see a circuit in which the D-list reduces from two elements to a single element which is detectable, yet the fault which initiated processing is not detectable by the given test pattern. The output of gate 5 S-A-0 is detectable by the pattern $(1, 0, 0)$ on inputs 1, 2 and 3. However, if we assign $\overline{D}$ to primary input 2, to test for a S-A-1 at that input, then the output of gate 4 is a $\overline{D}$ and the

4.8 FAULT SIMULATION

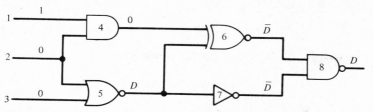

Figure 4.21 Recombining sensitized paths.

output of gate 5 is a D, so they are both placed in the D-list. We then choose gate 4 for processing. It goes to gate 6, which now has $(\bar{D}, D)$ on its inputs. That produces a 0 on the output of gate 6. We are now left with only gate 5 in the D-list, and that was previously determined to be detectable by the applied pattern, yet the S-A-1 at primary input 2 is not detectable because the 0 on the output of gate 6 prevents the D at gate 5 from reaching the output.

4.8.2 Deductive Fault Simulation

The failure to detect the fault on the output of gate 9, even though it drives gates on which faults are detected, is caused by the fact that a signal reconverges and two sensitized paths subsequently cancel each other. If there were no problems caused by reconverging logic, Testdetect could run quite rapidly and work straight from the outputs back to the inputs. However, reconvergent fan-out makes it necessary that any fan-out point be examined using D-lists.

The need to check for testability at every net where fan-out occurs can significantly slow down the Testdetect procedure. Deductive fault simulation[21] starts at the inputs and works toward the outputs. Simulation only takes place for the fault-free machine. The simulator deduces what faults are tested by an input vector and creates lists of those that are sensitized at any given node. As simulation proceeds, some faults cease to be sensitized, their symptoms become masked, and they are dropped by the simulator. Meanwhile, other faults become sensitized and are added to the list of sensitized faults.

To illustrate, consider the fault-propagating characteristics of a three-input OR gate. We associate with each input a list of faults from preceding logic that are sensitized to that point. If the present values on the OR gate inputs are all 0s, then the fault list on the output of the OR gate is the union of the fault lists on all the inputs. This follows from the fact that the fault list on any input is the set of faults that cause that input to assume a value that is opposite to its correct value. On the other hand, if the fault-free signals at all three nodes are 1s, then a fault symptom would be propagated only if it could cause all three inputs to assume an incorrect value. Therefore, the set of faults that propagates to the output of the OR gate is the set that results from the intersection of the fault lists at the three inputs. If one or two inputs are at 1 and the other is at 0, then the computations become slightly more complex.

EXAMPLE ■

Suppose we have an OR gate in which the fault lists are:

$$A = \{1, 2, 4, 7, 11\}$$
$$B = \{2, 5, 7, 8\}$$
$$C = \{1, 3, 7, 12\}$$

If all three inputs are at logic 0, then the output fault list D is the set $D = A \cup B \cup C \cup \{d_1\}$ where d_1 is the output of the OR gate S-A-1. In the present example $D = \{1, 2, 3, 4, 5, 7, 8, 11, 12, d_1\}$. If all three inputs are at logic 1, then the output fault list D is the set $D = A \cap B \cap C \cup \{d_0\}$, where $\cap$ denotes set intersection and $\{d_0\}$ denotes the output S-A-0. In this example $D = \{7, d_0\}$. If the upper two inputs are logic 1s and the lower input is 0, then an incorrect output will be obtained only if a fault f can change the values of the upper two inputs but does not change the lower output, that is, if $f \in A \cap B - C$. In our example, if fault 2 occurs, it will cause the wrong value to appear at the output. To that result we add the output fault which causes the gate to assume the opposite value and get $D = \{2, d_0\}$. If a single input is at 1, then that input S-A-0 will also propagate to the output and hence must be added to the list. ■ ■

A general rule for processing OR gates is as follows:

To the fault list at each input, add the fault corresponding to that input S-A-0 if the value on that input is a 1.

If all inputs are 0, then form the union of all these sets and add the fault corresponding to the output S-A-1.

If one or more inputs are 1, then
    form the intersection S of sets corresponding to inputs that have 1s,
    form the union T of sets corresponding to inputs that have 0 values,
    compute $S - T$,
    add the fault corresponding to the output S-A-0.

4.8.3 Concurrent Fault Simulation

Deductive fault simulation can require processing of enormous lists of faults using equations for manipulation of these lists which vary according to the values on the inputs of the gate being processed. In an event-driven environment, extensive list processing may be required even when no logic activity occurs. For example, if the three-input OR gate has values $(1, 1, 0)$ on its inputs, and if the inputs change to $(1, 0, 0)$ in response to a logic change, then the formula for computing the output fault list changes; hence the output fault list for the gate must be recomputed, even though no logic activity occurred on the output of the fault-free gate. If the fault list on the output of that gate changes, then the fault list must be recomputed forward for gates in the fan-out list of that gate, and this must be continued until fault list changes cease. Further complications occur when performing n-value simulation, $n \geq 3$, and when sequential circuit simulation is performed.

4.8 FAULT SIMULATION

The concurrent fault simulator avoids set union, intersection, and difference calculations by appending or linking new copies of a circuit element to the original element whenever faults cause faulted machine signals to differ from good machine signals. Furthermore, new circuit elements are added as long as the error signal continues to propagate. This is illustrated conceptually in Figure 4.22. In (a) the fault-free circuit is illustrated with correct logic values at each net. In (b) a modified version is illustrated in which each of the gates is replicated several times. In the following discussion, the element Y is followed by the subscript i, which is interpreted as follows:

$$
\begin{array}{ll}
0 & \text{— fault-free machine} \\
1 & \text{— input 1 S-A-}x \\
& \quad \ldots \\
n & \text{— input } n \text{ S-A-}x \\
n+1 & \text{— output S-A-0} \\
n+2 & \text{— output S-A-1}
\end{array}
$$

where the element Y is assumed to have n inputs and S-A-x denotes S-A-1 for an AND gate, S-A-0 for an OR gate.

The purpose of the multiple copies of the various gates is to represent, simultaneously, the fault-free gate as well as instances of the gate in which faults originate, and instances of the gate in which the logic value at an input is affected by faults occurring on other gates. The concurrent fault simulation algorithm recognizes two classes of faults: fault origins and fault effects. A *fault origin* (FO) is a gate at which a fault originates. An input fault origin (IFO) occurs on a gate input and an output fault origin (OFO) occurs on the output. Fault origins are linked together and attached to the unfaulted gate. A separate FO is used for each fault. If an FO causes a destination gate to have an input value that differs from the fault-free gate, then a *fault effect* (FE) is created or *diverged* and attached to the fault list of the destination gate. Whenever the output of an FO or FE is different from the corresponding unfaulted circuit, the FE or FO is said to be *visible*. When the output of an FE or FO is visible, then an FE is diverged at the destination gate. FEs continue to be diverged forward in the circuit until either the error signal is no longer visible or a primary output is encountered. When the error signal is no longer visible, the FE is *converged*[22].

These concepts are illustrated in Figure 4.22(b). Note first that there are five copies of gate G. The copy G_0, driven by inputs A and B, corresponds to the fault-free circuit. The remaining four copies are all IFOs. Copy G_1 (G_3) has one input S-A-1 (S-A-0) and the other input driven by input B. Copy G_2 (G_4) has one input S-A-1 (S-A-0) and the other input driven by input A. There are three copies of gate F, one corresponding to the fault-free circuit and two FOs corresponding to the output S-A-0 and S-A-1, respectively, Gate H has a fault-free copy H_0 and IFOs for stuck-at faults on each of its inputs as well as OFOs for S-A-O and S-A-1 faults on its output. It also has an FE, which consists of unfaulted copy H_0 driven by fault origin F_1. Gates J and K also have several copies which are interpreted similarly.

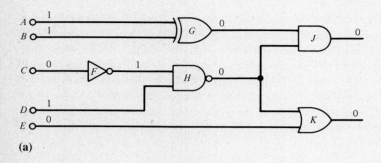

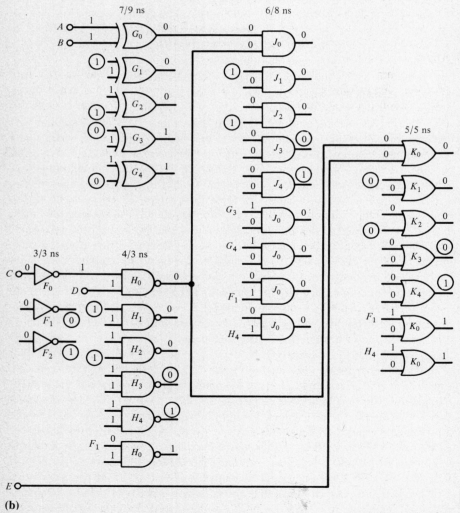

Figure 4.22 (a) Circuit for concurrent fault simulation. (b) Circuit with linked fault effects.

4.8 FAULT SIMULATION

The circled logic values in the figure are used to denote signals which are S-A-1 or S-A-0, hence the gates at which they occur are IFOs or OFOs. FEs are indicated by an unfaulted copy of a gate in which one or more inputs are sourced by an FO or FE. In the discussion which follows, we use the notation X_0/Y_i to represent a fault effect which originates at fault origin Y_i and is diverged to drive an unfaulted copy X_0 of X. The rise and fall delays for the elements are indicated above the unfaulted copy of the elements.

Before describing the rules for concurrent fault simulation, we informally describe what happens when a signal change occurs. Given the signal conditions and the attached fault effects indicated in Figure 4.22(b), suppose that primary input D changes to 0. It drives not only the unfaulted circuit H_0 but also some copies, including H_1 and the fault effect H_0/F_1. Fault origin H_2 is unaffected by the signal change because the gate input connected to primary input D is stuck-at-1. The OFOs H_3 and H_4 are unaffected by any input change. The gate H_0 in the unfaulted circuit must be simulated. The corresponding gates H_1 and H_0/F_1 in the faulted circuit must also be simulated.

When H_0 is simulated, its output switches from 0 to 1, therefore it must be scheduled for processing at time $t + 4$. Gate H_1 also changes but the value on H_0/F_1 does not change, therefore H_1 is scheduled but H_0/F_1 is dropped from further processing. Gates H_0 and H_1 are retrieved from the scheduler at time $t + 4$ and their outputs are updated. Then fault lists attached to gates in the fan-out of gate H_0 are processed. We describe here only the processing for gate J_0. Fault effects H_4 and H_0/F_1 no longer differ from H_0 so they are converged and dropped from the fault list attached to J_0. However, H_2 and H_3 now differ from H_0 so those fault signals must be linked to the fault list attached to J_0, that is, they are diverged at J_0. Also, the fault-free signal change on H_0 reaches the lower input of FEs J_0/G_3 and J_0/G_4, so those FEs must be simulated. Since the outputs of those FEs change, they must be placed on the scheduler.

The fault origin H_1 was also simulated. Its output is identical to the output of the unfaulted copy. A check of the fault list attached to J_0 shows that there is no fault effect labeled H_1 in the list, so no further processing need take place. Those fault effects which eventually reach a primary output, in this case J_4, J_0/G_3, and J_0/G_4, define a sensitized path from the fault origin to the output, hence they correspond to detected faults.

It is possible that the faulted copy could change and the unfaulted copy not change. For example, if the change on input D is followed by a change on input C, then H_2 will change while H_0 remains unchanged. In that case, it is necessary to trace the faulted output change to the destination gate(s) and perform divergence and convergence, as the situation warrants. It is also possible that the unfaulted copy may change in one direction and the faulted copy change in the opposite direction, as would be the case when primary input A changes. G_0 and G_2 change to 1, G_1 and G_3 are unaffected, and G_4 changes to 0. Furthermore, because the rise and fall times for G are different, G_0 and G_4 are placed in different time slots on the scheduler.

This model expands and contracts as input signals change. The basic fault-free circuit remains fixed but the remainder of the circuit is quite fluid. Gates with fault signals are added when fault effects cause the value on a gate input to differ from the corresponding value on the good machine. Gates in the fan-out of a faulted element continue to exist as long as the error signal persists. If the logic values on a gate change so that an error signal is no longer distinguishable from the fault-free signal, then that path terminates. When an error signal terminates, its forward propagation path must be deleted in its entirety.

An advantage of concurrent fault simulation is the fact that, apart from the processing required to cope with the emergence and disappearance of these circuit elements, in other respects the processing for these short-lived fault elements is the same as the processing for the more permanent good machine elements. Therefore the fault modeling capabilities are more flexible than for other methods of fault simulation because a faulted model may, in fact, represent a delay fault or virtually any other fault for which modeling code can be written. More complex functional models can also be fault-simulated with no more difficulty than can the nonfaulted models.

Implementation of the concurrent fault simulator does not require complete descriptor cells for each fault signal which differs from the good machine signal. Rather, an abbreviated descriptor cell (ADC) can be used for FEs and FOs since much of the information is identical for faulted and fault-free machines. A format for the ADC is illustrated in Figure 4.23. The fault-free cell and all related faulted cells are linked via pointers. Except for the ADC the FOs and FEs are similar to regular gates. They are scheduled on the scheduler and the zoom table can be used to evaluate them. However, signal changes on FOs and FEs can only affect FEs with the same identification number, while the signal from the good gate affects both the good circuit and all faulted circuits. To help expedite processing, ADCs are ordered by fault identification number when linked to a descriptor cell.

When a logic value change occurs on the output of a gate in the fault-free circuit, then processing for an FO or an FE depends on whether it is linked to the fault list for the source gate, called the emission list (ELIST), or the fault list for the destination gate, called the receive list (RLIST), or both. The rules are as follows:

If in ELIST only:
   Diverge a copy (an FE) of the destination gate with input states identical to those that existed on the unfaulted destination gate before the change arrived.

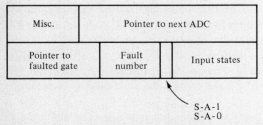

Figure 4.23 Abbreviated descriptor cell.

4.8 FAULT SIMULATION

If in RLIST only:
If it is an OFO, no action is taken. If it is an IFO, simulate unless the input change occurred on the faulted input. If an FE, simulate with the same change that occurred on the good gate.

If in both:
If the FE or FO output value in ELIST is x, then take the same action as when the FE or FO is in RLIST only. Otherwise, compare the input states of the FE in the RLIST to the states on the unfaulted gate and converge if they are identical.

EXAMPLE ■

We describe once again the events which occur when input D changes from 1 to 0. The signal change at primary input D is applied to the input of H_0 and it is simulated. Because its output changes, H_0 is scheduled for processing in time slot $t + 4$. Then, after it is scheduled, the fault list attached to H_0 is processed. No faults were attached to the primary input D. Therefore there is no ELIST, so we use the rule which defines actions taken when fault effects are in the RLIST only. H_1 and H_2 are IFOs, so H_1 is simulated but H_2 is not simulated. H_3 and H_4 are OFOs, therefore no action is taken. H_0/F_1 is an FE, therefore it is simulated with the same signal change that occurred on the unfaulted gate.

When H_0 is retrieved from the scheduler, gates J_0 and K_0 are simulated; however, we describe here only the processing for J_0. The output of gate J_0 did not change; nevertheless, it is necessary to process the fault list attached to J_0. J_1 is simulated and its output changes, so it must be scheduled. J_2 is faulted on the input that changed, so no processing is required. J_3 and J_4 are OFOs, so they are not processed. Fault effects G_3 and G_4 are in the RLIST but not in the ELIST for H_0, so they are simulated and placed on the scheduler. We have two FOs, H_2 and H_3, in the ELIST of H_0 which differ from H_0 and which are not in the RLIST, so it is necessary to diverge FEs J_0/H_2 and J_0/H_3 with input values identical to the values on J_0 before the change arrived. We have two FEs, J_0/F_1 and J_0/H_4, which are in both the ELIST and the RLIST. The logic values on the inputs of J_0/F_1 and J_0/H_4 are identical to the values that exist on the inputs of J_0 after the signal change arrived from H; therefore the two FEs are converged. ■ ■

The good circuit can affect good circuits in its fan-out and possibly all faulted circuits, according to the rules given above. However, the faulted circuit can only affect faulted circuits in its fan-out. Therefore, when the output of H_1 changes, the only gates that it will affect are FEs labeled H_1 in the fault list attached to J and K. Since there are none, and since the output of H_1 remains identical to the value on the unfaulted circuit H_0, no further processing is required.

Both deductive and concurrent fault simulation require large amounts of memory and efficient processing techniques. Memory management for a concurrent fault simulator in a virtual memory management environment can be accomplished simply by ignoring the issue and letting the virtual memory system provide memory as needed. This can cause many page faults, severely degrading performance. That can be avoided by breaking fault lists into smaller lists and fault-simulating in two or more passes. The number of passes is estimated based on the circuit size, the size of the fault list, the amount of available memory, and the amount of memory used to implement the descriptor cells and abbreviated descriptor cells. Since some of the numbers are dependent on the implementation, they must be derived empirically.

4.8.4 Fault Simulation in Sequential Circuits

When simulating sequential circuits, faults propagate through time as well as through logic gates. In the concurrent fault simulator, FEs are diverged and remain attached to a gate from one pattern to the next until converged, since a signal change on a primary input is no different from a signal change on any logic gate. If a fault causes a latch to assume a value other than the correct value, then the diverged gate(s) remains until some input change causes the latch to assume a value which matches the good machine value. Meanwhile, the fault effect at the latch continues to be propagated forward, via simulation, toward a primary output. The fault effect may therefore be latched up in two or more latches in the circuit. The deductive fault simulator attaches lists of faults to individual latch elements in a somewhat similar fashion.

When performing parallel fault simulation with a compiled simulator, faults may propagate to primary outputs or to pseudo-outputs. If a fault reaches a pseudo-output, then it is trapped in a feedback loop. In an actual sequential circuit, the fault causes the circuit to latch up in the wrong state. When simulating consecutive test stimuli, it is necessary that these error signals be simulated forward from the point where their symptoms presently exist. Consider the circuit of Figure 3.9. Three test patterns were needed to set up a test for the fault in question. Initially, the pseudo-input was set to x. The first pattern establishes a 0 on the pseudo-output of the first copy. When simulating the second pattern in this sequence, it is necessary that this 0 be applied to the pseudo-input of the second copy. At the conclusion of the second simulation an error signal, as indicated by the D, appears on the pseudo-output. Therefore, when simulating this circuit, the good machine simulation will produce a 1 and the faulted machine simulation for the S-A-1 fault on the inverter labeled E will have a 0 on the pseudo-output. When simulating the third test pattern, it is necessary to simulate with a 1 on the pseudo-input for the good machine and a 0 for the faulted machine.

In general, many faults could cause a different state on one or more stored state variables between the ith and $(i + 1)$st test pattern. The stored state for each fault must be correctly preserved between patterns. This can be done in either of two ways. The first approach is to select $W - 1$ faults, where W is the machine word width, and simulate the entire set of test patterns, or until all $W - 1$ faults have been detected. Then, repeat with another set of faults. Continue until all faults have been simulated. In the second approach all faults are simulated, $W - 1$ at a time, for each vector. The state of each stored state variable or pseudo-primary output for each faulted machine is saved between test vectors. On the $(i + 1)$st

4.8 FAULT SIMULATION

pattern the values on the stored state variables corresponding to the $W - 1$ selected faults are restored. This approach requires considerable storage, since

$$k \cdot (\text{no. of faults}) \cdot (\text{no. of stored state variables})/(W - 1)$$

words of storage must be used to hold the values on all machines for all stored state variables between test vectors, where the parameter k represents the number of binary bits needed to encode each logic value.

When simulation is completed for a test pattern, it is possible to use the simulation results in an interactive fashion with the ATPG. The ATPG, rather than picking a fault at random from a master fault list, inspects simulation results. The SOFTG (simulator-oriented fault test generator)[23] test generation system inspects the results of simulation to determine whether there is a fault trapped in a latch which is close to a primary output. If it finds a fault on a feedback line or in a flip-flop that appears to be easy to propagate to a primary output, and if the fault is not yet detected, then it is selected as the next candidate fault by the ATPG. Since the ATPG is using the values which already exist on the feedback lines, it does not create an initializing sequence; rather, it only needs to create a propagation sequence. This approach would retain the value on every feedback line in the circuit for each fault being considered, either for the entire simulation process or until it was decided that the fault had been detected and could be dropped from further consideration.

4.8.5 Fault Simulation Performance

The performance of fault simulation algorithms is of vital importance because it can consume many hours of CPU time. It was pointed out at the beginning of this chapter that a growing circuit size generally results in a growing fault list and a larger number of test vectors required to test the circuit. These three parameters suggest that simulation time may, as a worst case, increase in proportion to the third power of circuit size.

The various fault simulation algorithms have emerged in response to this rapidly escalating CPU time. Studies have been made comparing the efficiency of the various simulation methods. A comparison of the parallel fault simulator, using a table-driven model and event-driven simulation, with deductive fault simulation[24] indicated that, in general, the deductive fault simulator is faster but the parallel fault simulator has an advantage for small, highly sequential circuits of size up to about 600 gates. The deductive simulator generally requires more storage and its performance could be adversely affected by excessive paging in a host computer.

Another study[25] confirmed that deductive simulation is faster than parallel simulation. It also gave some results on deductive versus a variation of Testdetect that the authors called SFP (single fault propagation). The comparison indicates that the SFP approach is faster on circuits with high fan-in, often implemented in programmable logic arrays (PLAs), whereas deductive simulation is faster on random logic. The high fan-in circuits would appear to adversely affect the list processing method used in deductive fault simulation.

Numerous methods have been applied to speed up fault simulation. Some of them were previously discussed. These include event-driven simulation, fault dropping, and simulating only one fault from a set of equivalent faults. Another method for accelerating simulation is *states applied analysis*[26]. In this approach, a test sequence is first logic-simulated and the results are evaluated at each gate. Recall from Chap. 2 that the D-algorithm begins by applying the primitive D-cube of failure (PDCF) for a particular fault as the first step in creating a test. The PDCF is a necessary, but not sufficient, condition for a successful test. If the PDCF for a particular fault is not applied to a primitive element by any test pattern, there is no point in simulating the corresponding fault.

EXAMPLE ■

A three-input AND gate requires the input patterns 011, 101, and 110 in order to detect stuck-at faults on its inputs. If the pattern 011 never occurs on its inputs, then there is no point in simulating the S-A-1 fault on its first input. ■ ■

Other methods for improving performance include circuit ordering and statistical fault simulation. When a circuit is ordered, an element in memory is close to the element that drives it and the probability of page faults in a virtual memory environment decreases. Statistical sampling estimates the fault coverage, and hence the quality of a test, by simulating a small random sample of the faults. Sufficient faults can be simulated to give a high level of confidence that the fault coverage is within some range of the predicted value. The statistical fault simulation can be preceded by a states applied analysis[27]. If the analysis reveals that the percentage of potentially detectable faults is less than the required fault coverage, then there is no point in performing statistical fault simulation until the percentage of potentially detected faults is increased.

Despite efforts to enhance its performance, the parallel fault simulator suffers from the fact that it performs considerable redundant effort. For a particular fault list and a machine with a word size W, each gate is simulated (no. of faults)/$(W - 1)$ times when activity occurs.

For concurrent fault simulation, some performance estimates are provided by Ulrich et al.[28] for a simulator running on a VAX11/780, a machine with performance of about 1 million instructions per second (1.0 MIPS). Using zoom tables and other performance enhancements, approximately 6000–7000 signal changes per second have been achieved for basic logic simulation.

Given a circuit with 50,000 gates and 160,000 faults, the storage requirement for program and fault-free model is estimated at 2.8 Mbytes. This works out to about 50 bytes allocated per gate. A single-pass concurrent fault simulation would require about 28 Mbytes. Dividing this into five fault simulation passes, a peak of 8.4 Mbytes is required. This model assumes 1500 signal changes per test vector, which corresponds to 3% activity per vector, and a fault simulation, even with five passes, that takes only about eight times as much CPU time as does the fault-free logic simulation.

4.9 SUMMARY

The above estimates are conditional on the number of passes, hence on the amount of memory available. For some other amount of memory, the number of passes may change and will affect the performance numbers. The objective is to hold the number of passes to a minimum because the fault-free copy of a gate is simulated in each pass, resulting in redundant activity. Despite the multiple passes, the numbers compare favorably with parallel fault simulation, where each element is simulated $(160,000)/(W-1)$ times (about 5160 times for a machine with a 32-bit word).

It is possible to combine the features of parallel and concurrent fault simulation[29]. The parallel value (PV) list simulates all faults in one pass, as in concurrent fault simulation, but stores faulty values using individual bit positions in a word. Each fault is uniquely identified by a group number and bit position pair. Faults grouped together in a given parallel value word are chosen on the basis of their proximity to one another. If they are close together in the circuit, and if no activity is present in that area of the circuit, the fault word is dropped from forward propagation quickly. The evaluation techniques also differ, depending on whether the output activity occurred on the fault-free or the faulted copy of the gate. The PV techniques can save considerable amounts of memory and reduce the number of passes required for fault simulation in a large circuit.

Improvements to the concurrent fault simulation algorithm are possible through coding techniques[30]. A simulation program for a fast, pipelined architecture computer (CDC Cyber 176) which achieved simulation speeds of 30,000 events per second was reprogrammed to take greater advantage of the computer architecture. Short loops with many branches, which can be destructive of performance in a pipelined architecture, were modified via *loop unrolling*. The same series of operations is performed on several contiguous arguments. For example, the following FORTRAN software

```
      DO 100 I = 1, 32, 4
      A(I)     = B(I)     + CONST
      A(I + 1) = B(I + 1) + CONST
      A(I + 2) = B(I + 2) + CONST
      A(I + 3) = B(I + 3) + CONST
100   CONTINUE
```

increases the amount of code but reduces the number of branches that must be performed. Since most programs are characterized by the fact that a high percentage of the CPU time is spent in a very small part of the program, modifications to that part of the program can sometimes significantly increase the overall performance of the program. In the example cited here, events from the scheduler are retrieved and rearranged for optimized processing. This has led to a reported increase to a maximum of 90,000 events per second when performing gate-level simulation.

4.9 SUMMARY

Simulation is growing in importance because of the need for greater confidence in the design of complex ICs. The larger, more complex designs are more error-prone, and mistakes are both costly and time-consuming. Because it is not possible to repair design defects inside ICs, an incorrect design could set a project back many weeks. As a result, hierarchies of simulators are being developed that provide simulation capability at several levels of detail and features that permit ease of use.

The effective development and use of simulation tools requires an understanding of the design environment in which the tools will be used. Assumptions which hold in one design environment may not hold in another. Tools developed for use in a synchronous design environment may give incorrect results when applied to asynchronous designs. On the other hand, the synchronous design environment permits simplifying assumptions which can help to speed up simulation. Another factor that must be taken into account when considering the use of simulation tools is the amount of control that can be exercised over the design environment. The tools that one uses will depend to some extent on whether the user is designing his ICs in-house or buying LSI and VLSI chips from outside vendors. VLSI chips obtained from vendors rarely are accompanied by sufficient information to permit simulation at the logic gate or switch level.

The speed at which simulation is performed grows in importance as circuits grow larger. This is especially true for fault simulation. Four factors can be identified which contribute to simulation performance: choice of algorithm, implementation efficiencies, the basic speed of the CPU employed, and the architecture of the machine. Concurrent fault simulation is memory intensive but as memory costs decrease, the trade-offs increasingly favor the concurrent fault simulation algorithm. Implementation efficiencies can be enhanced by identifiying and eliminating every possible redundant or unnecessary calculation in the algorithms. This is especially true for something highly repetitive like simulation. Statistical bias can also be useful. When one of two or more outcomes is known to have occurred, check the most probable event first. In Chap. 10 we will look at some special architectures for simulation. It is interesting to note that, despite the advantages of newer algorithms, an old concept, the compiled simulator, has found its way into at least one hardware simulation machine, the Yorktown Simulation Engine.

Because of space limitations, we have presented an overly simplified view of MOS simulation in this chapter. The reader interested in a more detailed discussion of MOS simulation is referred to Hayes[31] for a discussion of lattice ordering techniques or Bryant[4] for a discussion of switch graphs.

PROBLEMS

4.1 What general rule can be stated regarding the minimum duration of the clock pulse applied to the latch in Figure 4.2(b) in order to avoid the glitch?

4.2 (a) Rewrite the Z-80 compiled simulator in the text to perform three-value simulation on the circuit of Figure 4.3, using the ordered pairs defined in Sec. 4.4.1. The Z-80 has registers B, C, D, and E, which can be loaded and stored (but not used in arithmetic or logic operations) as needed to swap bits.

(b) Rewrite the Z-80 three-value compiled simulator to perform parallel fault simulation of the seven faults corresponding to the inputs of G and H S-A-1 and the

PROBLEMS

inputs of I S-A-0. Use the JR LABEL (jump relative to LABEL) to jump to instructions which bug the circuit.

4.3 How would you simulate multiple faults in the compiled simulator?

4.4 When using a READ/WRITE array, what is the maximum number of passes before the compiled simulator has completely processed an input vector applied to
  (a) An n-level deep combinational circuit?
  (b) A sequential circuit with n_1 latches, a maximum of n_2 levels of combinational logic between feedback lines, and using the x value between input changes?

4.5 Describe the computations performed by the compiled simulator when the input transition $(0, 0) \rightarrow (1, 1)$ is simulated on the cross-coupled NAND latch model using READ and WRITE arrays. Repeat, but with $(0, 0) \rightarrow (x, x) \rightarrow (1, 1)$.

4.6 Prove lemmas 4.1 and 4.2 and theorems 4.1, 4.2, 4.3, and 4.4.

4.7 Create a nine-value simulation table capable of detecting hazards at an OR gate.

4.8 Given the following four combinations on the inputs to a three-input AND gate, what is the resulting output for each of the combinations?

input 1	M	/	W	M
input 2	W	∧	∧	*
input 3	1	\	∧	M

4.9 Using the figure shown:
  (a) Compute the timing of the paths from A, B, C, and D to the output for both 1 and 0, assume the rise time of the NAND gates is 8 ns and the fall time is 5 ns.
  (b) What maximum value would you get if you ignored the signal inversions and just used

average propagation delay?
maximum propagation delay?

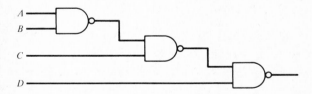

4.10 Perform a block-oriented analysis on the circuit of Figure 4.15 using separate rise and fall delays. Assume that all elements have 4 units of rise time and 6 units of fall time. Assume the required arrival time is 20 units. If there is a minimum arrival time of 4 units at the output, will it be satisfied?

4.11 Define a general rule for processing the AND gate and the exclusive-OR during deductive fault simulation.

4.12 Show the computations that both a deductive and a concurrent fault simulator would perform on the circuit in Figure 4.21 when applying the pattern $1, 2, 3 = (1, 0, 0)$. Initially set all nets to x. For the concurrent fault simulator, attach fault origins to each of the gates. Then apply the pattern. Assume that all elements have identical rise and fall times of 1 unit and that all inputs change at time $t = 0$.

4.13 Show the computations that a concurrent fault simulation would perform on the circuit of Figure 4.5(a). Assume S-A-1 faults on the inputs of gates 1 and 2, initially set all nets to x, then use the sequence $\overline{\text{Set}}, \overline{\text{Reset}} = \{(0, 1), (1, 1), (1, 0), (1, 1)\}$. Assume the elements have identical rise and fall times of 1 unit.

THE SIMULATOR: A Description

The remaining problems for this chapter are based on the simulator which follows. It is a three-valued $(0, 1, x)$, event-driven simulator written in Microsoft BASIC for a Radio Shack Model II computer with 64K memory. The simulator is barely usable as it exists and some of the problems will request elementary enhancements intended to make it more usable and to make its output readable.

The simulator recognizes seven element types. When coding a circuit description for input to the simulator, the function is specified by means of a one-letter code. The functions and their corresponding one-letter codes are: Primary Input (I), AND (A), OR (O), Invert (N), Primary Output (Y), Delay (D), Fan-out (F). When entering a circuit description into the program, the one-letter function code is always the first item on a line. It is followed by the function name, which is a decimal number between 1 and MAX, where MAX is simply an arbitrary upper limit on the number of elements in the circuit. The remaining items in a line of input are nominal delay values, the names of the inputs to this element, and the name of the output(s) driven by the named element.

The general format for the circuit description statements is as follows:

    Input: I, name, output
    Output: Y, name, input
    Invert: N, name, delay, input, output
    Delay: D, name, delay, input, output
    Fan-out: F, name, output1, output2, output3, output4
    OR: O, name, delay, input1, input2, input3, input4, output
    AND: A, name, delay, input1, input2, input3, input4, output

The third entry in a statement depends on the function. For a primary input, the third entry is the name of the gate to which the input is connected; for a primary output it is the name of the input which drives the output; and for the logic gates and the Delay element, the third entry specifies the amount of delay inherent in the element. The Fan-out element is used when an element drives more than one element. For instance, if an AND gate drives two or more other gates, then the AND gate in the circuit model drives a Fan-out element, which in turn drives the elements normally driven by the AND gate in the actual circuit. This is merely an artifice to simplify the simulation program. The third through sixth elements in the Fan-out entry list the gates driven by the original device which are now driven by the Fan-out element.

The remaining entries in the logic gate formats identify one or more inputs, and then an output. If an AND gate or an OR gate has fewer than four inputs, the unused positions must be indicated by consecutive commas in the input statement. If an AND (OR) used in a circuit has more than four inputs, then two or more four-input AND (OR) gates can be ANDed (ORed) together to model the gate. The last statement in a circuit description is ENDCKT.

THE SIMULATOR: A DESCRIPTION

After the circuit description has been processed, the remaining statements are used to specify simulation times, the inputs that are scheduled to change, and the new value they are to assume. The first number specifies the time at which the values are to be scheduled. Then the remaining entries on a statement are, alternately, the names of inputs and the logic value assigned. The entries are separated by commas and the last entry is an asterisk. The end of input stimuli is indicated by a statement which begins with the number 9999.

Simulation of an element is performed when a change occurs on any of its inputs. However, the output value is not updated immediately; rather, the element is scheduled for update at time $t + d$, where t is present time and d is the nominal delay of the element. At time $t + d$ the computed value is compared to the existing value on the output of the gate and, if it differs from the existing value, then the output value is updated and elements in the fan-out of the gate are simulated. The three values 0, 1, and x are encoded using two binary bits as follows:

$$0, 0 - 0$$
$$0, 1 - x$$
$$1, 1 - 1$$

The timing wheel, in the fashion of a clock, advances time by use of a counter, analogous to the minute hand on a clock. One complete rotation of the wheel represents W1 time units, which, in the simulator, are referred to as minutes (MI). After each complete rotation, MI is reset to zero and hours (HR) is incremented. The wheel is represented by two arrays, the usage bit (WH) and the link pointer (LP). If one or more elements must be processed at time i, then the usage bit WH(i) is turned on and link pointer LP(i) points to an entry in a queue.

The queue contains six items for each entry. The name of the element to be updated is placed in the queue array (QU). If additional elements are to be simulated at this time, then another link pointer (LK) points to the next element in the queue that is linked to this time slot. The link flag (LF) is on ($= 1$) if another element is linked to the element presently being processed, and it is off ($= 0$) if the present element is the last one in the linked list. The usage bit US(i) simply indicates whether position i in the queue is available or is presently in use. The scheduler searches through the queue to find an empty slot whenever an event must be scheduled. When an empty slot is found, the element is placed in the queue and the usage bit is turned on. After the element is processed the usage bit is turned off and the slot is free to be used again.

The value array (VL) stores the computed value on the element. This value is compared to the existing value stored in the (VA) array. The future time array (FT) is used to schedule remote events. These are events which occur too far in the future to permit scheduling within the next W1 time units. For example, if an element has a delay of 250 units, then it is scheduled for $t + [250(\bmod W1)]$ but FT(i) is set equal to 250/W1. When this element is encountered in the queue, if FT(i) > 0, then it is decremented but the element is left in the queue. If FT(i) = 0, then the element is processed and removed from the queue.

The program first reads in the circuit description and creates a circuit model. Then the input stimuli are read. When inputs and their logic values are read in, they are scheduled on the timing wheel. The number of lines of input stimuli read in at a given time is determined by the length of the timing wheel. Rather than read in all input stimuli and schedule them for future time via the FT array, the read process is temporarily suspended when a stimulus time exceeds present time plus maximum wheel time.

Simulation begins at statement 790. Each timing wheel slot is examined to determine if any entries are scheduled for that time. If not, then time is incremented. If the slot is nonempty, then processing is performed as described earlier. Two major subroutines exist at the end of the program. The first one puts an entry in the queue and either links it to the timing wheel slot, if the slot is empty, or links it to the last entry in the list presently linked to the wheel. When all processing at time i is complete, the second routine removes entries from the wheel. If queue position i is linked to the wheel, and if FT(i) is zero, then the entry is deleted. If FT(i) is nonzero, the entry is retained and FT(i) is decremented. When the end of the timing wheel is reached, the hours variable (HR) is incremented and the software which reads the input stimuli is revisited to again load the timing wheel. This continues until there are no additional input stimuli and the circuit is stable, that is, there are no more entries on any timing wheel slots.

Some of the later problems suggest program changes designed to make the simulator easier to use. You may prefer to implement some of those suggestions before working problems which make use of the simulator, in particular, those problems which simplify the input format and add some additional primitives such as the NAND, NOR, and exclusive-OR. Additional problems based on the simulator can be found at the end of Chap. 5.

Simulator Listing

```
10 DEFINT A-Z
20 CLS : PRINT @(2,0), "EVENT DRIVEN SIMULATOR":PRINT
22 PRINT "HR: MIN","VALUE":PRINT "-------","-----"
30 MAX = 400                    'MAXIMUM NUMBER OF LOGIC GATES
40 W1 = 100                     'MAX TIMING WHEEL SIZE
50 W2 = 200                     'MAX QUEUE SIZE
60 DIM F(MAX)                   'GATE FUNCTION
70 DIM NA(MAX)                  'GATE NAME
80 DIM D(MAX)                   'GATE DELAY
90 DIM IP(3,MAX)                'GATE INPUTS (MAX OF 4)
100 DIM DS(MAX)                 'DESTINATION OF SIGNAL
110 DIM VA(1,MAX)               'GATE VALUE (2 BITS)
120 DIM WH(W1)                  'TIMING WHEEL USAGE BIT
130 DIM LP(W1)                  'LINK POINTER TO QUEUE
140 DIM US(W2)                  'USAGE > 0 IF ENTRY IN THIS SLOT
150 DIM QU(W2)                  'QUEUE - CONTAINS NAME OF ENTRY
170 DIM FT(W2)                  'FUTURE TIME
180 DIM LK(W2)                  'LINK POINTER TO NEXT ENTRY IN STRING
190 DIM LF(W2)                  'LINK FLAG
200 DIM VL(1,W2)                'NEW SIMULATION VALUE
210 ZX = 0
230 FOR I=0 TO MAX       '-------- READ CIRCUIT DESCRIPTION --------
240 READ F$:FC=9
250 IF F$="I" THEN FC=1 ELSE IF F$="A" THEN FC=2 ELSE IF F$="O" THEN FC=3
260 IF F$="N" THEN FC=4 ELSE IF F$="Y" THEN FC=5 ELSE IF F$="D" THEN FC=6
270 IF F$="F" THEN FC=7 ELSE IF F$="ENDCKT" THEN FC=8
280 ON FC GOTO 290,330,330,310,300,310,320,400,340
290 GOSUB 370:READ DS(NM):GOTO 350
300 GOSUB 370:READ IP(0,NM):GOTO 350
310 GOSUB 370:READ D(NM),IP(0,NM),DS(NM):GOTO 350
320 GOSUB 370:READ IP(0,NM),IP(1,NM),IP(2,NM),IP(3,NM):GOTO 350
330 GOSUB 370:READ D(NM),IP(0,NM),IP(1,NM),IP(2,NM),IP(3,NM),DS(NM):GOTO 350
340 PRINT "ILLEGAL FUNCTION CODE",F$:STOP
350 NEXT I
360 PRINT "CIRCUIT SIZE EXCEEDS LIMIT":STOP
370 READ NM:IF NM > MAX THEN 390
375 VA(0,NM)=0:VA(1,NM)=1        'INITIAL VALUE OF ALL ELEMENTS SET TO "X"
380 F(NM)=FC:NA(NM)=NM:RETURN
390 PRINT "INVALID NAME",NM:STOP
400 ES=0                         'END OF SIMULATION FLAG ES = 0
405 FF=0                         'SET FANOUT FLAG=0
440 MI=0:HR=0
450 SZ=0:SR=0:SM=0               'INITIAL STIMULUS TIME / HOUR=0 / MINUTES=0
455 GOTO 490
460 MI=MI+1                      'UPDATE SIMULATION TIME
```

SIMULATOR LISTING

```
465 IF MI >= W1 THEN 470
467 IF WH(MI)=0 THEN 460 ELSE 790    'IF NO ACTIVITY -- UPDATE TIME
470 MI=0:HR=HR+1
480 IF SZ=9999 THEN STOP ELSE 510
490 READ SZ:IF SZ=9999 THEN 467
510 SM=SZ:SR=0
520 IF SM < W1 THEN 540
530 SM=SM-W1:SR=SR+1:GOTO 520
540 IF HR < SR THEN 467              'DONT READ STIMULUS
610 READ NM$:IF NM$="*" THEN 490
620 NM=VAL(NM$)
630 READ VA$:Z2(0)=0:Z2(1)=1         'INITIALIZE VALUE TO X
640 IF VA$="0" THEN Z2(1)=0 ELSE IF VA$="1" THEN Z2(0)=1
670 Z1=NM:Z3=SR:Z4=SM
680 GOSUB 1710
690 GOTO 610
715 REM ------------------------ BEGIN SIMULATION ------------------------
716 REM
785 REM         <<<<<< GET FANOUT FOR CHANGING GATES  >>>>>>
786 REM         <<<<<< PUT ON QUEUE AT FUTURE TIME    >>>>>>
787 REM
790 QP=LP(MI)                        'GET POINTER TO FIRST QUEUE ENTRY TO BE UPDATED
800 IF HR < FT(QP) THEN 879
810 NM=QU(QP)                        'GET NAME OF ELEMENT IN QUEUE
812 IF VA(0,NM)=VL(0,QP) AND VA(1,NM)=VL(1,QP) THEN GOTO 879
814 VA(0,NM)=VL(0,QP)                'UPDATE VALUE ON THE OUTPUT
816 VA(1,NM)=VL(1,QP)                ' -- OF ELEMENT IN QUEUE
820 FX=F(NM)                         'GET FUNCTION OF THIS ELEMENT
830 IF FX=7 THEN 1315                'FANOUT FUNCTION SHOULD NOT BE IN QUEUE
840 ND=DS(NM)                        'GET NAME OF DESTINATION GATE
850 FY=F(ND)                         'GET FUNCTION OF DESTINATION GATE
860 ON FY GOSUB 1000,1010,1100,1200,1290,1260,920   'COMPUTE ITS NEW VALUE
865 IF FY=5 OR FY=7 THEN 879         'DO NOT PUT PRIMARY OUTPUT ON WHEEL
870 Z1=ND
871 Z2(0)=NV(0):Z2(1)=NV(1)
873 Z3=HR
874 Z4=MI+D(ND)
876 IF Z4 < W1 THEN 878
877 Z4=Z4-W1:Z3=Z3+1:GOTO 876
878 GOSUB 1710
879 IF FF=1 THEN RETURN              'IF FF=1, CONTINUE FANOUT PROCESSING
890 IF LF(QP)=0 THEN GOSUB 1880:GOTO 460   'IF ZERO, GOTO PURGE ROUTINE
900 QP=LK(QP):GOTO 800               'ELSE, GET NEXT ELEMENT IN QUEUE LINK LIST
920 REM ---------------- COME HERE TO PROCESS FANOUT ELEMENT -------------
922 IF FF=1 THEN 1320
930 VA(0,ND)=VA(0,NM)
935 VA(1,ND)=VA(1,NM)
940 FF=1:NF=ND
943 FOR FO=0 TO 3
946 ND=IP(FO,NF)
947 IF ND=0 THEN 952
949 GOSUB 850
952 NEXT FO
955 FF=0
958 GOTO 890
960 REM ------------------------ COMPUTE NEW VALUES ON LOGIC ELEMENTS ---
1000 PRINT "CANNOT HAVE FANOUT TO PRIMARY INPUT":STOP
1010 NV(0)=1:NV(1)=1
1030 FOR IX=0 TO 3                   '*** AND ***
1040 TM=IP(IX,ND)
1050 IF TM=0 THEN 1080               'IF INPUT NOT CONNECTED
1060 NV(0)=NV(0) AND VA(0,TM)        ' -- THEN SKIP COMPUTATION
1070 NV(1)=NV(1) AND VA(1,TM)
1080 NEXT IX
1090 RETURN
1100 NV(0)=0:NV(1)=0                 '*** OR ***
1120 FOR IX=0 TO 3
1130 TM=IP(IX,ND)
1140 IF TM=0 THEN 1170
1150 NV(0)=NV(0) OR VA(0,TM)
1160 NV(1)=NV(1) OR VA(1,TM)
1170 NEXT IX
```

```
1180 RETURN
1200 REM                                 '*** INVERT ***
1210 TM=IP(O,ND)
1230 NV(O)=1-VA(1,TM):NV(1)=1-VA(O,TM)
1250 RETURN
1260 TM=IP(O,ND)                         '*** DELAY ***
1270 NV(O)=VA(O,TM):NV(1)=VA(1,TM)
1280 RETURN
1289                                     '*** PRIMARY OUTPUT ***
1290 IF VA(O,ND)=VA(O,NM) AND VA(1,ND)=VA(1,NM) THEN RETURN
1300 VA(O,ND)=VA(O,NM):VA(1,ND)=VA(1,NM)
1303 IF VA(O,ND)=1 THEN LO$="1" ELSE IF VA(1,ND)=O THEN LO$="O" ELSE LO$="X"
1305 PRINT HR;MI,LO$
1310 RETURN
1315 PRINT "FANOUT ELEMENT SHOULD NOT BE IN QUEUE":STOP
1320 PRINT "FANOUT NOT PERMITTED TO DRIVE FANOUT":STOP
1330 REM
1700 REM ----------------- PUT NEW ENTRY AT END OF LINKED LIST ------------
1701 REM          CALL WITH
1702 REM               Z1 = GATE NAME
1703 REM               Z2 = NEW VALUE
1704 REM               Z3 = STIMULUS HOUR
1705 REM               Z4 = STIMULUS MINUTE
1710 FOR Z5=ZX TO W2                   'FIND EMPTY SLOT IN QUEUE
1720 IF US(Z5)=O THEN 1735 ELSE NEXT Z5
1722 FOR Z5=O TO ZX-1
1724 IF US(Z5)=O THEN 1735 ELSE NEXT Z5
1730 PRINT "NO MORE ROOM IN EVENT QUEUE":STOP
1735 ZX=Z5
1740 IF WH(Z4) > O THEN 1780
1750 WH(Z4)=1:LP(Z4)=Z5                'SET LINK BIT AND LINK POINTER
1760 QU(Z5)=Z1:VL(O,Z5)=Z2(O):VL(1,Z5)=Z2(1):FT(Z5)=Z3 'PUT ENTRY IN QUEUE
1770 US(Z5)=1:LF(Z5)=O:RETURN           'SET USAGE BIT, ZERO LINK BIT
1780 SK=LP(Z4)                          'GET LINK POINTER FROM WHEEL
1790 FOR K1=1 TO W1
1810 IF LF(SK)=O THEN 1840      'IF LINK FLAG=O, REACHED END OF LINKED LIST
1815 SK=LK(SK)                          'GET LINK POINTER
1820 NEXT K1
1830 PRINT "CANNOT FIND END OF LINKED LIST":STOP
1840 LK(SK)=Z5                  'INSERT POINTER TO NEXT STRING ENTRY
1850 US(Z5)=1                           'SET USAGE BIT
1860 LF(SK)=1                           'LINK FLAG = 1
1870 GOTO 1760
1880 REM --------------------- PURGE WHEEL -----------------------------
1890 QP=LP(MI)
1900 IF HR=FT(QP) AND LF(QP)=O THEN WH(MI)=O:US(QP)=O:RETURN
1910 IF HR<FT(QP) AND LF(QP)=O THEN RETURN
1920 IF HR<FT(QP) THEN UX=QP:QP=LK(QP):GOTO 1950
1930 US(QP)=O:QP=LK(QP)
1940 LP(MI)=QP:GOTO 1900
1950 IF HR=FT(QP) AND LF(QP)=O THEN QP(UX)=O:RETURN
1960 IF HR<FT(QP) AND LF(QP)=O THEN RETURN
1970 IF HR<FT(QP) THEN UX=QP:QP=LK(QP):GOTO 1950
1980 US(QP)=O:LK(UX)=LK(QP):QP=LK(QP):GOTO 1950
1990 REM ------------------- END SIMULATOR -----------------------------
3000 DATA I,1,7
3010 DATA I,2,5
3020 DATA I,3,8
3030 DATA I,4,9
3040 DATA F,5,6,7,,
3050 DATA N,6,2,5,8
3060 DATA A,7,2,1,5,,,9
3070 DATA A,8,2,6,3,,,9
3071 DATA O,9,2,7,8,4,,10
3072 DATA Y,10,9
3080 DATA ENDCKT
3100 DATA 0,4,1,*
3101 DATA 3,2,0,4,0,*
3102 DATA 7,3,1,*
3103 DATA 14,2,1,*
3104 DATA 20,1,1,*
3106 DATA 30,2,0,*
3108 DATA 40,1,0,*
```

PROBLEMS

```
3110 DATA 50,4,1,*
3112 DATA 60,4,0,*
3118 DATA 90,2,1,*
3120 DATA 100,4,1,*
3122 DATA 110,1,1,*
3124 DATA 120,1,0,*
3126 DATA 130,4,0,*
3128 DATA 140,3,0,*
3130 DATA 150,1,1,*
3133 DATA 160,2,0,*
3135 DATA 398,4,1,*
3137 DATA 416,4,1,*
3139 DATA 440,4,0,*
3150 DATA 9999
```

Problems Based on the Simulator

4.14 The circuit encoded at the end of the simulator is a 2-to-1 multiplexer with data input ports connected to primary inputs 1 and 3. At time 34 the output goes to 0 even though both inputs are 1. Explain that. Redesign the circuit to prevent it from happening. Resimulate to insure that your design is correct.

4.15 Enter the following two-input AND gate to the simulator:

```
3000 DATA I, 1, 3
3010 DATA I, 2, 3
3020 DATA A, 3, 4, 1, 2,,,4
3030 DATA Y, 4, 3
3040 DATA ENDCKT
3100 DATA 0, 1, 0, 2, 1, *
3110 DATA 4, 2, 0, 1, 1, *
3200 DATA 9999
```

After running the circuit, replace statement 3110 with

```
3110 DATA 4, 1, 1, 2, 0, *
```

and rerun the circuit. If you encoded the simulator and data correctly, the output data will differ. Explain why. Alter the simulator to prevent this from happening.

4.16 If an input change occurs at time t on an element with delay d, the element is simulated immediately. However, the output is not checked until time $t + d$. Why not check immediately after simulation and skip the scheduling process if the output has not changed? Alter the simulator to try this approach. Delete line number 812 and insert a check at 872 before going to the scheduler, then repeat the experiment in the previous problem.

4.17 Add new primitive elements to the simulator. Include NAND, NOR, exclusive-OR.

4.18 Simulate all 32 input combinations on the circuit of Figure 4.20 and try to determine what functions are performed by the circuit. Which inputs appear to be the data lines and which appear to be the control lines?

4.19 In the exercises following Chap. 2, you were asked to redesign the circuit in Figure 2.5 in terms of NAND gates. Simulate your NAND gate model to verify that it is correct.

4.20 The simulator terminates at statement 480. It is possible that the simulator may terminate while residual activity remains scheduled on the timing wheel. Insert a check for activity on the wheel before allowing the program to terminate.

4.21 Simulate the delay flip-flop (DFF) of Figure 3.6. Assign 3 units of delay to each NAND. Set the Clock and Data inputs low and clear the DFF. Then raise the Data line and clock it (raise the clock line) 3 units later. Repeat this entire sequence while increasing the interval between Data and Clock to determine Setup time for the DFF. Then, preset the DFF while holding the Data input high and the Clock low. Clock it and drop the Data line 3 units later. Repeat the sequence while increasing the interval between Clock and Data to determine Hold time.

4.22 Simulate the four-state machine in Figure 3.20, using the synchronizing sequence (0, 1, 0, 1, 0). First simulate with both DFFs in the indeterminate state, and then start from each of the four known states in turn. Use the Preset and Clear lines to force the machine into each of the four start states. For the DFFs, use the circuit created in the previous problem. Repeat the experiment using the machine in Figure 3.21 and the synchronizing sequence (0, 1, 1, 0).

4.23 Add AND-OR primitives in two different configurations. One configuration consists of two three-input AND gates driving an OR gate. The other consists of three two-input AND gates driving an OR gate. Create two more primitives by adding an inverter to the output of each of the AND-OR primitives. Expand the IP array to allow six inputs per element and allow two or more alphanumeric characters for specifying the element function. Note that unused inputs may cause problems.

4.24 Encode the 74181 ALU, first as an interconnection strictly of AND, OR, and Inverter elements (obtain circuit descriptions from standard manufacturer TTL literature) and run several input vectors. Then, recode and rerun the circuit, using, wherever possible, the primitives created in the previous problem. Compare the amount of time required to simulate these two versions of the circuit. It may be necessary to connect several 4-bit stages and use many input vectors to detect a difference.

4.25 Implement a Zoom table for computing output response of logic elements. Create a five-dimensional array (AR) as a lookup table. Recode the simulator so that the values 0, x, and 1 are represented by integers 0, 1, and 2. The value stored in lookup table location $AR(i_1, i_2, i_3, i_4, i_5)$ is determined by the function code i_1 and the values i_2, i_3, i_4, and i_5 on the inputs to that function.

4.26 Provide an option in which users can create and enter up to four primitives and the corresponding zoom table without having to alter the simulator program.

4.27 Implement a second delay variable so that primitives can have both rise and fall delay. To test the circuit, use a two-input AND gate with rise time of 10 ns and fall time of 4 ns. At time 0 set input 1 to 0 and input 2 to 1. At time $t = 20$ set input 1 to 1 and at $t = 25$ set input 2 to 0. Did your simulator compute the correct answer?

4.28 When an input change occurs, the simulator evaluates the result and schedules the event on the timing wheel. At future time it checks for a change on the output. Rewrite the simulator so that it evaluates the change and schedules it on the wheel only if an element output change occurs. Provide exception handling for cases where an element is scheduled two or more times. Discuss the relative merits of these two approaches. Given a four-input AND gate, how many single input transitions exist? How many can cause a change on the output of the gate?

4.29 Add a delay flip-flop primitive to the simulator. Provide it with Data, Clock, and Clear inputs. The primitive will require memory; that is, it must remember the last Data value clocked in. Create a version which clocks in data on the rising edge of the clock and a second which clocks in data on the falling edge of the clock. Add software checks for setup and hold times. If these times are not met, set the internal memory and the output to x. Check for a minimum pulse duration on the Clear line.

PROBLEMS

4.30 Use the delay flip-flop primitive and some combinational logic to design a JK flip-flop. Simulate the circuit to verify that it works correctly.

4.31 The length of the timing wheel can affect performance. If all elements have unit delay, the wheel is not needed since all activity occurs in the present time frame or the next time frame and two lists are sufficient (analogous to a timing wheel with two slots). However, if elements have a wide range of delay, and wheel length is small relative to that range, then many events may occur in the distant future, and excessive computation occurs during scheduling. Program a counter that records the number of times that $FT(i)$ is set greater than 0. Use the 74181 encoded in an earlier problem, assign individual element delays to range from 1 to 50 units, and vary wheel size W1 from 20 to 100. Record the effect of wheel size on the amount of distant future activity.

4.32 The event-driven simulator only simulates when activity occurs. In a given time slot there is often very little activity; perhaps as few as 1 to 3% of the gates switch. However, when employing nominal delay simulation, a single input change can produce activity in many time frames. Write a routine which counts the number of

> Distinct time frames generated by one test vector
>
> Elements computed during each time frame
>
> Times each individual element is evaluated for a given input vector

To test your software, use a cross-coupled NAND latch in which both NAND gates have zero delay. Start by applying 0 to both Set and Reset. Then change both inputs to 1 at the same time. Repeat this test, but give both NAND gates nominal delays of 1 ns.

4.33 Using the program written in the previous problem, collect statistics on runs using the 74181 ALU. Create a sensitized path from the $A = B$ output back to one of the B_i, $0 \le i \le 3$, inputs and propagate a 0 and a 1 along the path. Space changes on primary inputs far enough apart so that activity resulting from a PI change settles out before the next PI change occurs.

4.34 The timing wheel scheduling software can consume a great deal of CPU time. If a logic element has zero delay, it is not necessary to schedule the element on the wheel. The alternative is to implement a stack and place zero delay elements on the stack when their inputs change. Then, after all of the elements linked to the wheel have been simulated, elements on the stack are processed before updating time. Implement a stack for zero delay elements.

4.35 Simulator results are unpredictable when a fan-out element drives another fan-out. Create a circuit in which an element drives five other logic elements. Encode it so that the element drives a fan-out. The fan-out drives three of the original elements and another fan-out, which drives the remaining two elements. Make changes to the simulator if necessary so that this circuit is correctly simulated.

4.36 The simulator does very little error checking. Encode statements containing the following errors. Add software which writes error messages and otherwise responds to these errors in some appropriate fashion.

> Illegal function
>
> Illegal name
>
> Missing comma(s)
>
> Too many or too few gate inputs
>
> Stimulus time less than previous stimulus time

4.37 An effective simulator must feature ease of use, data checking to catch and report user errors, and output formats that permit the user to easily identify relevant information. The number of data entry errors is reduced if the user is not required to enter fan-out elements and destination gates in the circuit description. Write a data input routine that does not require that information. After reading in the circuit, the routine will examine the inputs of each element and use that information to create the destination array DS and the fan-out elements.

4.38 Ease of use is further enhanced if arguments in the input data are not required to be separated by commas. Write a data entry routine that recognizes one or more consecutive blanks as delimiters. If a logic gate has fewer than four inputs, this will be detected by the subroutine; the user will not be required to indicate unused inputs via successive commas.

4.39 Create a *display formatter* routine as follows: read in a list of elements whose values are to be displayed during simulation. The elements may be any combination of primary inputs, primary outputs, and internal elements. Also read in a start time at which simulation results are to be displayed, and an interval n between each display update. Then create a display with a time tape at the top, the element names on the left, and the results of simulation printed horizontally across the screen. If the user requests the values on elements 4, 11, and 37 to be displayed every 4 time units, starting at time 150, then the display may appear as follows (this is sometimes referred to as a logic analyzer format):

```
           150   170   190   210   230   250   270   290   310   330
           V....V....V....V....V....V....V....V....V....V....V

4          ....11111...1111111111..1111.......11.11111111
11         xxx...111111...1111..11111.............111...
37         x11.......1111..11111111111.....1111...111111.
```

When implementing the display formatter, the display data must be updated every n units. Implement this by creating *bulletins*. The bulletins will be scheduled on the timing wheel like logic events. However, they will be assigned a unique function code so that, when a bulletin is encountered on the timing wheel, the simulator will update the display rather than simulate an element.

4.40 Note the bulletins may fall in the middle of a linked list, causing the display to be updated before all events are evaluated. To avoid this, expand the wheel to $2 * W1$ units. Schedule logic events at even positions and bulletins in the immediately following odd-numbered slot so that the display is updated only after all elements linked to a wheel slot have been processed.

4.41 Bulletins can also be used to schedule the read operation for input stimuli. Rewrite the input stimulus read routine to be initiated by a bulletin rather than when the end of the wheel is encountered.

4.42 A commercial quality simulator permits users to enter signal names alphanumerically rather than as numbers. Implement such a capability in the routine that reads the circuit description. Permit alphanumeric names up to six characters in length. When a name is read in, convert it to a number for internal processing but save the name in a table so it can be used with the display formatter.

4.43 Some signals, such as clock signals, switch periodically. Implement a periodic function in which the keyword PERIODIC appears at the beginning of the line, followed by

PROBLEMS

the name of the signal. This is followed by a positive integer >0 which specifies the duration of each signal and a binary number specifying the initial value of the signal.

4.44 Successive input changes on an element may cause a pulse on the output whose duration is less than the inertial delay of the element. Incorporate the ability to detect inertial delays. The user must be able to specify inertial delay of elements when entering a circuit description.

4.45 Write a serial fault simulator based on the following outline and run the fault simulator on the circuit of Figure 4.20:

```
Create test vectors
Create stuck-at fault list for logic gate inputs
FOR J=1 TO LAST_TEST_VECTOR
  Simulate Good Machine with VECTOR(J):save output response
  FOR I=1 TO LAST_FAULT
    IF FAULT(I) = DETECTED THEN NEXT I
    INSERT FAULT(I):perform simulation of VECTOR(J)
    IF steady-state output ≠ good machine response
    THEN SET FAULT(I) = DETECTED
  NEXT I
NEXT J
```

4.46 When performing serial fault simulation, the simulation time can be reduced by pre-screening faults. If an n-input AND (OR) gate has two or more 0s (1s) on its inputs after good machine simulation of a test vector, then the test vector cannot possibly detect any of the input stuck-at faults on that gate (why?). Write a routine that pre-screens the circuit after each good machine simulation and marks the testable faults so that serial fault simulation only occurs on *potentially detectable* faults.

4.47 Many functions made up of logic gates, such as latches and flip-flops, are used repeatedly in a circuit design. It is tedious and error-prone to encode them every time they are used. A simulator, like an assembler, can use macros. In the simulator, the macro is a description of the function in terms of simulator primitives, using dummy variables. Write a function expander that accepts a gate-level macro description of a function. This macro is replaced by its expanded version when the circuit model is being created. We illustrate with a cross-coupled NAND latch. It uses the freeform input developed in earlier exercises.

```
FUNCTION
LATCH &1,&2,&3,&4,&5,&6
NAND &5 &3 &6 &1
NAND &6 &4 &5 &2
END
```

The latch, when used in a circuit, would appear as follows:

```
LATCH SET RESET DELAY1 DELAY2 OUT1 OUT2
```

The function expander, upon encountering function LATCH, replaces it with the expanded version and replaces dummy variables with arguments present in the function call. The macro functions can be greatly simplified if some of the other simulator capabilities have been first programmed.

4.48 Recode the simulator in a compiled language and compare the amount of run time required for typical circuits of 50 to 100 elements and some fixed number of input stimuli.

4.49 When using descriptor cells, a signal change on the output of a gate is stored in the descriptor cell of the destination gate as one of several contiguous signals. Then, for evaluation, all input values are accessed as a single entity to index into the zoom table and provide a rapid calculation. Furthermore, the manner in which inputs and outputs are linked together in descriptor cells is rather straightforward to implement and does not require special fan-out tables. Implement these concepts in your compiled simulator.

REFERENCES

1. Druian, R. L., "Functional Models for VLSI Design," *Proc. 20th Design Automation Conf.*, 1983, pp. 506–514.
2. Rappaport, A., "Capable Digital-Circuit Simulators Promise Breadboard Obsolescence," *EDN*, March 17, 1983, pp. 105–126.
3. Johnson, W. A., "Behavioral-Level Test Development," *Proc. 16th Design Automation Conf.*, 1979, pp. 171–179.
4. Bryant, R. E., "A Switch-level Model and Simulator for MOS Digital Systems," *IEEE Trans. Comput.*, vol. C-33, no. 2, Feb. 1984, pp. 160–177.
5. Ulrich, E., and D. Hebert, "Speed and Accuracy in Digital Network Simulation Based on Structural Modeling," *Proc. 19th Design Automation Conf.*, 1982, pp. 587–593.
6. Eichelberger, E. B., "Hazard Detection in Combinational and Sequential Switching Circuits," *IBM J. Res. Dev.*, vol. 9, no. 2, March 1965, pp. 90–99.
7. Hardie, F. H. and R. J. Suhocki, "Design and Use of Fault Simulation for Saturn Computer Design," *IEEE Trans. on Elec. Comput.*, vol. EC-16, no. 4, Aug. 1967, pp. 412–429.
8. Fantauzzi, G., "An Algebraic Model for the Analysis of Logical Circuits," *IEEE Trans. Comput.*, vol. C-23, no. 6, June 1974, pp. 576–581.
9. Phillips, N. D., and J. G. Tellier, "Efficient Event Manipulation: The Key to Large Scale Simulation," *Proc. 1978 IEEE Int. Test Conf.*, pp. 266–273.
10. Ulrich, E. G., "Exclusive Simulation of Activity in Digital Networks," *Commun. ACM*, vol. 12, no. 2, Feb. 1969, pp. 102–110.
11. Ulrich, E. G., "Non-Integral Event Timing for Digital Logic Simulation, "*Proc. 14th Design Automation Conf.*, 1976, pp. 61–67.
12. Bowden, K. R., "Design Goals and Implementation Techniques for Time-Based Digital Simulation and Hazard Detection," *Proc. 1982 Int. Test Conf.*, pp. 147–152.
13. Hitchcock, R. B., "Timing Verification and the Timing Analysis Program," *Proc. 19th Design Automation Conf.*, 1982, pp. 594–604.
14. Wold, M. A., "Design Verification and Performance Analysis," *Proc. 15th Design Automation Conf.*, 1978, pp. 264–270.
15. Ng, P., et al., "A Timing Verification System Based on Extracted MOS/VLSI Circuit Parameters," *Proc. 18th Design Automation Conf.*, 1981, pp. 288–292.
16. Levin, H., "Enhanced Simulator Takes on Bus-Structured Logic." Electron. Des., Oct. 29, 1981.
17. Holt, D., and D. Hutchings, "A MOS/LSI Oriented Logic Simulator," *Proc. 18th Design Automation Conf.*, 1981, pp. 280–287.

REFERENCES

18. McDermott, R. M., "Transmission Gate Modelling in an Existing Three-Value Simulator," *Proc. 19th Design Automation Conf.*, 1982, pp. 678–681.
19. Roth, J. P., et al., "Programmed Algorithms to Compute Tests to Detect and Distinguish between Failures in Logic Circuits," *IEEE Trans. Comput.*, vol. EC-16, no. 5, Oct. 1967, pp. 567–580.
20. Roth, J. P., *Computer Logic, Testing, and Verification,* chap. 3, Computer Science Press, Rockville, Md., 1980.
21. Armstrong, D. B., "A Deductive Method for Simulating Faults in Logic Circuits," *IEEE Trans. Comput.*, vol. C-21, no. 5, May 1972, pp. 464–471.
22. Schuler, D. M., and R. K. Cleghorn, "An Efficient Method of Fault Simulation for Digital Circuits Modeled from Boolean Gates and Memories," *Proc. 14th Design Automation Conf.*, 1977, pp. 230–238.
23. Snethen, T. J., "Simulator-Oriented Fault Test Generator," *Proc. 14th Design Automation Conf.*, 1977, pp. 88–93.
24. Chang, H. Y., et al., "Comparison of Parallel and Deductive Fault Simulation Methods," *IEEE Trans. Comput.*, vol. 23, no. 11, Nov. 1974, pp. 1132–1138.
25. Ozguner, F., et al., "On Fault Simulation Techniques," *J. Des. Auto. and Fault Tol. Computing*, vol. 3, no. 2, pp. 83–92.
26. Case, G. R., "SALOGS-IV, A Program to Perform Logic Simulation and Fault Diagnosis," *Proc. 15th Design Automation Conf.*, 1978, pp. 392–397.
27. Case, G. R., "A Statistical Method for Test Sequence Generation," *Proc. 12th Design Automation Conf.*, 1975, pp. 257–260.
28. Ulrich, E., "High Speed Concurrent Fault Simulation with Vectors and Scalars," *Proc. 17th Design Automation Conf.*, 1980, pp. 374–380.
29. Moorby, P. R., "Fault Simulation Using Parallel Value Lists," *Proc. ICCAD*, 1983, pp. 101–102.
30. Krohn, H. E., "Vector Coding Techniques for High Speed Digital Simulation," *Proc. 18th Design Automation Conf.*, 1981, pp. 525–529.
31. Hayes, J. P., "A Unified Switching Theory with Applications to VLSI Design," *Proc. IEEE*, vol. 70, Oct. 1982, pp. 1140–1151.

Chapter 5

The Automatic Test Pattern Generator

5.1 INTRODUCTION

In previous chapters we looked at methods for creating and evaluating test stimuli for digital logic circuits. In this chapter we turn our attention to implementation details. We will examine the entire automatic test pattern generator (ATPG) environment including the devices that must be tested and we will see how the growing complexity of these devices affects the problem of creating test stimuli.

Test stimuli are created more efficiently if features are available which accommodate user interaction with the simulation and test pattern generation software. We will look at some user features and examine the ways in which these can help the user to work more effectively and more accurately.

The concepts described in previous chapters have evolved as a means of creating and evaluating test stimuli for digital logic circuits, whether they are single ICs or logic boards with 200 or more ICs, while minimizing the amount of resources, both CPU and engineering time, required. As with many engineering endeavors, the state of the art at any given time represents a trade-off between cost and benefit aimed at obtaining maximum results for a reasonable investment of resources. As the product to be tested changes, so also do the trade-offs change. Therefore, the overview presented here represents a snapshot in time of an art that is undergoing rapid change.

5.2 OVERVIEW OF THE ATPG

The flowchart in Figure 5.1 illustrates the complete process of creating a test program for a digital product. Given a circuit for which test stimuli must be created,

5.2 OVERVIEW OF THE ATPG

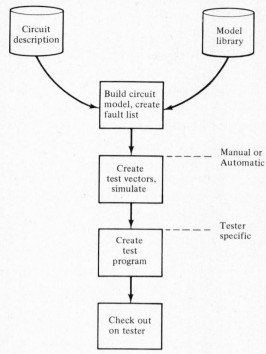

Figure 5.1 Test program creation.

whether it is an IC or a digital logic board, it is necessary to create a computer model of the circuit expressed as an interconnection of primitives for which the ATPG has processing capability. If the ATPG employs the D-algorithm, the circuit description is an interconnection of primitives for which a singular cover, propagation D-cubes, and primitive D-cubes of failure (PDCF) exist.

After the circuit model is created, a fault list is compiled. Then the ATPG can begin to create tests. A fault is selected from the fault list and an attempt is made to generate a test for that fault. If the pattern generator succeeds in creating an input pattern, or sequence of patterns, control is passed to the simulator. The simulator determines which faults, if any, have been detected by the input sequence. A test pattern generator often makes simplifying assumptions, some of which may be invalid, causing the simulator to reject the test. If the test is acceptable, and if one or more faults are detected by the test, the simulator checks off the faults that were detected. Then, if more faults remain and if termination criteria have not been satisfied, it passes control back to the pattern generator.

Postprocessing consists of issuing reports on faults tested, formatting the test stimuli and response for a particular tester, and perhaps creating diagnostic data. Other data may be required, depending on the particular test equipment used. That data will be discussed in more detail in the next chapter.

5.3 CREATING THE CIRCUIT MODEL

The circuit model is the data upon which the ATPG operates. It includes a description of the elements that make up a circuit, their interconnections, and the faults for which tests must be generated. The circuit description can be entered into the ATPG as a list of components or as a list of nets.

5.3.1 Circuit Description by Components

When entering a circuit description into an ATPG or simulator as a list of components, each entry typically includes the name of a component and its function, a list of its inputs, and possibly some additional information such as propagation delay time. For the basic switching elements such as AND, OR, Invert and simple combinations of these elements, the inputs can be listed in any order since the logic operations are commutative. For more complex devices such as flip-flops, the list of inputs is position-dependent. The processing program will associate the input in a given position with a specific action to be performed.

EXAMPLE ■

We will create a circuit description for the circuit in Figure 5.2.

```
A       PI
B       PI
C       PI
D       PI
E       PI
F       PI
G       PI
H       PI
I       AND 8 A B
J       NAND 6 C D
K       OR 8 I J E
L       DFF 12 - 6 F G - H
M       EXOR 7 K L
N       PO M
ENDCKT
```
■ ■

Each entry in this circuit description consists of the name of an element followed by its function. For primary inputs (PI) there is no additional information. For primary outputs (PO) there will be an entry listing the name of the device that drives the PO. For a switching element there are additional entries including propagation delay and a list of elements connected to its inputs. The DFF (delay flip-flop) is followed by two numbers and an intervening dash. The first number represents the delay from Data input to the output following the arrival of a clock pulse. The second (unused in the example, as indicated by the dash) and third numbers represent delay from the occurrence of a Set or Reset input to the appearance of its effect on the output. The inputs are positional. The processing program assumes that the first input is connected to the Data input, the second is connected to the Clock input, the third is connected to the Set input, and the fourth is connected to the Reset input. The second dash in this list indicates that the Set input is not used.

5.3 CREATING THE CIRCUIT MODEL

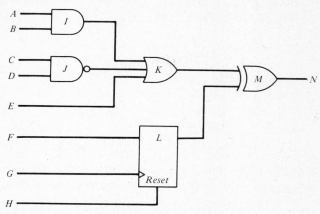

Figure 5.2 Gate-level interconnection.

This circuit description can easily be converted into a computer model. As each element is read into the computer, a descriptor cell (see Chap. 4) of the appropriate size is created and pointers to other elements are established. The descriptor cell contains an entry defining the function of the element represented by the cell. This is used as an index by the ATPG or simulator to access the singular cover, propagation D-cubes, and PDCFs or other processing information.

5.3.2 Circuit Description by Wire List

The gate- or element-oriented circuit description is convenient and is easily implemented in a design environment where all elements have just one or two outputs. However, in the board test environment, where many ICs, each having numerous outputs, are interconnected, and where each net may interconnect multiple inputs and outputs, it is more convenient to use a net-oriented circuit description.

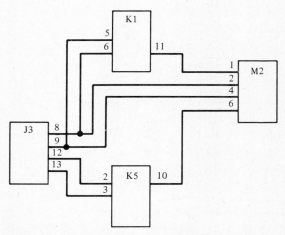

Figure 5.3 Interconnection of ICs.

Consider the partially illustrated circuit of Figure 5.3. Several interconnections exist between the four ICs labeled J3, K1, K5, and M2. This circuit could be entered into an ATPG preprocessor program in the following format[1]:

```
J3-8,K1-6,M2-2
K1-11,M2-1
J3-9,K1-5,M2-4
J3-12,K5-2
J3-13,K5-3
K5-10,M2-6
```

Each line of the description describes one net which connects two or more component pins. The IC names may be names assigned by the manufacturer, arbitrary names assigned by the designer, or codes used to designate the physical position

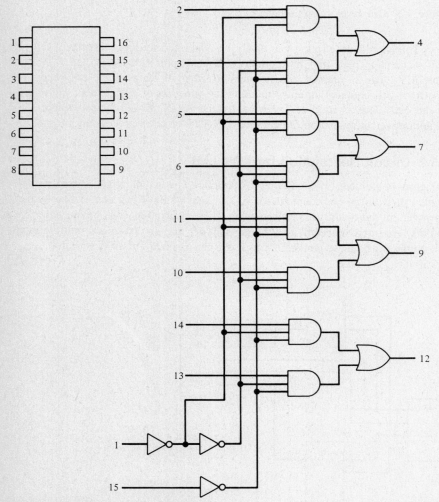

Figure 5.4 Gate-level description of IC.

5.3 CREATING THE CIRCUIT MODEL

of the device on the board. The number following the dash is the pin number. If the name is a physical location or arbitrarily assigned code, then another file is provided which contains a cross reference between the IC location or code name and the manufacturer's name assigned to the part. The IC part number is used to access a library where the IC is described in terms of an expansion model containing primitives which the ATPG is capable of processing. Figure 5.4 illustrates this concept.

A quad two-input multiplexer is illustrated in Figure 5.4. The library entry describes the device as an interconnection of AND, OR, and Invert elements, using the previously described gate-level description format. For each such device encountered in the wire list, the ATPG retrieves a corresponding description from the library and substitutes it for the IC. When finished, the circuit has been expanded into an interconnection of basic primitives.

5.3.3 Modeling the IC

The individual IC members in the library can often be modeled in several different ways, depending on the types of primitives implemented in the ATPG or simulator. The quad multiplexer of Figure 5.4 could exist in the library as an interconnection of AND gates, OR gates, and inverters. This gate-level description would require 15 elements to represent the four multiplexers: four OR gates, eight AND gates, and three inverters.

The D-algorithm does not require that a circuit be modeled solely in terms of basic logic gates. The IC could be modeled as four 2-to-1 multiplexers. Each of these four primitives is then the logical equivalent of the circuit illustrated in Figure 5.5(a), which we arbitrarily label as primitive P22. The IC then appears, in terms of the primitive P22, as in Figure 5.5(b). The P22 primitive has associated with it a singular cover, primitive D-cubes of failure, and propagation D-cubes. The PDCFs are determined by creating a cover corresponding to each of the failures to be modeled and using that, together with the cover for the unfaulted circuit, as described in Chap. 2, to form PDCFs.

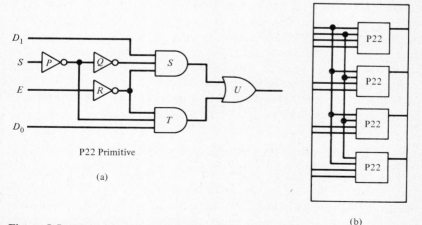

Figure 5.5 Multiplexer expressed as a primitive.

There are advantages to incorporating several gates into a single primitive. First, it requires less computer memory to contain the circuit image because the ICs are represented in terms of fewer primitives. Second, less CPU time is required to process the circuit since the propagations, implications, and justifications are performed on fewer primitives. This is important in a CPU-intensive operation such as automatic test pattern generation. However, a more complex primitive usually has larger tables for the cover, PDCF, and propagation D-cubes. This may increase the CPU time required by the ATPG to search through the tables and thus offset some of the CPU savings.

Another advantage to the more complex primitive, not so obvious, is the fact that it may be easier for an ATPG to justify a signal required on the output of the P22 primitive than to justify the same signal on a gate-level model. Suppose a 1 is required on the output of P22. Using gate-level primitives, the ATPG would likely assign values to the Select input and one of the Data inputs and then attempt to justify those assignments. However, if the Select line comes from control logic that is difficult to control, it might be easier to justify 1s on both Data inputs. A disadvantage, in an asynchronous environment, to the use of more complex primitives is the fact that it may be more difficult to detect switching hazards. This can be seen in Figure 5.5(a), where, if inputs D_0 and D_1 are at 1, and E is at 0, and S switches from 0 to 1, there will be a negative pulse on the output, equal in duration to the propagation time through inverter Q.

The modeling effort for sequential circuits is complicated by the added dimension of time. Edge-triggered and level-sensitive devices are especially prone to modeling errors and their behavior must be carefully analyzed. On some edge-triggered flip-flops new data appears on the output very shortly after the active edge, while on others configured as master/slave flip-flops the new data may not appear until the following clock edge. An incorrect model can invalidate the results of a test, causing the tester to come up with miscompares and causing several logic boards to be "repaired" before an incorrect model is discovered.

5.4 FAULT MODELING

The fault list is determined, at least in part, by the primitives employed. However, other factors must also be considered. There are several MOS and bipolar technologies in use. Failures in these different technologies do not always produce similar error behavior. Another important consideration is the amount of CPU time required to process a circuit. This is affected by the size of the fault list. Methods have been developed to reduce the size of the fault list.

5.4.1 Fault Equivalence and Dominance

The stuck-at fault model has been used for many years as a basic model for switching elements. It has the attraction that fault list compilation and processing are easy to automate. Even when more complex primitives such as the 2-to-1 multiplexer (MUX) are used in the system, the stuck-at faults can be used as a

5.4 FAULT MODELING

guide to creating the PDCFs. For example, we may create a PDCF for an S-A-1 fault on the output of inverter Q in the 2-to-1 MUX primitive of Figure 5.5(a) by first creating a cover corresponding to the circuit with the S-A-1 fault on the inverter output and then using that, in conjunction with the cover for the fault-free circuit, to create a PDCF for the fault.

In building fault lists, it is often found that some faults are indistinguishable from others. In Figure 5.5(a), if we create a truth table describing the behavior of the primitive when the output of gate S is S-A-0, and another describing the behavior when the upper input to the OR gate is S-A-0, we find that they are identical. This is not surprising since they are tied to the same net. We say that the two faults are *equivalent*. There is no logic test that can distinguish between them. More precisely, if T_a is the set of tests which detect fault a and T_b is the set of tests which detect fault b, and if $T_a = T_b$, then it is not possible to distinguish a from b.

In Chap. 2 we remarked that the fault list for an n-input AND gate consisted of $n + 2$ entries. However, when we tested for input i S-A-1, we simultaneously tested for the output S-A-1. The converse does not hold; a test for an S-A-1 on the output does not necessarily test for any of the input S-A-1 faults. We say that the output S-A-1 fault dominates the input S-A-1 fault. In general, fault a *dominates* fault b if $T_b \subseteq T_a$. From this definition it follows that if fault a dominates fault b, then any test which detects fault b will detect fault a.

We define a function F to be *unate* in variable x_i if the variable x_i appears in the sum-of-products expression for F in its true or complement form, but not both. The concept of fault dominance for logic elements can now be characterized[2]:

Theorem 5.1. Given a combinational circuit $F(x_1, x_2, \ldots, x_n)$, a dominance relation exists between faults on the output and the input x_i if and only if F is unate in x_i.

We define a function to be *partially symmetric* in variables x_i and x_j if $F(x_i, x_j) = F(x_j, x_i)$. A function is *symmetric* if it is partially symmetric for all input variable pairs x_i, x_j. With those definitions we have:

Theorem 5.2. If a logic gate is partially symmetric for inputs i and j, then either faults on those inputs are equivalent or no dominance relation holds.

Theorem 5.3. In a fan-out free logic circuit realized by symmetric, unate gates, tests designed to detect stuck-at faults on primary inputs will detect all stuck-at faults in the circuit.

Fault equivalence and dominance relations are used to reduce the size of fault lists. Since computer run time is affected by the size of the fault list, the reduction of the fault list, a process called *fault collapsing*, can reduce test generation and simulation time. Consider the primitive of Figure 5.5(a). An S-A-1 fault on the output of gate U is equivalent to an S-A-1 fault on any of its inputs and an S-A-0 fault on the output of gate U dominates an S-A-0 fault on any of its inputs.

S-A-x faults on the inputs to gate U, in turn, are equivalent to S-A-x faults on the outputs of gates S and T. Therefore, for the purpose of detection, if faults on the outputs of gates S and T are detected, faults on gate U can be ignored.

5.4.2 Checkpoint Faults

The process of fault collapsing via equivalence and dominance can be continued. The output S-A-0 faults on S and T are equivalent to input S-A-0 faults on U, and output S-A-1 faults dominate input S-A-1 faults. However, the process can be performed more efficiently in combinational circuits.

The Stuck-At-Faults We first define g/e to denote an S-A-e fault at g, where $e \in \{0, 1\}$. Then[2]:

Theorem 5.4. At any fan-out point in combinational logic, we label the driving gate output as y, and the fan-out paths as $p_1, p_2, \ldots, p_n$; then y/e dominates p_i/e, $1 \leq i \leq n$ and

$$\bigcup_{i=1}^{n} T_{p_i/e} \subseteq T_{y/e}$$

This dominance relation is the reverse of the previous relation defined relative to primary inputs, so the set T of tests for primary inputs will not detect all faults. However, it suggests the following strategy:

1. Assign S-A-0, S-A-1 faults to the primary inputs.
2. Assign S-A-0, S-A-1 faults to the paths p_i emanating from a signal source y with fan-out greater than 1.
3. If any of the stuck-at faults are equivalent to faults on the outputs of symmetric logic gates, translate them to the output.
4. If faults occur at the input of an inverter, translate them to the output.
5. Apply Theorem 5.4 wherever it applies.

By translating faults, we mean delete the input stuck-at fault and add the equivalent stuck-at fault on the output. As an example, if an AND gate has two inputs connected to primary inputs, the S-A-0 faults on these two inputs are deleted and are replaced on the fault list by a single S-A-0 fault at the AND output. Application of Theorem 5.4 allows us to delete faults at primary inputs which fan out immediately or through an inverter. The signal path that originates at a primary input or at one of the fan-out paths of a signal with multiple fan-out is called a *checkpoint arc*. The faults obtained by use of checkpoint arcs are called *checkpoint faults*[3].

EXAMPLE ■

In the circuit of Figure 5.5(a), there are eight checkpoint arcs, the four primary inputs and two fan-out paths from each of P and R. Therefore, there are initially 16 faults. The faults on the inputs to P and R are translated to their outputs, where they dominate faults on the fan-out paths; hence they

5.4 FAULT MODELING

are deleted. The faults on the input to Q are translated to its output. We now have 12 faults, namely S-A-0 and S-A-1 faults on each of the three inputs to S and T. The three S-A-0 faults at the inputs of each of gates S and T are translated to the outputs, where they are represented by a single equivalent S-A-0 fault, resulting in a total of 8 faults. ∎∎

The checkpoint faults establish a correspondence between signal path faults and a subset of the set of all stuck-at faults. If all checkpoint faults have been tested, then all stuck-at faults have been tested, or, put another way, the ability to propagate a 0 and 1 along the entire length of each checkpoint arc or signal path has been verified. The value of checkpoint faults lies in the fact that they provide an automated means to identify these paths, starting with stuck-at faults on primary inputs and nets with multiple fan-out. It is interesting to note that the sequential path sensitizer (SPS) test generation system described in Chap. 3 creates tests by propagating 0 and 1 along each signal path, or transmission path, as they were called[4]. In this section we started with checkpoint faults, then minimized that set wherever possible on the basis of fault dominance, to arrive at a set of stuck-at faults which identified the transmission paths.

The checkpoint faults provide a useful means of helping the test engineer when he is manually writing test patterns for a circuit. Rather than provide him with a list of all untested faults, a more effective strategy may be to have the fault simulator provide him with a list of just the untested checkpoint faults. Then he is not as likely to write test vectors for faults that will be detected by some other vector. Furthermore, the smaller list of undetected checkpoint faults is probably a better gauge of how much work remains to be done. Note that checkpoint faults do not represent equivalent numbers of stuck-at faults. In Figure 5.5(a), the path from P to the output through Q, S and U is longer than the path through T and U. This suggests, assuming all faults are equiprobable, that checkpoint faults can be weighted based on the number of gates in the path. When creating test vectors, automatically or manually, the checkpoint faults of highest weight are then given highest priority.

It is important to note that dominant faults can be detected by test patterns that do not detect dominated faults. This is important if fault resolution is required. Also note that the theory was not established for sequential circuits. In Sec. 3.3.1 we discussed the problem of testing for an S-A-0 fault on the output of an inverter connected to the Enable input of a latch. We concluded that we could not confirm a test for the fault because it causes the latch state to remain permanently indeterminate. As a result, based on our definitions, the S-A-0 fault does not dominate faults on the inputs of the NAND gates in its fan-out path, both of which are testable. However, from a practical standpoint, it is interesting to note that in the given circuit, successful tests for S-A-0 on both fan-out paths from the inverter are equivalent to a successful test for S-A-0 on the output of the inverter. This follows from the fact that if both fan-out paths are fault-free, then the latch can be both set and reset, which implies that the inverter output is fault-free, and conversely, if the inverter is faulty, or if one of the two fan-out paths is faulty, then one of the two latch operations, either set or reset, will fail.

The Delay Faults It is possible that a circuit is free of structural defects but the delay along one or more signal paths is excessive. Simply propagating 1 and 0 along the paths, while sufficient to detect stuck-at faults, is not sufficient to detect delay faults since the signal being propagated to a flip-flop or a primary output may have the same value that existed on the preceding clock period. It cannot then be determined whether the signal clocked into the flip-flop or observed at a primary output is the new signal or the old signal.

To detect delay faults it is necessary to propagate rising and falling edges along signal paths. The existence of checkpoint faults as identifiers of unique signal paths for propagation of 1 and 0 suggests the following extended strategy to detect both the stuck-at faults and delay faults:

1. Identify all unique signal paths.
2. Select a path, apply a 0 to the input, then propagate through the entire path.
3. Repeat the signal propagation with a 1 on the input, and then again with a 0 on the input.
4. Continue until all signal paths have been exercised.

A complete signal path may include several flip-flops (see Sec. 3.4.5). The test strategy described will check the delay relative to clock pulse width along paths where both source and destination are flip-flops. The strategy may also be effective in detecting stuck-open faults in CMOS circuits (see Sec. 5.4.4). The number of unique signal paths will normally be significantly less than the number of checkpoint faults since several such faults will often lie along a single signal path.

5.4.3 Redundant Faults

Redundant connections can cause a fault to become untestable. A connection is defined as *redundant* if it can be cut without altering the output functions of a circuit[5]. If a circuit has no redundant connections, then it is irredundant. The following theorem follows directly from the definition of redundancy.

Theorem 5.5. All S-A-1 and S-A-0 faults in a combinational circuit are detectable if and only if the circuit is irredundant.

The simplest kind of redundancy, and one seen often in practice, is to tie two or more signal pins together at the input to an AND or an OR gate. This is often the practice when an n-input gate is available in an IC package and a particular application requires fewer than n inputs. When two inputs are tied together, it is impossible to establish an input condition wherein the two inputs differ; therefore any test that requires the two inputs to differ cannot be applied.

For the purpose of fault analysis, there are two fault possibilities when inputs are tied together (see Figure 5.6):

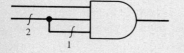

Figure 5.6 AND gate with redundant input.

5.4 FAULT MODELING

1. An open occurs somewhere between the common connection point and one of the inputs.
2. An open occurs prior to the common connection point.

If an open exists between the common connection and the gate input, the fault cannot be detected. If the open occurs prior to the common connection, then the open affects both gate inputs and the behavior appears the same as if there were a single input with an S-A-1 on the input. Therefore, it is a common practice in industry to check gates for inputs with identical signal names. If matching signal names are found, then all but one signal can be deleted. Other kinds of redundancy can be more difficult to detect. Redundancy designed into logic solely to prevent a hazard will create an untestable fault. If the fault occurs, it may or may not produce an error symptom since a hazard represents only the possibility of a spurious signal. No general method exists for spotting redundancies in logic circuits.

5.4.4 Technology-Related Faults

There is growing concern over the adequacy of the stuck-at model. Some faults are technology-dependent and cause behavior unlike the traditional S-A-x faults. Circuits are modeled with the commonly used logic symbols for test purposes, but in practice it is quite difficult to correlate faults in the actual circuit with faults in the behaviorally equivalent circuit represented by logic gates.

ISL A circuit implemented in integrated Schottky logic (ISL) is shown in Figure 5.7. If diode D_1 is shorted, then signal A becomes a wired-AND of A and B. If A fans out to another gate, then the input to the other gate is no longer signal A but, rather, the signal $A \cdot B$. To test for diode D_1 shorted, it is necessary to set signals

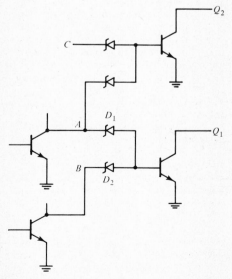

Figure 5.7 Integrated Schottky logic (ISL) gate.

A and C high, and signal B low. Then, if D_1 is shorted, the signal $A \cdot B$ will go low and the output of Q_2 will go high where, normally, setting inputs A and C both high is sufficient to cause Q_2 to go low.

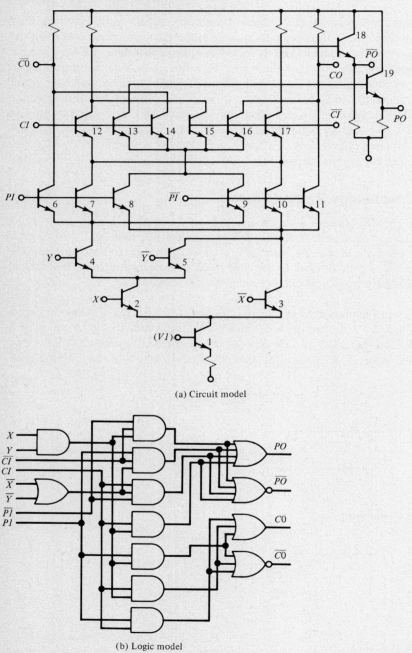

Figure 5.8 ECL full adder. (© 1982 IEEE.)

5.4 FAULT MODELING

ECL Emitter-coupled logic (ECL) fault effects have been analyzed in depth by Beh el al.[6]. A full adder is shown in Figure 5.8, both at the circuit level and at the logic level. Test patterns were generated to test all defects in the circuit model. Then, independently, a set was generated to test for stuck-at faults in the logic model. The two sets obtained were:

	Circuit model				Logic model		
CI	PI	Y	X	CI	PI	Y	X
0	0	1	0	0	0	1	1
0	0	1	1	0	1	1	1
0	1	1	1	1	0	0	1
1	0	0	1	1	0	1	0
1	1	0	1	1	0	1	1
				1	1	1	1

Two of the patterns derived from circuit analysis, 0010 and 1101, were not generated by the logic model. Yet, circuit level analysis shows that the two patterns are needed. The pattern 0010 detects a collector-to-emitter short on transistor 2, base-to-emitter shorts on 3 and 8, and open contacts at the emitter, base, and collector of 3. The pattern 1101 detects emitter, base, and collector contact opens on transistor 8. None of these faults are detected by the test patterns generated from the logic model.

MOS Metal oxide semiconductor (MOS) circuits can also be implemented in ways which make it difficult to characterize faults. The circuit in Figure 5.9 is intended to implement the function

$$F = (A + C)(B + D)$$

With the indicated open it implements

$$F = A \cdot B + C \cdot D$$

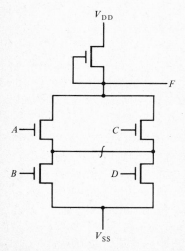

Figure 5.9 MOS circuit with open.

It is not immediately obvious how to implement this MOS circuit as an interconnection of logic gates so as to conveniently represent both the fault-free and faulted versions (although it can be done).

CMOS The complementary metal oxide semiconductor (CMOS) NOR circuit is illustrated in Figure 5.10. When A and B are low, both p-channel transistors are on, and both n-channel transistors are off. This causes the output to go high. If either A or B goes high, the corresponding upper transistor(s) is cut off, the corresponding lower transistor(s) is turned on, and the output goes low.

The conventional S-A-x faults occur when the output shorts to V_{SS} or V_{DD} or when opens occur at the input terminals to the NOR circuit. The opens cause S-A-0 faults on the inputs since the input signal cannot turn off the corresponding p-channel transistor and cannot turn on the corresponding n-channel transistor. Another type of fault which can occur in the CMOS circuit is the open, which occurs in a transistor or at the connection to a transistor. Three such faults can be identified in the two-input NOR gate of Figure 5.10. These faults, usually referred to as *stuck-open* faults, include a defective pull-down transistor connected to A or B or an open pull-up transistor anywhere between the output channel and V_{DD}[7].

If Q_4 is open, the signal A can cut off the path to V_{DD} but it cannot turn on the path to V_{SS}. Therefore, the value at F will depend on the electrical charge trapped at that point when signal A goes high. The equation for the faulted circuit is

$$F_{n+1} = \overline{A} \cdot \overline{B} + A \cdot \overline{B} \cdot F_n$$

Table 5.1 illustrates the effect of all seven faults. In this table F represents the fault-free circuit. F_1 and F_2 represent the output S-A-0 and S-A-1, respectively. F_3 and F_4 represent open inputs at A and B. Faults F_5 and F_6 correspond to opens

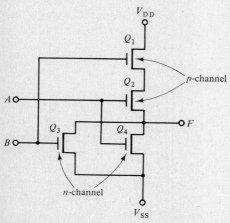

Figure 5.10 CMOS circuit.

5.4 FAULT MODELING

TABLE 5.1 FAULT BEHAVIOR FOR CMOS NOR

A	B	F	F_1	F_2	F_3	F_4	F_5	F_6	F_7
0	0	1	0	1	1	1	1	1	F_n
0	1	0	0	1	0	1	0	F_n	0
1	0	0	0	1	1	0	F_n	0	0
1	1	0	0	1	0	0	0	0	0

in the pull-down transistors connected to A or B or the leads connected to them. F_7 is the function corresponding to an open anywhere in the pull-up circuit.

Some circuit output values become dependent on previous values held by circuit elements when the circuit is faulted, so that in effect the faulted circuit exhibits sequential circuit behavior. For example, we note from Table 5.1 that F_5 differs from F, the fault-free machine, only in row 3, and then only when F has value 0 and F_5 had a 1 at the output on the previous pattern. To detect this fault, it is necessary to establish the values $(0,0)$ on the inputs A and B. This produces the value 1 at the output of the gate. Then, the values $(1,0)$ are applied to the inputs and the sensitized value is propagated to an output.

A suggested approach for testing stuck-open faults[8] develops tests for the conventional faults first. When simulating faults, the previous pattern is checked to see if the value F_n from the previous pattern, in conjunction with the present value, will cause the output of the gate to be sensitized on the present pattern. In the situation cited in the previous paragraph, if the previous pattern causes a $(0,0)$ to appear on the inputs of the NOR, and if the present pattern applies a $(0,1)$ or $(1,0)$ to the NOR, then one of the two stuck-opens on the pull-down transistors is sensitized at the output of the NOR and it simply remains to simulate it to determine whether it is sensitized to an output.

If stuck-open faults remain undetected after all stuck-at faults have been processed, then it is necessary to explicitly sensitize them using a two-pattern sequence. The first pattern need only set up the initial conditions on the gate being tested. The second pattern must be propagated to an output. Note that during these patterns, it is also possible to check, during simulation, for detection of other stuck-open faults.

5.4.5 Bridging Faults

Faults can occur because of either shorts or opens. In general, the opens are easier to model. An open on an input to an AND gate inhibits that input from pulling the gate down to 0, hence the input is S-A-1. Shorts can be more difficult to characterize. If a signal line is shorted to ground or to a voltage source, we can model it as S-A-0 or S-A-1, but signal lines can become shorted to each other. In any reasonably sized circuit, it is impractical to model all pairs of shorted nets. However, it is possible to identify and model shorts which have a high probability of occurrence.

Adjacent Pin Shorts We define a function F to be *elementary* in variable x if it can be expressed in the form

$$F = x^* \cdot F_1$$

or

$$F = x^* + F_2$$

where x^* represents x or $\bar{x}$ and F_1, F_2 are independent of x. An *elementary gate* is a logic gate whose function is elementary. An *input-bridging fault* of an elementary gate is a bridging fault between two inputs, neither of which fans out to another circuit. With these definitions, we have[9]:

Theorem 5.6. A test set which detects all single input stuck-at faults on an elementary gate also detects all input-bridging faults at the gate.

The theorem tells us that tests for stuck-at faults on inputs to elementary gates, such as the AND gates and the OR gates, will detect many of the adjacent pin shorts that can occur. However, because of the random nature of pin assignment in IC packages (relative to test strategies), the theorem rarely applies to IC packages. It is common in industry to model shorts between adjacent pins on these packages because the shorts have a high probability of occurrence, due to the manufacturing methods used to solder ICs to printed circuit boards. Adjacent pin shorts may cause a signal on a pin to alter the value present on the other pin. To test for the presence of such faults it is necessary to establish a sensitized signal on one pin and establish a signal on the other pin which will pull the sensitized pin to the failing value. If the sensitized value D is established on one of the pins, then a 0 is required on the adjacent pin. Given a pair of pins p_1 and p_2, the following signal combinations on a pair of pins will completely test for all possibilities wherein one pin may pull another to a 1 or 0.

$$p_1 - D\,\bar{D}\,0\,1$$
$$p_2 - 0\,1\,D\,\bar{D}$$

It is possible to take advantage of an existing test to create, at the same time, a test for adjacent pin shorts. If a path is sensitized from an input pin to an output pin during test pattern generation, and if a pin adjacent to the input pin has an x value assigned, then that x value can be converted to a 1 or 0 to test for the adjacent pin short. The value chosen will depend on whether the pin on the sensitized path has a D or $\bar{D}$.

Programmable Logic Arrays Shorts created by commercial soldering techniques are easily modeled because the necessary physical information is available. Recall that IC models are stored in a library and are described as an interconnection of primitives. That same library entry can identify the pins most susceptible to solder shorts, namely, the adjacent I/O pins.

Structural information is also available for programmable logic arrays (PLA) and can be used to derive tests for faults which have a high probability of occur-

5.4 FAULT MODELING

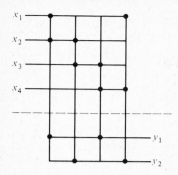

Figure 5.11 Programmable logic array.

rence. Logically, the PLA is a pair of arrays, the AND array and the OR array. The upper array in Figure 5.11 is the AND array. Each vertical line selects some subset of the input variables, as indicated by dots at the intersections or crosspoints, to create a product term. The lower array is the OR array. Each horizontal line selects some subset of the product terms, again indicated by dots, to create a sum-of-products term at the outputs.

The PLA is susceptible to bridging faults and crosspoint faults[10]. The *crosspoint fault* is a physical defect caused by an unconnected diode at a crosspoint or a diode which is connected at a crosspoint where it should not have been connected. In the AND array, the product term logically expands if a device is disconnected and the product term logically shrinks if an additional input variable is connected to the vertical line. In the OR array, a product term is added if an additional column is connected into the circuit and a product term will disappear from the circuit output if a column is not connected where required.

Bridging faults can occur where lines cross. The symptom is not necessarily the same as when an additional device is connected into a circuit. For example, the bridging fault may cause an AND operation whereas the crosspoint fault may cause an OR operation. The crosspoint open is similar in behavior to opens in conventional gates. The bridging fault, like shorts between signal lines in any logic, is complicated by the fact that a signal is affected by a logically unrelated signal. However, the regular structure of the PLA makes it possible to identify potential sources of bridging faults and to perform fault simulation if necessary to determine which of the possible bridging faults are detected by a given set of test patterns.

The Integrated Design Environment When particular bridging faults, such as those that occur between IC pins or PLA signal runs, can be identified based on physical structure, then it is possible to specifically check for those faults when performing fault simulation. The MOS faults previously described are more difficult because there is no regular physical structure in the circuit. However, in many design environments physical design macros exist in a system library (a physical macro is a description of the physical geometry of a function). The designer creates his design by connecting together standard functions which exist in the macro library, much as he would if he were designing a printed circuit board using standard SSI and MSI parts. The physical macro may have a corresponding logic macro which describes it as an interconnection of logic gates. The logic function attempts to describe the circuit in such a way as to permit modeling of as many real faults as possible. From the physical macro, shorts with high probability of occurrence can be identified. This requires a careful analysis of the physical macro, based on an understanding of the particular fault mechanisms of the technology and an understanding of design practices.

5.4.6 Selecting Fault Classes for Test

Creation of test stimuli and their validation through fault simulation can be a very CPU-intensive activity. Therefore, it has been the practice to direct test pattern generation and fault simulation at fault classes having the highest probability of occurrence. In the manufacture of printed circuit boards, two major fault classes include *manufacturing faults* and *field faults*. Manufacturing faults are those that occur during the manufacturing process, including opens at pins and shorts between pins. Field faults occur during service and include opens at IC pins, but also include internal shorts and opens. Testing in a manufacturing environment is often restricted to the manufacturing faults because it is assumed that the individual ICs have been thoroughly tested for internal faults before being mounted on the board. Although this can significantly reduce CPU time, the test so generated suffers from the drawback that it may be inadequate for detecting faults which occur while the device is in service. Studies of fault coverage for printed circuit boards comprised mainly of SSI and MSI parts show that tests which provide coverage for about 95% of the manufacturing faults often provide only about 70 to 75% fault coverage for field faults.[11,12]

5.5 THE PATTERN GENERATOR

Most pattern generators employ the D-algorithm[13]. It can create tests for all testable faults in combinational logic and it can be modified to operate on sequential logic, as was demonstrated with the iterative test generator (ITG). However, it is not uncommon for the D-algorithm to be preceded by random test pattern generation in order to reduce the amount of computation that must be performed.

5.5.1 Random Patterns

The use of random patterns is motivated by the efficiency curve shown in Figure 5.12. The first dozen or so patterns applied to a combinational logic circuit typically detect anywhere from 35 to 60% of the faults selected for testing, after which the rate of detection falls off. To see why this curve holds, consider a simple n-input, 1-output circuit. It can implement any of 2^{2^m} functions. A test pattern in which all inputs have known values, 0 or 1, partitions the functions into two equal-sized equivalence classes, based on whether the output response is a 1 or 0. A second input will further partition the functions so that there are now four equivalence classes. The functions in three of the classes will disagree with the correct circuit in either one or both of the output responses. In general, for a combinational circuit

5.5 THE PATTERN GENERATOR

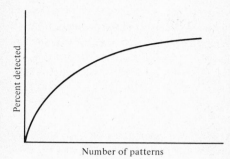

Number of patterns

Figure 5.12 Test efficiency curve.

with n inputs, and assuming all inputs are assigned a 1 or 0, the percentage of functions distinguished from the correct function after m patterns is given by the following formula:

$$P_D = \frac{1}{2^{2^n} - 1} \cdot \sum_{i=1}^{m} 2^{2^n - i} \times 100\%$$

The object of a test is to partition functions into equivalence classes such that the fault-free circuit is in a singleton set relative to functions that represent faults of interest. Since a complete partition of all functions is usually impractical, a fault model, such as the stuck-at model, defines the subset of interest so that the only functions in the equivalence class with the fault-free machine are functions corresponding to faults with very low probability of occurrence. A diagnostic test can also be defined in terms of partitions. The diagnostic test attempts to partition the set of functions such that as many functions as practical, representing faults with high probability of occurrence, are in singleton sets.

EXAMPLE ■

The 16 possible functions that can be represented by a two-input circuit are listed below. The two-input EXOR circuit is represented by F_6. Its output is 1 whenever A and B differ.

A	B	F_0	F_1	F_2	F_3	F_4	F_5	F_6	F_7	F_8	F_9	F_{10}	F_{11}	F_{12}	F_{13}	F_{14}	F_{15}
0	0	0	0	0	0	0	0	0	0	1	1	1	1	1	1	1	1
0	1	0	0	0	0	1	1	1	1	0	0	0	0	1	1	1	1
1	0	0	0	1	1	0	0	1	1	0	0	1	1	0	0	1	1
1	1	0	1	0	1	0	1	0	1	0	1	0	1	0	1	0	1

■ ■

Application of any single pattern to inputs A and B will distinguish between F_6 and 8 of the other 15 functions. Application of a second pattern will further distinguish F_6 from another four functions. Hence, after two patterns, the correct function is distinguished from 80% of the remaining functions. The formula expresses percentage tested for these single-output combinational functions strictly on the basis of the number of unique input patterns applied and makes no distinction concerning the values assigned to the inputs. It is a measure of test effectiveness for all kinds of faults, single and multiple, and suggests why there is a high initial percentage of faults detected. However, it does not reveal any information about particular classes of faults and, in fact, simulation of single stuck-at faults generally reveals a somewhat slower rise in percentage of faults detected. This should not be surprising, however, since there are many more multiple faults than single faults and no evidence to suggest that detection of single and multiple faults will occur at the same rate. In fact, it was pointed out in Sec. 5.4.6 that studies show differing rates of detection for manufacturing and field faults.

Random patterns are significantly less effective when applied to sequential circuits. They are also ineffective, after the first few patterns, against specific fault classes with high probability of occurrence such as the S-A-x faults in combinational circuits. At that point the problem has shifted. Initially, we want to detect large numbers of faults. Then, after some threshold is reached, we want to detect specific faults. Therefore, when random patterns are used, their use is normally followed by deterministic calculation of test patterns for specific faults.

5.5.2 The Imply Operation

In his original paper on the D-algorithm[14], Roth propagated sensitized signals along one or more signal paths to outputs before performing justification. In a subsequent paper[15] Roth described a modified D-algorithm, called DALG-II, in which the full implication of every assignment is carried out at every step of the propagation or justification phase. In general, an implication exists if, as a result of existing assignments on the inputs and output(s) of a primitive, only one entry in the cover exists which does not conflict with the existing assignments. If no entry exists, then a conflict has occurred.

EXAMPLE ■

In Figure 5.13 we want to derive a test for an S-A-0 on the upper input of gate J. We start by assigning a $(1, 0)$ to its inputs. The 0 on the lower input implies 1s on D and E. On gate I, a 1 on the output and a 1 on the input from D implies a 0 on the output of G. That implies 1s on inputs B and C. We finally propagate a D through J. That requires a 1 on the upper input to K. Input B was previously assigned a 1. A 0 is implied on input A and the test is complete. ■ ■

When decisions are encountered, they can frequently be postponed. Gate-level test pattern generation is one endeavor where it is desirable to postpone decisions as often as possible. We avoided a decision in the example just described by starting with the lower input to gate J. If the upper input had been selected first for processing, then a decision would be required as to which input to gate I would be assigned a 0. That could have caused a 0 to be assigned to input D, resulting in a conflict. By postponing the decision, it was ultimately avoided. The general rule is to avoid making decisions as long as any alternative activity can be performed. When decisions are made, it is necessary to record enough information so that if a decision leads to a conflict, it is possible to restore the machine to the

5.5 THE PATTERN GENERATOR

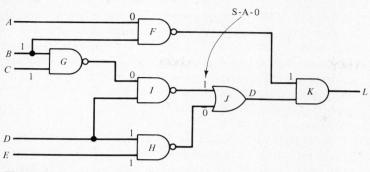

Figure 5.13 The implication operation.

state that existed when the decision was made. This permits an alternative decision to be made and evaluated.

5.5.3 Comprehension Versus Resolution

When creating test stimuli for digital circuits it is possible to bias the algorithm for either maximum or minimum fault detection with each pattern. If it is only necessary to determine whether an IC is good or bad, and there is no requirement to diagnose the cause of a failure, then we may want to make that determination with a minimum number of vectors; that is, we want maximum fault coverage or comprehension with each test vector. Minimizing the number of test vectors will reduce the number of passes through fault simulation, resulting in less CPU usage. Furthermore, it can reduce the amount of storage space required to store stimulus and response data at the test station. On the other hand, when testing a printed circuit board that may contain up to 200 IC packages, it is desirable to locate a failed IC so that the board can be repaired. This can usually be done more easily if fewer failures are detected by a given test pattern.

The algorithm can be biased by applying propagating or nonpropagating input values to primitives during the justification phase. This is illustrated in the circuit of Figure 5.14. When testing the output of gate 10 S-A-0, we may select $(0, 0)$ for the inputs or we may select either of $(0, 1)$ or $(1, 0)$. If we select $(0, 0)$, then no fault on preceding logic will propagate through the NAND gate and the only fault detected is the output of gate 10 S-A-0. If $(1, 0)$ or $(0, 1)$ is selected, then other faults can propagate through gate 8 or 9 to the output.

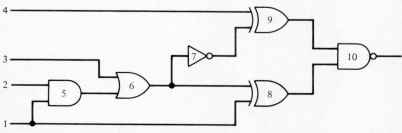

Figure 5.14 Extending a sensitized path.

This concept of desensitizing propagation paths in order to increase resolution can be enhanced by initially selecting faults at or near primary outputs and desensitizing signal paths at every opportunity. Maximizing comprehension when using the D-algorithm may be achieved in combinational circuits by initially selecting faults at or near the inputs and selecting propagating values whenever possible. It can also be achieved by using dynamic compaction, as explained in the next section, or the subscripted D-algorithm (AALG).

Another feature proposed by Roth for DALG-II is the "fast plunge." Normally, at fan-out points, the next gate selected for propagation is the lowest numbered gate in the fan-out list. In the circuit of Figure 5.14, a D on input 1 would be propagated through gate 5. However, the fast plunge selects the highest numbered gate, in this instance gate 8, and propagates through it rather than through gate 5. Since the ordering scheme generally assigns higher numbers to the primary outputs, the algorithm will often get to an output in a smaller number of steps, and with fewer gate assignments requiring justification. This provides additional opportunities for desensitizing signal paths. Another motive for selecting a gate other than the lowest numbered gate in the fan-out list is that, because of reconvergent fan-out, it may not be possible to propagate a test through a lower numbered element such as gate 5 in Figure 5.14.

5.5.4 Test Pattern Compaction

Quite often a test for a given fault requires assigning values to relatively few of the primary inputs. If there are many patterns with just a small number of input values assigned, then pairs of these test patterns can be merged, provided that none of the input positions conflict. The general rule for merging is:

> If one vector has a 1 in position i and the other vector has a 0 in position i, they cannot be merged.
>
> If one vector has $e \in \{0, 1, x\}$ in position i and the other has x, then position i is assigned the value e.

Merging can also be performed on sequences of vectors. When self-initializing sequences of test patterns are created for sequential circuits, as is done when employing the iterative test generator, a sequence could be placed, in its entirety, immediately following the previous sequence. However, the number of test patterns can sometimes be significantly reduced by merging sequences.

EXAMPLE ■

We will attempt to merge the following two sequences.

	1	2	3	4	5	6
1:	1	x	0	0	1	1
2:	0	0	x	0	1	0
3:	1	1	1	0	x	0
4:	0	1	0	x	1	x
5:	x	x	1	1	x	1

(1)

	1	2	3	4	5	6
1:	x	1	0	1	x	1
2:	0	0	x	1	0	x
3:	0	0	x	x	1	1
4:	1	x	x	1	0	0

(2)

5.5 THE PATTERN GENERATOR

We start with the first pattern of the second sequence and compare it with the last pattern of the first sequence. There is a conflict in the third bit position. We then compare it to the fourth pattern of the first sequence. This time there is no conflict. However, we cannot simply merge the patterns because the sequences are chronologically dependent. The successful application of the second sequence depends on applying all four patterns in sequence. Therefore, it is necessary to compare the second pattern of the second sequence with the last pattern of the first sequence. If they conflict the sequences cannot be merged. In this case there is no conflict so the two sequences can be merged by combining the last two patterns of the first sequence with the first two of the second sequence. We get:

```
1:  1  x  0  0  1  1
2:  0  0  x  0  1  0
3:  1  1  1  0  x  0
4:  0  1  0  1  1  1
5:  0  0  1  1  0  1
6:  0  0  x  x  1  1
7:  1  x  x  1  0  0
```
■ ■

Test pattern reduction can be accomplished dynamically while patterns are being created[16]. In this approach the ATPG attempts to create tests for additional faults after a test has already been successfully created for a fault. In Figure 5.15 a test was created for the upper input of gate Q S-A-1. This test was extended as far back as possible toward the inputs to maximize fault comprehension. In doing

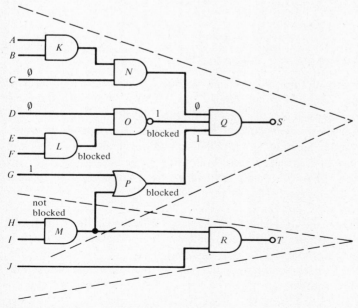

Figure 5.15 Dynamic compaction.

so, several gates were assigned values. However, in the circuit shown, gate M has fan-out which leads to another primary output. It is possible that additional faults can be selected and sensitized to the other output. To do so would require selecting a fault and sensitizing a path to the other output, subject to the constraint that values not be altered on gate inputs already assigned. Values on those inputs are fixed and are not permitted to be changed. In addition, gate outputs become "blocked" because of assignments on the inputs which make it impossible to propagate a sensitized path through the gate.

We attempt to propagate additional tests through gates which are not blocked. To increase the likelihood of selecting faults which can be successfully tested, cones are created from the outputs (see Sec. 4.4.5). Two cones are illustrated in Figure 5.15 by means of the dashed lines. Cones generally overlap since signals, especially control signals, affect many areas of logic. If a given fault is only contained in cones whose outputs already have assigned values, then it is pointless to select that fault during the dynamic compaction process.

In the circuit of Figure 5.15 a test on gate K could not be propagated to output S because it is blocked from the output. It cannot be propagated to output T because it is not contained in the cone of T. If an output has not yet had a value assigned, then a fault contained in the cone of that output is a candidate for test creation. If the test attempt fails because of excessive numbers of blocked gates, then continue either until a fault is found in that cone for which a test can be achieved or until no more untested faults exist for which a test has not been attempted. At some point in the creation of any one test pattern it becomes impractical to try to continue to create tests. Obviously, if all outputs in the circuit are assigned values, no additional faults can be propagated to these outputs. It also becomes difficult when nearly all of the inputs, $\geq 85\%$, have already been assigned values.

5.5.5 The Asynchronous Environment

When clock signals for a flip-flop are created by logic combinations within a circuit, the flip-flop becomes vulnerable to unintended clock pulses caused by hazards and races. Asynchronous Set and Reset inputs on clocked flip-flops as well as asynchronous latch inputs are also vulnerable to random pulses. To solve this problem, vulnerable inputs can be identified in an ATPG such as the iterative test generator and requirements can be imposed on these lines that they be hazard-free[17].

In the circuit of Figure 5.16, the latch inputs are required to remain fixed from time t_n to t_{n+1}. However, there is the possibility of a negative-going pulse from the OR gate of sufficient duration to cause the latch to switch. If the pulse occurs, the latch may make a permanent state transition. If the circuit is simulated using three-value simulation, an x value would be interposed between the two occurrences of the logic 1 on the $\overline{\text{Set}}$ line and this would put the latch into the x state. The subsequent transition on the input from x back to 1 would not be able to restore the latch to a known state. This is a conservative response to the problem and is intended to prevent the simulator from predicting the wrong value. Unfortunately, it does not address the problem of trying to avoid the hazard.

5.5 THE PATTERN GENERATOR

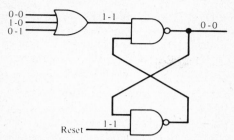

Figure 5.16 Occurrence of hazard.

To prevent the adverse effects of hazards, it is necessary to identify circuits, such as latches and flip-flops, where hazards can cause erroneous operation. Then it is necessary to identify internal states and input combinations during which they are vulnerable to hazards. Not all states or input conditions cause problems. For example, a hazard on an input to the OR gate in the circuit of Figure 5.16 will not cause an error if the latch output is at 1.

The hazard in Figure 5.16 may occur because of the manner in which justification is performed. If the ATPG simply requires that any single input to the OR gate be at 1, then establishment of a 1 on the lower input to the OR gate would be deemed sufficient by the ATPG to satisfy the logic conditions imposed by the justification process. However, when it is also required that the net be hazard-free, the ATPG must take into account the previous assignments on that gate and determine whether a hazard is created as a result of the signal change. Furthermore, it may be required that the circuit be free from exposure to dynamic as well as static hazards since a dynamic hazard on some circuits, such as a counter, can also cause erratic operation.

A delay flip-flop (DFF) must not be exposed to hazards on its Clock, Set, or Reset lines. It is assumed that the Data line will have numerous changes during most clock cycles but that the data will stabilize before the clock is applied. For the cross-coupled NAND latch, the following requirements must hold:

$\overline{Set}$	$\overline{Reset}$	Q
1–1*	x–1	0–0
x–1	1–1*	1–1

In this table the first entry states that when the $\overline{Reset}$ goes from x to 1, and $\overline{Set}$ is at 1, and the latch is in state $Q = 0$, then the $\overline{Set}$ line must be free of hazards (an asterisk denotes the hazard-free requirement). In the second case, when in the state $Q = 1$, the $\overline{Reset}$ line must be hazard-free. Basically, any combination of internal state and input combination which could cause a state change in response to an unwanted pulse on an input line requires that the input line which is vulnerable to the pulse be hazard-free.

When an input to a latch or flip-flop must be hazard-free, then the inputs to any combinational logic element which drives the sequential element must be selected so as to avoid creating hazards on the output. Hazard conditions corresponding to various input conditions depend on the type of gate. The table in Sec. 4.5.1 defines the conditions for which hazards may exist at an AND gate. For more general primitives used in the D-algorithm, detection of hazards requires examining all vertices contained in the cube formed by setting to x all input positions which change.

EXAMPLE ∎

Consider the P22 primitive of Figure 5.5. If the values on the inputs S, D_1, D_0, E change from $(0, 0, 1, 1)$ to $(1, 1, 0, 1)$, then it is necessary to look at all vertices in the cube $(x, x, x, 1)$. Four of the eight vertices have zero values, hence a static hazard exists. ∎ ∎

5.6 THE SIMULATOR

The simulator is used to verify that a test is not invalidated by races or hazards. It is also used to determine which faults are tested by a given test pattern. The simulator may be required to simulate all faults in the fault list for each applied pattern or it may be required to simulate only faults which have not yet been detected.

5.6.1 Fault Dictionaries

During fault simulation it is common for several faults to be detected by each test pattern. When testing a printed circuit board it is desirable to isolate the cause of an erroneous output to as small a group of candidate faults as is practical. Therefore, rather than stop on the first detection of an output error and attempt to diagnose the cause of the error, a tester may continue to apply patterns and record the pattern number for each failing test. At the conclusion of the test, the list of failed patterns is used to retrieve diagnostic data which identifies the potential faults detected by each applied pattern. If one or more faults are common to all failed patterns, the common faults are high-probability candidates.

To assist in identification of the cause of an error, a fault dictionary can be employed. A *fault dictionary* is a data file which defines a correspondence between faults and symptoms. It can be prepared in several ways, depending on the amount of data generated by the ATPG[18]. If we denote the ith fault in a circuit as F_i, then we can create a set of binary pass-fail vectors $F_i = (f_{i1}, f_{i2}, \ldots, f_{in})$ where

$$f_{ik} = 1 \quad \text{if } f_i \text{ is detected by test } T_k$$
$$= 0 \quad \text{otherwise}$$

These vectors can be sorted in ascending or descending order and stored for fast retrieval during testing.

During testing, if errors are detected, a pass-fail vector is created in which position i contains a 1 if an error is detected on that pattern and a 0 if no error is detected. This vector is compared to the pass-fail vectors created from the simulation output. If one, and only one, vector is found to match the pass-fail vector

5.6 THE SIMULATOR

resulting from the test, then the fault corresponding to that pass-fail vector is a high-probability fault candidate. It is possible, of course, that two or more nonequivalent faults have the same pass-fail vector, in which case it is possible to distinguish between them only if they have different symptoms, that is, they fail the same test pattern numbers but produce different failing responses at the output pins.

EXAMPLE ■

The following table lists four tests and pass-fail vectors corresponding to five failing machines, f_1 through f_5.

	T_1	T_2	T_3	T_4
F_1	0	1	0	0
F_2	1	1	0	1
F_3	0	0	1	0
F_4	0	1	0	0
F_5	1	0	0	1

Faults f_2 and f_5 are both detected by test T_1. If tests T_2 and T_4 also fail then the vector F_2 matches the pass-fail vector. If T_4 is the only additional test to fail, then F_5 is a match. Faults f_1 and f_4 have identical pass-fail vectors. The only hope for distinguishing between them during testing is to compare the actual output response to the predicted response for fault machines f_2 and f_4. ■ ■

Because the matrices are quite sparse, it is generally more compact to simply create a list of the failing test numbers for each fault. The fault number then serves as an index into the list of failing test numbers for that fault. Then, when one or more tests fail at the tester, the fault simulator output indicates which faults are potential causes of each test pattern failure. These faults are used to access the fault dictionary to find that fault for which the failing test numbers most closely match the actual test failures observed at the tester.

Test generation and fault simulation are based on the single-fault assumption, hence the fault list for a failing test can be inaccurate. This is especially true on the first few patterns applied to a circuit since that is when gross defects are most frequently detected. However, after about the first half-dozen patterns, gross defects have usually all been detected and the likelihood increases that the error is a single-fault error. In that case the fault data recorded by the simulator for each pattern becomes more reliable as a source of diagnostic data. However, even without the presence of gross physical defects, unmodeled faults such as noise, crosstalk, or parametric faults cause error symptoms that are not always detectable by fault dictionaries.

5.6.2 Fault Dropping

Fault dictionaries are not used as frequently as in the past. Circuits are growing to the point where the amount of diagnostic data needed for dictionaries has become unmanageable and the amount of CPU time required to simulate every fault on every pattern is unacceptably high. More practical means for locating faults on printed circuit boards include automatic test equipment which has the ability to isolate faults by means of circuit probing algorithms (see Chap. 6). In such cases diagnostic data is not required; therefore the fault is deleted from the fault list after it has been detected. This process, called *fault dropping*, can significantly speed up simulation. If full fault simulation is impractical, but diagnostic data is required, then a compromise between full fault simulation and fault dropping is to maintain a count of the number of times each fault has been detected. After the fault has been detected some specified number of times, it is dropped from further simulation.

The criterion for determining when to drop a fault is a function of circuit size and the number of faults detected with each pattern. The objective is to reduce simulation time while obtaining enough information to minimize the number of components that must be replaced on a board in order to restore it to proper operation. The problem is complicated by the fact that equivalent faults will always appear together if they have not been reduced to a single equivalent fault. Interestingly, the amount of CPU time may sometimes be reduced if the ATPG is required to create patterns for maximum resolution rather than maximum comprehension. More test patterns are created but fewer faults are detected by each pattern; thus fault resolution is achieved more quickly and faults can be dropped sooner.

If a fault contained in a list of faults for the nth test pattern is the only previously undetected fault in that list, it can be dropped from further simulation. The reasoning here is that if any of the other faults actually exist in the device being tested, then during testing they will cause an output error on an earlier pattern. If the nth pattern is the first to fail, then the lone previously undetected fault is the likeliest candidate for replacement. If two or more previously undetected faults are detected, and if they can be distinguished from one another by virtue of unique responses, then they can also be dropped.

5.7 MODELING LARGE DEVICES

The growing use of LSI and VLSI parts, including microprocessors and related peripheral devices, and large random access memories (RAMs) on printed circuit boards presents two problems. The devices must be modeled at a level that can be processed by the ATPG or fault simulator and the ATPG must be effective enough to create meaningful test stimuli for these devices. For the moment we confine our attention to the modeling problem.

One of the major testing problems associated with LSI and VLSI components is the virtual nonexistence of structural information. The intent of the ATPG is to create tests for structural defects and yet, for many of the VLSI parts in use, the only available information is a few sheets of functional data. VLSI designs put increasingly more logic into the design and correspondingly less information into the documentation. It becomes necessary to guess at the structure in order to test them or, worse still, try to work around them.

5.7 MODELING LARGE DEVICES

5.7.1 Behaviorally Equivalent Circuits

An approach sometimes used for LSI and VLSI circuits is to design a gate-level circuit which is behaviorally equivalent to the circuit being tested. Such a circuit, if designed correctly, is logically indistinguishable from the original circuit, except possibly for minor timing variations, when analyzed at the terminals. Tests can then be developed for structural faults in the behaviorally equivalent structural model on the assumption that an effective set of tests for the equivalent circuit will effectively test a similar percentage of faults in the original structure. That this is not always a correct assumption is illustrated by the two circuits in Figure 5.17, which have behaviorally identical terminal behavior[19]. There is no way to tell them apart based on stimuli applied at the inputs. However, the set of six test vectors listed below will test all S-A-1 and S-A-0 faults in the NAND model but only 50% of the faults in the NOR model!

	Test set			
	X_1	X_2	X_3	X_4
1:	1	1	1	1
2:	0	0	0	0
3:	1	0	0	0
4:	0	1	0	0
5:	0	0	1	0
6:	0	0	0	1

Fortunately, circuits to be tested are rarely that small. If such a circuit were embedded in a much larger circuit, it is more likely that all or most of the faults would be detected incidentally when testing for other faults.

The rationale for constructing gate-level equivalent models stems from the fact that ATPGs operate on structural models. Such models provide a means by which the ATPG can create, and the fault simulator can evaluate, stimuli at the device inputs and correctly determine what value should appear at the device outputs. For circuits of reasonable size, the fault coverage estimate for the structurally equivalent circuit is close enough to the fault coverage for the real circuit to permit realistic estimates of product defect levels. Furthermore, because the two circuits are behaviorally identical at their terminals, the stuck fault coverage for faults at the terminals, the so-called manufacturing faults, will be an accurate measure of fault coverage for those faults in the real circuit.

5.7.2 RAM Model

Perhaps a bigger problem than fault coverage for models of LSI and VLSI circuits is the availability of such models. When a logic board includes a microprocessor and a number of other chips of comparable complexity which perform serial I/O, parallel I/O, direct memory address (DMA), priority interrupt control, floppy disk control, and so on, and when some or all of these peripheral chips have programmable control registers that permit them to operate in one of several modes, then the task of creating structural models for all of these devices becomes prohibitively expensive.

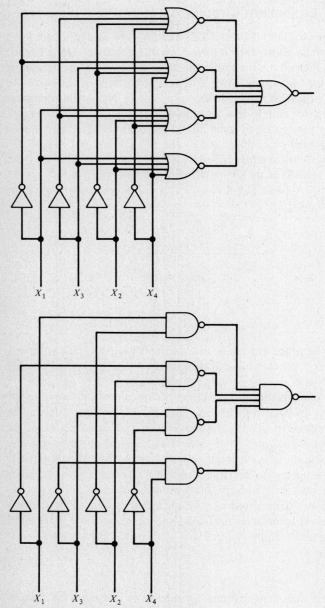

Figure 5.17 Two equivalent circuits.

Some devices can be modeled hierarchically, using MSI parts to design the device, and then reducing the MSI design to a gate level, in effect designing the LSI chip as a board. However, to design logically equivalent devices, just to be able to test them, is an expensive operation. When large devices are used on a logic board and structural models do not exist for the devices, it becomes necessary

5.7 MODELING LARGE DEVICES

to develop test strategies that accommodate the presence of these devices without letting their presence compromise testability of other circuits on the board. One method is to use sockets for the more complex parts. Parts are normally soldered onto boards in order to obtain acceptable levels of reliability. The socket, although less reliable, permits the removal of the chip from the board and the use of its pin connections as additional inputs and outputs for the tester.

Another strategy is to use partial models. A partial structural model is one which correctly reflects the behavior of a selected subset of a device's capabilities. The capabilities chosen to be modeled are determined in part by testing needs and in part by ease or difficulty of modeling the capability. Since only part of the device capabilities are modeled, there must be a response to an attempt to exercise functional capabilities which have not been modeled. For example, a RAM may be modeled to have only cells corresponding to addresses $0, 1, 2, 4, \ldots, 2^{n-1}$, $2^n - 1$ to permit detection of faults on the address lines. If we put different values into each of these memory cells, then an S-A-0 on address line A_1 would cause a write into location 0 instead of location 1. When the contents of memory location 0 are later read, it will be found to contain the data that was intended to be stored in memory location 1. This model thus permits testing of stuck-at pin faults. However, it must be possible to handle the case where an attempt is made to address an unmodeled memory cell. This is done via the "x-generator" illustrated in Figure 5.18.

The x-generator is a cross-coupled latch with no inputs. The NANDs are initialized to x at the beginning of test generation and there is no way to get out of the x-state. The address decoding circuits in the RAM model are designed so that an attempt to access any RAM address other than the modeled addresses causes the x-generator to be selected, hence the value read out of memory is indeterminate.

5.7.3 Microprocessors

Some microprocessors, coprocessors, and peripherals are so complex that it simply is not practical to accurately model them if, indeed, it is even possible. Microprocessors are appearing with more and more asynchronous subsystems. For example, the Intel 8086 has two functionally separate units, the bus interface unit (BIU) and the execution unit (EU). The BIU maintains a queue of instructions so that the EU seldom has to wait for instructions to be retrieved from memory. The BIU, running asynchronous to the EU, performs a memory fetch whenever there is room in its queue for an instruction. Other newly emerging microprocessors have

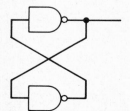

Figure 5.18 The x-generator.

on-chip cache memory for instructions and data, pipelined decoding and execution of instructions, and on-chip memory management. It is not enough to model individual functional units correctly. A model must correctly describe asynchronous behavior between any pair of units. Without information from the manufacturer detailing the register transfer characteristics, this may be impossible[20].

As was the case with the RAM, a partial model of the microprocessor can be used to achieve specific objectives. Such a model for the 8088 microprocessor appears in Figure 5.19. The Hold signal normally puts the microprocessor Data and Control lines into the high-impedance state while another device, such as a DMA controller, controls the memory lines without interference from the microprocessor. In the configuration of Figure 5.19, the Hold signal either causes Data and Control lines to be flooded with *xs*, or it puts them into the high-impedance state and allows the test inputs to furnish control signals which normally originate in the microprocessor. Although this configuration will not permit testing of the microprocessor, it will permit testing of other components on the board when the microprocessor is soldered in place and cannot be removed. It does require the addition of pins solely for testing purposes. Care must be exercised when using this strategy because a microprocessor may not go immediately to the Hold state following arrival of a Hold signal. It may continue running for a few clock states to complete an instruction.

An alternative approach is to model some of the elementary instructions in a microprocessor[21]. This may include instructions which load and store registers, the NOP (no operation), branch instructions, and logic operations. This will permit testing for the manufacturing faults and it will permit propagation of tests for other devices through the microprocessor. However, an attempt to propagate unmodeled

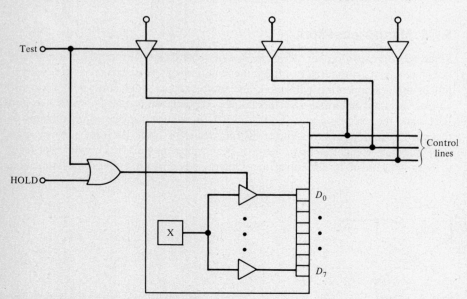

Figure 5.19 Dummy microprocessor model.

5.8 USER FEATURES

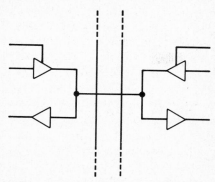

Figure 5.20 Net with bidirectional pins.

Op Codes through the microprocessor may cause the contents of internal registers and flip-flops to become indeterminate.

5.7.4 Bidirectional Signals

Bidirectional signals are common on microprocessors and peripheral chips. These permit sharing of physical pins, sometimes by signals which are totally unrelated. A pair of bidirectional pins connected together is illustrated in Figure 5.20. The configuration is used to connect a microprocessor to a memory data port or to a bidirectional data port of a peripheral. The driver and receiver share the same physical pin, but otherwise the test problem is similar to any other in which two or more drivers are connected to a single net. There must not be more than one driver actively driving the net at any given time.

Circuits are normally designed so that only one driver can be active on a given net. When a microprocessor driver is active, the microprocessor simultaneously issues a signal to the peripherals which forces them to deactivate their drivers. In other situations a 2-to-4 or 3-to-8 decoder is used to select one of several drivers. However, the tri-state capability can be a serious problem on circuits where the bus crosses two or more circuit boards. In that case, the tri-state controls are also distributed across two or more boards and control combinations which are impossible in the system environment become possible in the board test environment. It is important, if this possibility exists, to have an ATPG and simulator capability that checks for conflicts among the drivers connected to the nets.

5.8 USER FEATURES

The growing complexity of circuits makes them more difficult to test, and for precisely that reason it becomes more desirable to have software programs that can analyze digital circuits and create effective tests. Much of what goes for test pattern creation involves tedious analysis of gate-level interconnections, devising logic values to propagate a test through the gates, and keeping track of the logic levels on a myriad of gates while being careful not to overlook a key signal which, if set to the wrong value, could negate the entire test. This is precisely the area where computers excel. Unfortunately, successful test generation programs which can perform these operations still remain more a wish than a reality.

5.8.1 User-Specified Inputs

To solve, at least in part, the problem of automating test stimulus creation, an ATPG can be provided with features that permit the user to guide it in its search for test patterns. On early ATPGs the test engineer would try to "steer" the programs with clever applications of the x-generator. Assume, in Figure 5.21, that the signals A and B are not permitted to be identical. Then we can modify the circuit model so that the circuit inputs which originally came from primary inputs A and B now come from 2-to-1 multiplexers. If $A = B$ the multiplexers select their input from the x-generator, otherwise they select their input from primary inputs A and B. This circuit is based on the assumption that the ATPG will not select input conditions which cause xs to appear if a choice exists which can produce known values. Other similar configurations can inhibit other illegal input combinations. Note that this is only used in an ATPG circuit model; the original unaltered circuit model is used during simulation.

The user can work much more efficiently if these capabilities are incorporated into the ATPG. If a combination of input values causes two or more tri-state devices to simultaneously drive a bus, it should be possible to inhibit that combination simply by specifying the combination as being illegal. Input combinations should also be prohibited if they cause transitions into illegal states, or simultaneous toggling of either the clock and data inputs of a flip-flop or load and clock lines of a serial/parallel register.

When inputs are held at fixed values, these assignments are implicated by the ATPG and cause other logic gates to become blocked, just as during dynamic compaction. The same can be done for logic combinations on inputs. If two inputs are inhibited from being high simultaneously, then whenever one of them is set

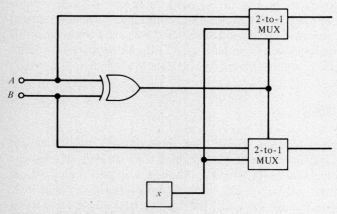

Figure 5.21 Inhibiting a choice.

5.8 USER FEATURES

high by the ATPG, the other is immediately set low and all possible implications are performed.

This concept of input assignment or constraint by the user can be extended to permit guiding an ATPG through complex control logic, such as state machines, which are otherwise difficult to control. Entire sequences of inputs can be specified to the ATPG. These assignments can be implicated on an individual pattern basis and then, as in the process used in dynamic compaction, faults are selected for processing within the constraints imposed by the preassigned inputs. This approach has the attraction that the user can solve difficult state transition problems and let the ATPG take care of the tedious bookkeeping involved in propagating individual faults to outputs.

5.8.2 Test Pattern Languages

The complexity of present-day circuits has outstripped the ability of ATPGs to keep pace. On some circuits techniques to circumvent problem areas, so called "workarounds," permit use of an ATPG in applications where it might not otherwise be feasible. The partial models previously described fall in the category of workarounds. However, numerous circuits exist where the ATPG simply will not work and the logic designer or a test engineer must write test stimuli. These stimuli can then be fault-simulated to evaluate their effectiveness.

The task of writing test stimuli is greatly simplified through the use of a test pattern language. When writing test stimuli for a small circuit board with just a few inputs, it is not difficult to write patterns as sequences of 1s and 0s. However, when dealing with large boards, having as many as 200–300 inputs, the process becomes both tedious and error-prone. Good test pattern languages have been developed which permit specification of test stimuli by name. They also provide constructs such as macros and loops similar to those found in high-level languages to deal with repetitive input sequences. The language compiler converts the test specification into a file containing 1s and 0s suitable for processing by the simulator. The language has the advantage that it is easier to read and understand, and thus easier to write and to alter if changes are made to a board. Furthermore, if two architecturally similar designs, implemented in different technologies, must be tested, then the same pattern set can often be employed with only minor revisions[22].

A test pattern language obviously must be able to specify high and low values on input pins. It should also be able to specify an initial default value on pins. Other desirable features include:

1. A pulse command which sets a pin to its opposite value and then back to the original value
2. A bus command which permits grouping a number of input pins together and specifying their value in binary, octal, hex, or decimal
3. Subroutines and/or macros so that frequently used sequences of inputs need only be coded once and then invoked as needed
4. Change command which sets a signal to its opposite value irrespective of present value

EXAMPLE ■

The following test is expressed in a hypothetical test language to illustrate some of the desirable features of such a language. ■ ■

```
NAME MEM=PIN16,
DATA1=PIN44,
DATA2=PIN47,
DATA3=PIN34
CLOCK=PIN18,
WE=PIN51,
ENPPI=PIN31
BUS ADDRESS=PIN7,PIN8,PIN9,PIN10,PIN11,PIN12,PIN13,PIN14
BUS DATA=PIN29,PIN30,PIN31,PIN32,PIN37,PIN38,PIN39,PIN40
;
MACRO WRITE DEVICE
LOW DEVICE                  ;SET CHIP SELECT LOW
PULSE WE                    ;PULSE THE WRITE ENABLE LINE
HIGH DEVICE                 ;DESELECT THE CHIP
ENDM
;
DEFAULT ALL HIGH
LOW DATA1 DATA2
PULSE CLOCK
PATTERN HIGH DATA1
        LOW DATA3
ADDRESS=B'11010011
DATA=H'00
WRITE ENPPI
CHANGE DATA3
ADDRESS=H'3A
;
DO 8
WALK DATA
INCR ADDRESS
WRITE MEM
ENDO
END
```

In this hypothetical example several reserved words are used. These include: NAME, BUS, LOW, HIGH, PULSE, PATTERN, MACRO, ENDM, CALL, CHANGE, DO, ENDO, WALK, INCR, END. It is assumed that PIN*nn* identifies a primary input pin that also exists in the circuit model. A comma is used to indicate that more arguments follow for a command. A semicolon terminates an input line. Any text following the semicolon is treated as a comment.

When PIN*nn* is assigned a value, the corresponding value in the circuit model is assigned that value during simulation. The NAME command allows substitution of a mnemonic name for the original name. A group of pins can be referenced as a bus by identifying them in a BUS statement. The MACRO and ENDM reserved words identify a group of commands which are inserted into the command stream whenever the name following the reserved word MACRO is invoked. In this example, the word WRITE causes the set of commands to be inserted. The word

5.8 USER FEATURES

DEVICE is a dummy variable which is replaced by an argument when the macro is invoked. The HIGH and LOW commands cause all pins following the command to switch on the same pattern.

When it is desired to switch some inputs high and others low on the same pattern, then PATTERN is specified. It causes all pin changes specified in the next two HIGH and/or LOW commands to be grouped as part of one pattern. If the command immediately following PATTERN, or the subsequent command, is not a HIGH or LOW, then PATTERN is ignored. DO and ENDO delimit a group of commands that are to be duplicated some specified number of times. WALK assigns a 1 to a single pin in a group of pins and 0s to all others. On successive occurrences of the WALK command, the 1 is rotated to the next pin on the right in the group. The INCR command increments the binary value on a group of pins.

The test pattern editor converts this set of instructions into a matrix of 1s and 0s to be processed by the simulator. The editor can check for syntax errors and, with access to a list of circuit primary input names, it can check for nonexistent or misspelled names. As a means of qualitatively evaluating a set of test vectors, an audit[23] function can be employed to evaluate test patterns and determine which inputs were never toggled. If tests are written as arrays of 1s and 0s, large files can become invalidated because of a circuit change which adds or deletes one or more primary inputs. The test written in symbolic language can often be updated very quickly to reflect these circuit changes. Furthermore, the simulation language is effective for design verification simulation as well as fault simulation.

5.8.3 The Simulator Output Display

When writing test patterns the diagnostic engineer usually has a specific purpose in mind. For example, he may want to drive a circuit into some specific state in order to exercise a certain functional unit. When writing stimuli he gauges his progress according to whether or not the fault comprehension has increased. Alternatively, he may look at the list of untested faults provided by the simulator. If results are not as expected, he will want to inspect simulation results to determine why a discrepancy exists between what he expected and what he actually obtained.

A convenient way to display simulator output on a video display is in a format similar in appearance to that of a logic analyzer or oscilloscope. The user can call for values on specific nets to be displayed and specify a range of patterns for which the values are to be displayed. If the simulator performs nominal delay simulation, then the range can be specified either in time intervals or in pattern numbers. By specifying a time interval, the display can show simulation results over a longer or shorter interval, like adjusting the sweep frequency on an oscilloscope.

When setting up a test for logic that requires a sequence of signals, or for logic that employs critical timing, the simulator output display can become an indispensable tool for debugging the test. In Figure 5.22, DATA2 goes to x at time 550. It is noted that CLOCK switches low at the same time. If DATA2 were the output of a flip-flop, then it would be necessary to examine the waveform at the input of the flip-flop. If the flip-flop input were at x, it would be necessary to trace

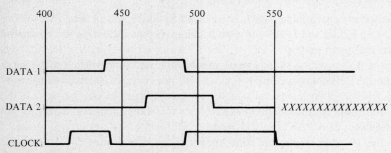

Figure 5.22 Simulator output.

back to the circuit that generated the signal for the flip-flop input. However, if the flip-flop input were at a known value, then it is possible that the Clock and Data inputs switched too close together in time, causing the simulator to decide that a race existed which was too close to call, resulting in the output being set to x.

A good simulator contains a great deal of information which can be useful to both the diagnostic engineer and the design engineer. However, the fact that there is a great deal of information makes it imperative that the information be usable by the engineer. The oscilloscope-like display provides the designer or test engineer with a visual window into circuit behavior which is usually easy to interpret.

5.9 OUTPUT FILES

The purpose of the ATPG is to create test stimuli. It must also provide response data, which the tester uses to determine whether the board under test responded correctly to the applied stimuli. Response at internal nodes is also useful when the user is debugging manually generated test patterns or probing a faulty board to isolate a defect. For this purpose a file is required which lists response at internal nodes for all applied stimuli.

For stimulus and response files, it is necessary to provide corresponding files which identify the order in which data appears in these files. This is simply a list of the input and output pins ordered according to the corresponding stimulus and response data. If response of internal nodes is recorded in a file, then a file is required which lists the names of the internal nodes, in the same order in which the data is recorded.

Physical information is required which describes the printed circuit board. The physical information includes a list of devices on the board, their physical location, and their interconnections. This information is used for probing defective boards to isolate faults. This will be discussed in more detail in the next chapter.

Various statistics are employed to gauge the effectiveness of an ATPG when applied to a particular circuit. The test percentage, a measure of number of faults for which tests were successfully generated versus total number of faults, is one such statistic. It may be broken down to indicate number or percentage of faults detected in various categories: S-A-1, S-A-0, stuck-open, shorts, and so on. The total number of circuit parameters is useful because, if a circuit has more inputs than the tester can drive, then obviously some action must be taken. Also, if fault collapsing is employed, the ATPG may indicate total number of faults and total number of equivalent faults.

5.10 SUMMARY

Automatic test pattern generators have been used since the early 1960s as a means of generating tests for digital logic. When used to test devices designed with SSI and MSI parts, or when used to test ICs for which the complete design is available, creating a circuit model is not a difficult task. Now, however, more devices are being used in designs for which no internal data is available. This represents a major impediment to structural level testing.

Automatic test pattern generators have been notably unsuccessful at testing sequential circuits. More logic is being placed on single boards and within individual ICs and, as a result, ATPGs are growing less successful at generating test stimuli at a time when the growing complexity of circuits demands more accurate tests. One result of the decreasing effectiveness is an increasing emphasis on tools to assist in writing and debugging manually written test patterns.

New fault types have been introduced as a result of new VLSI technologies. The limited number of I/O pins has resulted in a proliferation of bidirectional pins and tri-state gates. Another effect of VLSI is the fact that they are often configured to use only part of their functionality. When that happens, statistics on fault coverage become less meaningful because defects in unused circuits normally have no effect on circuit operation.

The difficulty in testing LSI and VLSI components, and the printed circuit boards on which they reside, has stimulated intense research into methods both to develop more effective algorithms to test circuits and to design circuits so as to be testable. Those topics will be the subject of subsequent chapters.

PROBLEMS

5.1 Prove the dominance and equivalence theorems.

5.2 Given an n-input AND gate, with an input S-A-0, can you devise a test to determine which input is S-A-0?

5.3 List all equivalent fault classes for the circuit of Figure 5.13. Identify all dominant faults.

5.4 Collapse the faults on the circuit in Figure 4.20. Translate input stuck-at faults to the output whenever possible; for instance, an input S-A-0 on the input of the AND is equivalent to the output S-A-0.

5.5 Define a procedure for detecting the ISL faults described in Sec. 5.4.4.

5.6 Find a sequence of four tests that will detect all seven CMOS NOR gate faults.

5.7 Create a table of stuck-at and stuck-open faults for the CMOS NAND similar to the NOR circuit of Table 5.1.

5.8 Is it possible to detect any faults in a circuit without testing any checkpoint faults?

5.9 Identify all checkpoint faults in the quad 2-to-1 multiplexer of Figure 5.4.

5.10 Using the circuit in Figure 5.13, create a complete test set for
  (a) Maximum resolution
  (b) Maximum comprehension

5.11 Complete the derivation of the equation for P_D.

5.12 Merge the following three sequences of patterns:

$$111xx0 \quad x1xx10 \quad 110011$$
$$x1x11x \quad x0x1x1 \quad 1x1110$$
$$00xx01 \quad x00001 \quad 000xx0$$
$$1xx0x1 \quad 111x10 \quad 1xxx10$$
$$\qquad\qquad\qquad\qquad\qquad\quad 111010$$

5.13 For the P22 primitive of Figure 5.5(a), find one or more hazard-free transitions.

5.14 Determine all faults in the NOR circuit that are tested by the six test vectors developed for the NAND circuit of Figure 5.17. Create a pass-fail vector for each fault and use that to create a fault dictionary. Two of the NOR gates in Figure 5.17 could be missing and their absence would not be detected by the test vectors created for the NAND gate circuit. Identify them.

5.15 Find the smallest set of vectors that will test for all stuck-at faults in the NOR circuit of Figure 5.17. How many faults in the NAND gate circuit will they detect?

5.16 Given the following two-input logic gates, with values on inputs and output as indicated, which of these assignments imply additional values?

	IN1	IN2	OUT
OR	x	x	0
NOR	x	x	1
NAND	1	x	1
AND	0	x	0

5.17 Prove: In a combinational circuit with a single output, if there is no reconvergent fan-out, then S-A-0 and S-A-1 faults on primary inputs will detect all single stuck-at faults in the circuit. Find a counterexample to show that it is not true for multiple-output circuits.

5.18 Given the following matrix of test patterns versus faults detected, if pattern 4 is the only failure, which fault is most likely to have occurred?

		\multicolumn{8}{c}{Fault number}
		1	2	3	4	5	6	7	8
Pattern number	1	1		1			1		
	2		1	1	1				
	3			1		1		1	
	4			1				1	1

5.19 From the above matrix, if all four tests fail, which fault is most likely to have occurred?

PROBLEMS

5.20 Using the circuit in Figure 4.20 and the D-algorithm, create tests for all inputs to gate 13 which can be tested without making any decisions. For each fault, proceed as follows: make the initial assignment of a PDCF and satisfy all implications. Continue until you either achieve a successful test or cannot proceed further without making a decision.

5.21 When propagating a D or $\overline{D}$ through an exclusive OR, a decision is made to propagate with a 1 or 0 on the other input. This occasionally leads to contradictions and the necessity for remaking decisions. How could you modify the D-algorithm to avoid decisions at the exclusive ORs?

5.22 During dynamic compaction, an attempt to create a test for a fault is unsuccessful. Under what conditions would another attempt be made to create a test for the fault?

5.23 Analyze the following test strategy: Given a fault which makes it impossible to drive a latch into a known state, simulate the unknown state until the x reaches an observable output. When the output of a physically defective circuit is measured on a tester, the x must either agree or disagree with the expected output from the good machine. If it disagrees, the fault is detected. If it agrees, the latch must be in the same state as the unfaulted circuit. Therefore, in the simulator, when the x reaches an output, set the faulted circuit to the same state as the good circuit and if the fault causes them to disagree again at a later pattern, mark the fault as detected.

The remaining exercises require the use of the simulator found at the end of Chap. 4.

5.24 Simulate all input combinations on the two circuits of Figure 5.17. Verify that the circuits are identical.

5.25 Perform serial fault simulation on the circuits of Figure 5.17, using the six vectors that test all stuck-at faults on the NAND circuit.

5.26 Write a routine which locates and compiles a list of checkpoint faults. Collapse the list by translating input S-A-0 (S-A-1) faults to the output of AND (OR) gates. Perform serial fault simulation on the circuit of Figure 4.20. Simulate only the potentially detectable faults from the collapsed checkpoint fault list and drop the fault from further simulation after it is detected.

5.27 Given a signal path originating at some arbitrary point in a circuit and terminating on the output of a gate G, such that the signal path does not at any point fan out to two or more destination gates, then we say that any faults along that signal path are *locally detectable* at the output of gate G. As an example, in Figure 5.14, faults along the signal path from primary input 3 or 2 to the output of gate 6 are locally detectable at the output of gate 6. Faults occurring at PI 1 are not locally detectable because PI 1 fans out to two destination gates. However, an open at the input to gate 5 that does not affect the signal path from PI 1 to gate 8 is locally detectable at gate 6.
Implement a Testdetect as follows:

> FOR I = 1 TO LAST_VECTOR
>
> Perform logic simulation of VECTOR(I).
>
> Perform serial fault simulation of potentially detectable output faults on gates with fan-out of 2 or more.
>
> For each gate output fault that is detected, use Testdetect to check locally detectable faults.

If all faults that are locally detectable at gate G have been detected, drop gate G from further serial fault simulation.

NEXT I

5.28 Using the serial fault simulator, apply random patterns to the circuit in Figure 4.20 and chart the fault coverage for stuck-at faults on logic gate inputs as a function of the number of test patterns applied.

REFERENCES

1. Dacier, W. C., "Software Generates Tests for Complex Digital Circuits," *Electron. Des.*, Oct. 28, 1982, pp. 137–146.
2. Mei, K. C. Y., "Fault Dominance in Combinational Circuits," Digital Systems Lab. Rep. 2, Stanford Univ., Aug. 1970.
3. Goel, P., J. Grason, and D. Siewiorek, "Structural Factors in Fault Dominance for Combinational Logic Circuits," *Proc. Fault Tolerant Computing Symp.*, 1973, p. 174.
4. Kriz, T. A., "Machine Identification Concepts of Path Sensitizing Fault Diagnosis," *Proc. 10th Symp. on Switching and Automata Theory*, Oct. 1969, pp. 174–181.
5. Armstrong, D. B., "On Finding a Nearly Minimal Set of Fault Detection Tests for Combinational Logic Nets," *IEEE Trans. Elecron. Comput.*, vol. EC-15, no. 1, Feb. 1966, pp. 66–73.
6. Beh, C. C., et al., "Do Stuck Fault Models Reflect Manufacturing Defects?", *IEEE Test Conf.*, 1982, pp. 35–42.
7. Wadsack, R. L., "Fault Modeling and Logic Simulation of CMOS and MOS Integrated Circuits," *Bell Syst. Tech. J.*, vol. 57, no. 5, May–June 1978, pp. 1449–1474.
8. El Ziq, Y. M., "Automatic Test Generation for Stuck-Open Faults in CMOS VLSI," *Proc. 18th Design Automation Conf.*, 1981, pp. 347–354.
9. Mei, K. C. Y., "Bridging and Stuck-at Faults," *IEEE Trans. Comput.*, vol. C-23, no. 7, July 1974, pp. 720–727.
10. Son, K., and D. K. Pradhan, "Design of Programmable Logic Arrays for Testability," *IEEE Test Conf.*, 1980, pp. 163–166.
11. Szygenda, S. A., and A. A. Lekkos, "Integrated Techniques for Functional and Gate-Level Digital Logic Simulation," *Proc. 10th Design Automation Conf.*, pp. 159–172.
12. Thomas, J. J., "Common Misconceptions in Digital Test Generation," *Comput. Des.*, Jan. 1977, pp. 89–94.
13. Breuer, M. A., et al., "A Survey of the State of the Art of Design Automation," *Computer*, vol. 14, no. 10, Oct. 1981, pp. 58–75.
14. Roth, J. P., "Diagnosis of Automata Failures: A Calculus and a Method," *IBM J. Res. Dev.*, vol. 10, no. 4, July 1966, pp. 278–291.
15. Roth, J. P., et al., "Programmed Algorithms to Compute Tests to Detect and Distinguish between Failures in Logic Circuits," *IEEE Trans. Electron. Comput.*, vol. EC-16, no. 5, Oct. 1967, pp. 567–580.
16. Goel, P., and Barry C. Rosales, "Test Generation and Dynamic Compaction of Tests," *Proc. 1979 Int. Test Conf.*, pp. 189–192.
17. Breuer, M. A., and R. Lloyd Harrison, "Procedures for Eliminating Static and Dynamic Hazards in Test Generation," *IEEE Trans. Comput.*, vol. C-23, no. 10, Oct. 1974, pp. 1069–1078.
18. Chang, H. Y., E. Manning, and G. Metze, "Fault Dictionaries," *Fault Diagnosis of Digital Systems*, chap. 5, Wiley, New York, 1970.

19. Miczo, A., "Fault Modelling for Functional Primitives," *IEEE Test Conf.,* 1982, pp. 43–49.
20. Fong, J. Y. O., "Microprocessor Modeling for Logic Simulation," *Proc. 1981 IEEE Test Conf.,* 1981, pp. 458–460.
21. Zimmer, B. A., "Test Techniques for Circuit Boards Containing Large Memories and Microprocessors," *Proc. 1976 Semiconductor Test Conf.,* pp. 16–21.
22. Wickham, G. G., "Concepts of a Tester Independent Functional Test Language System," *Proc. 1978 Semiconductor Test Conf.,* pp. 206–208.
23. Gelman, Stacey J., "VEEP-A Vector Editor and Preparer," *Proc. 19th Design Automation Conf.,* 1982, pp. 771–776.

Chapter 6
Automatic Test Equipment

6.1 INTRODUCTION

Detection and diagnosis of faults in digital products can be a very costly and time-consuming operation. The test equipment is often as complex as the product being tested and may require highly trained operators. The increasing complexity of individual components used in digital products requires that more thought be given to detecting faults as early as possible in the manufacturing process. Early fault detection can reduce the need for expensive equipment, it can reduce the skill levels required on the part of operators, and it can lower the work-in-process inventory levels.

Despite the great amount of thought, planning, and strategy employed, failed units continue to elude detection until far into the manufacturing process. Some inevitably reach the shipping dock. To catch a greater number of these defective units, it is necessary to have a highly effective test, one that can

- Provide high fault coverage
- Provide good diagnosis
- Run on the tester

In previous chapters we looked at methods for developing test sets aimed at achieving high fault coverage and resolution. We now examine the test equipment and the test strategies that are being used to detect failed units while minimizing cost.

6.2 THE MANUFACTURING TEST ENVIRONMENT

The "rule of ten" guideline introduced in Chap. 1 asserts that the cost impact of a defective component escalates rapidly if it eludes early detection and progresses

6.2 THE MANUFACTURING TEST ENVIRONMENT

further into the manufacturing process. Consequently, the guideline serves as a motivation to detect defective components as early as possible in the manufacturing cycle.

Manufacturers of complex digital equipment acknowledge the validity of the rule of ten by instituting comprehensive test strategies which distribute test resources throughout the manufacturing process. Testing begins, as shown in Figure 6.1, with incoming inspection. At this station components from vendors are tested to insure that they comply with some minimum set of specifications. This is frequently necessary because vendors may not test all of the components that they ship but, rather, may only test random samples. Components may also be exposed to environmental hazards or physical abuse during shipping. A second purpose of incoming inspection is to selectively sort parts. If two or more products use the same part type but one product uses the part in a signal path that requires tighter tolerances, then it may be necessary to sort the parts at incoming inspection and route the components with preferred characteristics to the design where they are most needed. A thorough screening may, as a beneficial side effect, influence a vendor to improve his quality control.

Bare-board testing is employed to detect faulty printed circuit boards. The object of the test is to verify point-to-point continuity and to check isolation, including high-resistance leakage, between each point and every other point on the board. The bare-board testers generally use self-learning wherein the tester takes readings between pairs of points on a known good board and stores the results in a file which becomes the test. The points between which readings are taken are usually separated by 0.10-inch centers. Multiple-layer boards, such as the one illustrated in Figure 6.2[1], which has 23 metal interconnection layers sandwiched between insulating material and connected together by means of through-holes in the insulating material, are usually tested after each metal layer is deposited. If defects exist, it is still possible to correct them. The contacts for the measurements are usually made by means of a "bed-of-nails" fixture, shown in Figure 6.3. Some manufacturers are starting to use visual recognition systems to detect opens and shorts; however, visual techniques, although capable of higher throughput, cannot quantify resistance and are not as effective at verifying conductivity of through-hole plating[2].

The boards which pass bare-board test are populated with components. The board can then be tested with an in-circuit tester. The in-circuit tester (ICT) also uses the bed-of-nails fixture shown in Figure 6.3 to make contact with electrical points on the board. The printed circuit board to be tested is placed on the perimeter

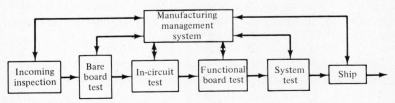

Figure 6.1 The manufacturing test process.

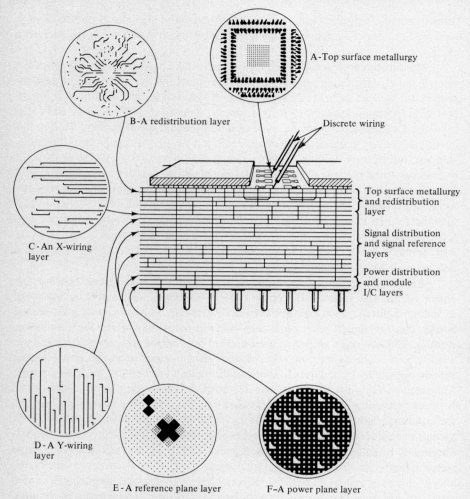

Figure 6.2 Multilayer ceramic module. (Reprinted with permission from *Electronic Design,* vol. 28, no. 1; © Hayden Publishing Co., Inc., 1978.)

gasket and then a vacuum is used to pull the board down onto the fixture and into contact with the spring-loaded nails or contacts. A wiring harness connects these nails to the tester. When the nails are brought into contact with the board, the tester, under program control, selectively applies signals to some of these nails and monitors others. In this way the tester can individually test components, including ICs, resistors, and inductors used within a circuit. The primary purpose is to identify defects introduced during manufacturing. These defects include missing components, wrong components, components inserted with wrong orientation, solder shorts between adjacent pins, and opens resulting from bent pins or cold solder joints. Since the in-circuit tester applies functional tests to integrated circuits, it is capable of detecting failed ICs which, although checked at incoming inspection,

6.2 THE MANUFACTURING TEST ENVIRONMENT

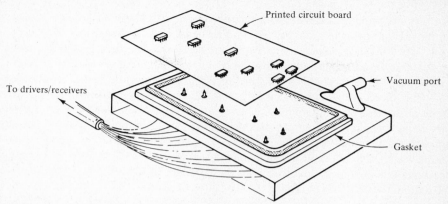

Figure 6.3 Bed-of-nails fixture for in-circuit tester.

can still fail during the manufacturing process from such things as electrostatic discharge or excessive heat.

From the in-circuit tester the board goes to a functional tester, which applies signals to edge pins and exercises the board as a complete functional entity. Since it is testing the board on a global scale, it can detect faults which an in-circuit tester may not detect, including faulty behavior caused by incorrect timing relationships between signals. Components may be functionally correct and individually respond correctly to stimuli, but one or more of them may respond too slowly as a result of parametric faults. The cumulative delays can alter the chronological order in which signal changes arrive at a device. A slow arriving data or clock at a flip-flop will eventually cause an incorrect value to be clocked in. The dynamic or high-speed functional tester can also detect signals which are too slow in arriving at board edge pins. The functional tester has special facilities for diagnosing fault locations, as well as provisions for margin testing of clock frequency and voltage ranges, features which are useful for detecting intermittents.

After a board has passed board test, either with or without one or more trips to a repair station, it must then be checked out as part of a system. A complete system is assembled and exercised in an operational environment. The problems now encountered include defects resulting from cabling problems, bent pins, high-resistance contacts, and erroneous behavior resulting from cumulative delays over two or more boards.

A feature which is growing in importance in modern-day manufacturing environments is the manufacturing management system (MMS). The MMS is capable of recording the manufacturing history of a board during its passage through the production cycle. Information collected on the board includes a history of test results. If a board fails at a particular test station, the cause is diagnosed, it is repaired, and then it is retested. If a board repeatedly fails and is tying up excessive resources, a decision must eventually be made, based on its history, either to continue retesting and repairing it or to scrap it. Information from the MMS can help in making the decision. By compiling statistics on types of defects, the MMS can also help to correct manufacturing processes which are error-prone. In addition, if excessive numbers of boards are incorrectly diagnosed, the MMS may be able to provide an indication that the test for that board must be upgraded.

The MMS can also be used to optimize the overall test strategy. As a product matures, it frequently becomes less prone to manufacturing defects. If statistics indicate that a board is rarely failing the in-circuit test, it may become cost-effective to bypass the in-circuit test and take the board directly to the functional test station. If, at a later date, the failure rate increases and exceeds some threshold, the MMS can issue a message noting this fact and recommend that boards be routed back through the in-circuit tester. The strategy may, of course, be modified to execute the in-circuit test and omit the functional test unless a threshold at the functional test station is exceeded. In either case, the optimum strategy must be to use feedback from the MMS to minimize the cost of testing. That may mean reducing the amount of capital tied up in expensive test equipment or reducing the skill level required to operate the equipment. The data from the MMS must be periodically reviewed to determine if additional test equipment should be purchased or if it might be more cost-effective to move some mature boards away from a particular test station in order to make it available for new products that must be tested.

6.3 THE FUNCTIONAL BOARD TESTER

The basic purpose of a functional tester is to apply test stimuli to a digital logic circuit board or to an integrated circuit and to monitor response. Obviously, the quality of the test applied to a given circuit can be no better than the quality of the test stimuli. However, even with effective test stimuli which exercise a circuit thoroughly and uncover a high percentage of physical defects, in a circuit with critical timing dependencies test effectiveness may depend on the ability to apply tests and measure response at the correct time in order to detect parametric faults. The ideal tester, then, may be defined as one which accurately emulates the operating environment in which a board must ultimately function. This permits detection not only of physical defects but of parametric defects which affect timing as well.

The kinds of automatic test equipment (ATE) available range from quite simple to extremely complex. In the past it has been the practice in many companies to build testers in-house. This was motivated in part by special problems related to speed and/or packaging (pin counts, etc.) which were not addressed by commercially available testers. It was also motivated by cost. The in-house tester is usually a relatively simple device, custom tailored to a particular problem, which merely applies high and low signals at the board under test edge pins and monitors response at output pins after first waiting some fixed time interval to permit circuits to reach a stable state.

Because of the ever-growing pervasiveness of digital logic products and their growing complexity, as well as the increasing cost of testing and the need to reduce this cost, it has, ironically, become necessary to invest more capital in test equipment in order to reduce the overall cost of testing. The objective of improved test equipment is to increase throughput by providing a better test, one which can both

6.3 THE FUNCTIONAL BOARD TESTER

identify more kinds of defects and help to more rapidly diagnose them. The preferred goal is to identify a specific failed component. However, even identifying the nature of the problem, such as a signal path with excessive timing, can save time because it eliminates the need to isolate a problem to a specific board later in a more complex system. To achieve this objective, the kinds of test equipment that are appearing include complex distributed processing systems in which a computer controls several subsystems, some of which are themselves controlled by microprocessors. In addition to hardware for applying stimuli to ICs and digital logic boards, they permit simultaneous access to test files via time share terminals so that one engineer may be testing a board while another is creating or updating a test file and a third engineer may be simultaneously downloading a newly created test file from a central computer complex where it was originally created.

6.3.1 The Reference Tester

Test stimuli for automatic test equipment can be obtained either from test patterns written by circuit designers and/or diagnostics engineers, or from an automatic test pattern generator (ATPG), or from some combination of these sources. The test response can be obtained either by simulating the test stimuli or by running the test stimuli on a reference board and monitoring response. The responses from the reference board, also called the known good board (KGB) or "golden" board, are recorded in a data file and then used as a standard of comparison for production boards. An alternative approach is to use a tester which can run a test simultaneously on two boards, one of them being the KGB. Then, if there is a miscompare during the test, it is assumed that the production board is faulty.

The KGB approach has the advantage that a test can be written very quickly, a test for a logic board sometimes being operational within one or two days. However, the approach has some pitfalls, the most obvious being the need to insure that the KGB is initially free of defects. If running a comparison test on two boards simultaneously, the KGB must be maintained in fault-free condition. It may be difficult to hang onto a KGB used for comparison purposes if a complex system, representing a large source of revenue, cannot be shipped to a customer for lack of a circuit board.

When using the known good board it is necessary to initialize all memory elements on a board to a known value at the start of test and to keep the board in a known state during the test. Random patterns used as test stimuli can create races and hazards, causing unpredictable state transitions, and result in miscompares on boards that are actually good. The failure to initialize a single memory element may go unnoticed for several months if the element is biased to come up in the same state every time. Then, a subtle manufacturing process change such as rerouting a wire may change the outcome of a critical race several months after a test was thought to be stable and thus produce disagreeable results.

It is difficult to provide a qualitative measure of a test since the estimate of test quality is usually derived from fault simulation. One solution to this problem is to use two KGBs, insert a fault in one of them, and then run the tests to determine whether the inserted fault was actually detected. After this is performed for some sufficiently large and representative sample of faults, a fairly accurate measure of fault coverage can be obtained. It is, however, time-consuming and could cause permanent damage to a KGB. Opens are usually harmless to insert, and excessive delays can be emulated with capacitive loading, but inserted shorts could cause a KGB to no longer be a KGB. Furthermore, it is usually not known how the results are affected by engineering change orders. It is also difficult or impossible, when using VLSI components, to emulate faults that occur inside the chip.

6.3.2 Architecture of a Stored Response Tester

Automatic test equipment is coming to rely more on the stored response concept wherein the response is obtained from a simulator and maintained in a storage device accessible to the ATE. The simulator can analyze the test patterns against a circuit model as discussed in a previous chapter. It can detect potential problems caused by races and hazards and replace unknown signals, or those where the outcome of a race is too close to call with certainty, with an x value. The user then has the option to modify the test stimuli and resimulate or send the stimuli to the ATE with some output pins and/or internal nets in the x state. This, of course, requires an ability on the part of the tester to selectively sample outputs and ignore those with unknown values.

A board tester which incorporates many of the features found in present-day complex ATE is the Teradyne L200 series of testers, depicted in the block diagram in Figure 6.4. The basic system is made up of three main units, the computer group, the digital subsystem, and the analog subsystem.

The computer group includes the system's supervisory processor and computer peripherals. It controls the digital subsystem, the analog subsystem, and communications with the external environment. The minicomputer has all of the functions normally ascribed to a computer system, including a CPU and main memory, I/O ports for peripherals, and an operating system. The standard minicomputer operating system, which supports multiuser multitasking, is enhanced by the addition of custom software designed to interface with the digital subsystem and the analog subsystem as well as to create a software environment conducive to writing, debugging, and maintaining test programs.

The commands which have been added to facilitate the editing and debugging of test data files are necessary because test plans for complex circuit boards do not always work the first time. The problem is compounded by the fact that, if the tester finds a miscompare between the expected response and the measured response from a circuit board, the problem may be a faulty board, but it may also be a bad test or a misconception about the interface between board and tester. A bad test can result even when test stimuli are simulated if the computer model is inaccurate or the simulator does not correctly predict the effects of hazards and critical races. Problems at the tester/board interface can result from skew in the switching time between individual pins on the tester, as well as insufficient drive capability or slow rise time at board input pins which have large fan-out. Tracking down the cause of miscompares can be less frustrating if the engineer has effective software support tools. These can help him to set up conditions leading to miscompares or

6.3 THE FUNCTIONAL BOARD TESTER

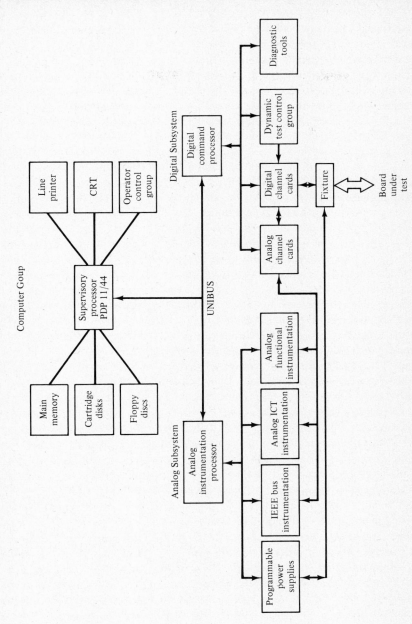

Figure 6.4 The L200 series test system. (© 1984 Teradyne, Inc.)

to put the tester into a program loop that permits him to probe the board with additional equipment. Software which features ease of use can minimize distractions and help the engineer to focus his attention on the problem to be solved.

The analog subsystem is controlled by its own dedicated microprocessor. It contains several programmable AC and DC power supplies. The analog functional instrumentation includes devices for measuring voltage, current, frequency, and period. A high-resolution time measurement option permits measurement of single-shot time interval, period, and frequency. A function generator generates sine, triangle, square, and arbitrary waveforms with programmable frequency, amplitude, and duty cycles. The ICT instrumentation and an IEEE-488 bus interface are also part of the analog subsystem.

The digital subsystem includes the test electronics (drivers/detectors) and a fixture which interfaces with the board under test. The digital subsystem in turn is controlled by the computer group through its digital command processor (DCP). A more detailed drawing of the digital subsystem is shown in Figure 6.5. As many as 1152 digital channels can be controlled by the DCP. The DCP fetches digital test stimuli and response data directly from main memory and stores the data in pattern memory.

Three measures are used to specify the speed at which a tester can operate; these are called the *pattern rate*, the *clock rate*, and the *effective test rate*. Pattern rate is the maximum speed at which a series of test patterns can be applied to a board under test. Clock rate is the maximum speed at which a channel can toggle; it is usually a multiple of pattern rate. The effective test rate is the average rate of test pattern application and takes into account the time required to update pattern memory. In the L200 series, tests can be applied to the board under test from pattern memory at bursts up to 10 MHz. However, if the number of test patterns exceeds the size of pattern memory, then pattern memory must be updated with a new set of test patterns and expected values (or both, if the channel is connected to a bidirectional pin). The L200 series, which has a maximum pattern memory size of 4096 words, uses a direct memory access (DMA) capability to update pattern memory from main memory at a typical effective test rate of 100,000 tests per second.

The channel drivers are tri-state devices and are disabled when the channel is connected to an output pin of the board under test. Provisions exist for connecting a load to the channel in order to permit the tester to more closely emulate the system environment in which the board will eventually operate.

Dynamic test controllers include the pattern controller, timing controller, and data acquisition controller. The pattern controller exercises control over the digital channels. It sequences memory addresses in test pattern memory and provides a capability for hardware looping within test pattern memory during testing to assist in debugging board problems. The timing controller determines edge placement. There are eight test phases for controlling the drivers and eight test windows for controlling the receivers. The timing generator provides a programmable cycle time ranging from 100 nanoseconds to 65 microseconds and permits the timing to change from pattern to pattern. (These features will be discussed in detail in Sec. 6.5.) The data acquisition controller provides facilities to assist in diagnosing board faults. Data can be acquired on various events and various trigger modes exist.

6.3 THE FUNCTIONAL BOARD TESTER

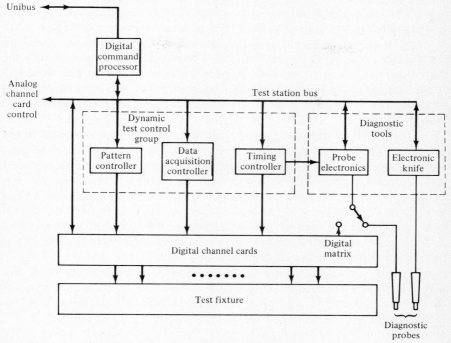

Figure 6.5 The digital subsystem. (© 1984 Teradyne, Inc.)

6.3.3 Diagnostic Tools

The functional test diagnostic tools include a guided probe and an electronic knife. The guided probe is used when an error is detected at a board edge pin or internal net being monitored. Upon detection of an error the guided probe is used to isolate the source of the error. This is done by either manually or automatically probing selected points on the circuit board. When probing is performed manually a display device instructs (guides) the operator to contact specified points on the circuit board with a hand-held probe. Automatic probing is accomplished by means of a bed-of-nails fixture. The automatic probe requires that the tester have a data file with information on the x, y coordinate of each pin of each chip on the board relative to some reference point.

The probing operation starts with the board edge pin or internal net at which the tester detects an erroneous signal. From the data base which describes the physical makeup of the board, the tester determines which IC drives the output pin. The tester then

1. Determines which inputs on that IC control the value on the erroneous output
2. Directs the guided probe to an input of the IC
3. Runs the entire test while monitoring the values on the input
4. Repeats steps 2 and 3 for all inputs that affect that output

If the tester detects an error signal on the output of an IC but does not detect an error signal on any of its inputs, the IC is identified as being potentially at fault. If an erroneous signal is detected on an input at any point during application of the test, then it is assumed that the error occurred at some device between the device presently being probed and the board inputs. Therefore, it is necessary to again back up to the IC that is driving the input pin on the IC currently being checked. This is repeated until an IC is found with an incorrect output but no incorrect inputs.

The guided probe can be very efficient at locating faulty components. It can help to substantially reduce the skill level required to detect and diagnose most faults on a circuit board because, in theory at least, the operator places the probe on IC pins in response to directions from the tester and then, when the tester detects an IC with a wrong output but correct inputs, it instructs the operator to replace that IC. However, it is not foolproof. Consider the circuit of Figure 6.6. Two tri-state registers are tied together at their outputs and are connected to the inputs of a third register. Register R2 is held in the high-impedance state. Register R1 is enabled for a brief time during the middle of a clock period. While it is enabled, data from R1 is clocked into R3. If erroneous data is found in R3 by the guided probe, it examines the inputs. If it examines the inputs at the end of the clock period, when R1 and R2 are both at high impedance, it will likely conclude that R3 is faulty when, in fact, R3 may have received faulty data from R1.

Notice in the previous paragraphs that a device was declared to be faulty if its output had an error signal but its inputs were correct. However, in practice, it is not quite that simple. If an IC is driving another IC, and the net which interconnects them is S-A-0 or S-A-1, it is possible that one of several equivalent faults may have caused the erroneous signal. A fault may exist in the IC which drives

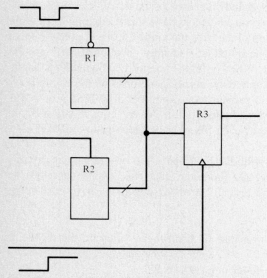

Figure 6.6 Time-dependent data transfer.

6.3 THE FUNCTIONAL BOARD TESTER

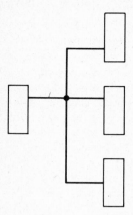

Figure 6.7 Isolating the failing IC.

the net or a fault may exist in the IC whose input is connected to the net. With three or more devices connected to a single net, as in Figure 6.7, the problem becomes more critical, because if we simply replace devices until the board passes the test, a large number of devices may have to be replaced before the failing device is replaced. This entails not only several trips to the repair station, but also several passes through the tester, and the entire process of debug and diagnosis may be repeated each time. In the meantime, each device removed and replaced increases the possibility of irreparable damage to the board.

To help resolve this problem, the L200 series employs an electronic knife. Its purpose is to locate internal device faults after the guided probe has located the net with an erroneous signal. The knife is a three-pronged instrument in which all three prongs contact a single DIP (dual in-line package) pin. This is done statically after the test is run up to the pattern at which the guided probe isolated the failing net. The knife first measures the DC voltage to insure that the measurement agrees with the voltage detected by the guided probe. If there is disagreement, the engineer must resolve that discrepancy before proceeding to the next step.

Once the tester is satisfied that probe and knife are reading identical voltages, the knife measures node resistance by forcing a DC current and measuring the change in DC voltage. Several different current levels can be tried until a significant measurable change is detected. If the measured resistance is greater than some specified threshold, for example, 400 ohms for TTL circuits, the net is assumed not to be controlled by an internal short and the device whose output is supposed to be driving the node is assumed to be at fault. If the measured resistance is less than the threshold, then the net is assumed to be controlled by a fault.

If DC tests do not reveal the cause of the problem, then AC ratio measurements are applied. In Figure 6.8(a), an AC current is injected at A. The knife measures the voltage drop V_1 from point B to point C caused by the fraction of injected current flowing away from the device. Then, as indicated in Figure 6.8(b), an AC current is injected at point C and the voltage drop from B to A, caused by the fraction of current flowing into the IC, is measured. Assuming equal IC lead resistances, the voltage ratio V_1/V_2 is equal to the ratio of internal device impedance to the impedance of the rest of the net. This ratio is measured for each IC pin

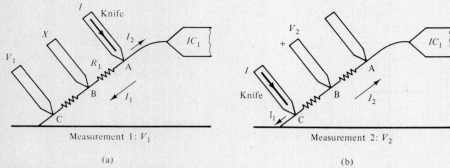

Figure 6.8 AC ratio measurement. (© 1984 Teradyne, Inc.)

connected to the net and the device with the lowest impedance is diagnosed as being at fault. The success of this method rests on the validity of two assumptions, namely

1. The voltage on a node is controlled by the lowest impedance.
2. The device controlling a failing net is bad.

The success of this method also rests on the accuracy of the voltage measurements V_1 and V_2, which in turn is determined by the probe's physical geometry. This requires that the probe not be abused, which could disturb its physical configuration and produce inaccurate measurements.

6.4 THE TEST PLAN

The functional board tester requires several files in order to test a circuit board. The data in these files can be classified as test data or diagnostic data. The test data is that data applied to every board and can be broken down into data that defines the board test environment and data that describes the actual stimuli to be applied. Diagnostic data is that data which is accessed in response to detection of an error.

Part of the task of establishing a test environment includes defining a mapping between I/O pins on the board under test and digital channels on the tester to insure that drivers and receivers in the tester drive or monitor the correct signals on the board under test. When test plans are written with symbolic names, this file will consist of a list of channel numbers and corresponding symbolic names. It is also necessary to define voltage levels for logic 1 and logic 0, as well as voltage ranges or tolerances, since these values will vary depending on the technology used. In addition, they may vary if it is required that a board be tested at operating margins. A board that normally operates at 5.0 V may be tested at 4.5 and 5.5 V to determine if it can operate at these voltage extremes without error. Intermittent errors can sometimes be induced at these marginal voltages.

6.4 THE TEST PLAN 241

If debug facilities such as the guided probe and electronic knife are available, then effective use of these resources requires that the tester have knowledge of each physically accessible IC pin, including the physical location of the pin and the expected logic values for each input vector applied. As with edge pins, the tester may require information defining the probe voltage levels corresponding to logic 1 and 0.

A circuit interconnection file is necessary when the guided probe is used to trace an error signal from an output pin back toward board inputs. The interconnection file describes all connections between ICs. A second file which is useful in conjunction with the guided probe is one which lists all inputs that affect each output of each IC on the board. In a circuit such as a quad two-input NAND gate IC, as depicted in Figure 6.9, if the guided probe traces back to NAND gate B, there is no point in observing inputs 1, 2, 5, 6, 7, and 8 since they do not affect output pin 10. This file reduces the number of measurements required and thus cuts down on the number of probe errors. This is particularly important when probing is done with a hand-held probe on a densely populated board, since such boards are especially susceptible to misprobes.

Fault isolation can be performed by means of a fault dictionary if the tester contains a computer or if a general-purpose computer is available for that purpose (see Sec. 5.6.1). This approach to diagnosis requires running a test completely from beginning to end, in contrast to the guided probe technique, which requires stopping on first fail and diagnosing the cause of the error at that point in time where the error was first detected.

The file which contains the test stimuli and response data must not only specify the output response, but must also specify those output pins which should not be observed, either because the output is indeterminate or because it does not add any additional information concerning circuit failures. The test data to be applied to the circuit can be written using ordered sets of 1s and 0s or symbolic

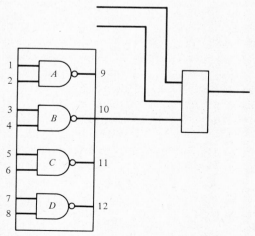

Figure 6.9 Optimizing the guided probe operation.

instructions from a test language having many constructs similar to those described in Sec. 5.8.2. We illustrate some features of the L200 test language by means of an example which includes a number of commands whose purpose is to define the test environment.

EXAMPLE ■

Given a circuit with input signals CLK and DATA and outputs named Q1 and Q2, with behavior for eight test patterns specified in the following table:

	1	2	3	4	5	6	7	8
DATA	0	0	0	1	1	1	0	0
CLK	0	1	0	0	1	0	0	1
Q1	x	x	x	x	x	x	x	0
Q2	x	1	1	1	0	0	0	1

A test plan to implement this sequence on the L200 tester appears as follows:

```
!PINMAP
DATA = (1,1,P);
CLK = (1,2,P);
Q1 = (1,3,P);
Q2 = (1,4,P);

SET DIGITAL (DATA,CLK,Q1,Q2)
       LEVEL = $A
       MODE = $STATIC
       LOAD = $OFF;

SET PIN (DATA,CLK,Q1,Q2)
       STATE = $DIGITAL;

SET ALEVELS VIH = 3.0 V
            VIL = 0.6 V
            VOL = 3.0 V
            VOL = 0.6 V
            VCOM = 1.5 V
            IOH = 16 mA
            IOL = -1.6 mA;

SET DCPOWER 1 FORCE = 5.0 V
              FORCE = 5.0 V
              LIMIT = 1.0 A
              MODE = $SUPPLY;

ML DATA, CLK, IOX Q1, Q2 FTEST;
MH CLK, OH Q1, FTEST;
ML CLK, FTEST;
MH DATA, FTEST;
MH CLK, OL Q2, FTEST;
ML CLK, FTEST;
ML DATA, FTEST;
MH CLK, OH Q2, OL Q1, FTEST;
```

In this example the PINMAP associates symbolic names used by the test engineer with physical pin numbers on the tester. The signal names are defined to be digital signals and the voltage and current high and low levels are specified for the drivers and receivers. After the tester has been configured, the remaining commands apply inputs and instruct the tester to monitor response. The first of these applies a low signal on the DATA and CLK inputs and, using the IOX command, states that the outputs Q1 and Q2 will respond with indeterminate values. On the next command the CLK is set high and the tester is told, via the OH command, to expect a high on Q2. On the next three commands the CLK is first set low, then the DATA is set high, and then the CLK input is raised. The tester is then told to expect a low on output Q2. The remaining signals are interpreted in a similar fashion.

The command processor compiles the symbolic test plan into logic 1s and 0s, some of which are applied to the inputs and others are monitored at the outputs. This has the obvious advantages provided by any symbolic language, namely that it is easy to read, understand, and remember. It permits the test engineer to focus his attention on writing the test plan. ■ ■

6.5 THE DYNAMIC FUNCTIONAL TESTER

Functional testers can be characterized as static or dynamic. The *static* tester applies stimuli and then waits for some extended time period until all internal activity in the circuit has settled. It then measures the outputs and determines whether they have responded correctly. The static tester does not attempt to accurately measure *when* events occur. Therefore, if a signal responds correctly but has excessive propagation delay along one or more signal paths, that fact may not be detected by the static tester. Excessive delays can be a problem when a board must plug into a complex system and function correctly with several other boards. Isolating problems caused by excessive propagation delays is especially difficult when the board has passed a functional test and is assumed to be good. Other problems not handled well by static testers include detection of pulses on outputs which are present only briefly, and application of tests to devices such as dynamic MOS parts which have minimum operating frequencies.

To detect timing problems, and to exercise devices at the clock frequency for which the devices were designed to operate, the *dynamic* tester is employed. It is also sometimes called a *high-speed functional tester* or a *clock rate tester*. It can be programmed to apply input signals and sample outputs at any time in a clock cycle. It is more complex than the static tester since considerably more electronics is required. Whereas many functions in the static tester are controlled by software, in the dynamic tester they must be built into hardware in order to provide resolution in the picosecond range. There are also added complexities for the user. Whereas the static tester employs low slew rates (the rate at which the tester changes signal values at the circuit inputs), the dynamic tester must employ high slew rates to avoid introducing timing errors. However, high slew rates increase the risk of overshoot, ringing, and crosstalk[3]. Programming the tester also requires more effort on the part of the test engineer since he must now be concerned with both the logic signals on the circuit being tested and the time at which they occur.

The general architecture of a dynamic tester is illustrated in Figure 6.10. The test pattern source is the same set of patterns that are used by the static tester. However, they are now controlled by timing generators and wave formatters. The test patterns specify the logic value of the stimulus and response and the remaining circuits specify when the stimulus is to be applied or when the response is to be sampled. The system is controlled by a master clock which determines the overall operating frequency of the board and controls a number of timing generators. Each of the timing generators employs delay elements and other pulse-shaping electronics to generate a waveform with programmable placement of leading and trailing edges. The placement of these edges is determined by the user and can be specified to within a fraction of a nanosecond.

The number of timing generators used in a functional tester depends in part on whether it is a shared resource or tester-per-pin architecture. A *shared resource* tester contains fewer timing generators than pins and employs a switching matrix to distribute the timing signal to tester pins, whereas the *tester-per-pin* architecture employs a timing generator for each tester pin. Advocates of the tester-per-pin believe that the switching matrix contributes greatly to skewing problems, and so eliminating the switching matrix makes it easier to deskew and thus improve the accuracy of the tester[4].

The L200 is a shared resource tester which employs eight timing generators for drivers and eight for strobing the device outputs. The wave formatters specify whether the output signal is to be Return to Zero, Return to One, Return to Complement, Return to High Impedance, or Non Return. This is illustrated in Figure 6.11 for the Return to One and the Non Return formats. Waveforms for two timing generators are shown for four periods. The master clock has a 500-ns cycle. Timing

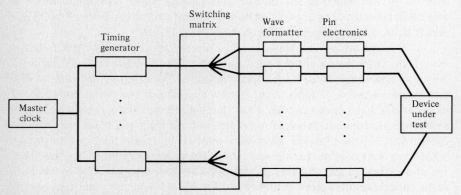

Figure 6.10 Architecture of dynamic tester.

6.5 THE DYNAMIC FUNCTIONAL TESTER

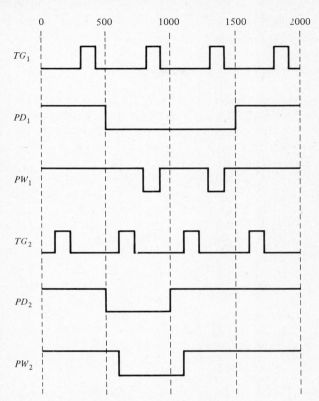

Figure 6.11 Formatted waveforms.

generator TG_1 is programmed to go high at 300 ns and low at 400 ns, while the second timing generator, TG_2, is programmed to go high at 100 ns and low again at 220 ns. Program data for two pins, called PD_1 and PD_2, are shown. Signal PD_1 is programmed with a Return to One format; therefore the resulting waveform PW_1, after going through the formatter, is normally high and assumes a low state only when PD_1 is low and TG_1 is high. Signal PD_2 is programmed as a Non Return so the resulting formatted signal PW_2 coming out of the formatter follows PD_2 with a delay of 100 ns.

With eight timing generators for drivers, it is possible to apply signals at eight different times to a board under test. Since bus structured printed circuit boards have many I/O pins with identical behavior, eight timing generators is usually sufficient. However, a board may have several modes of operation, each having unique timing characteristics. Therefore, the timing generators are usually capable of being programmed on-the-fly from one of several timing sets (TSETs). Each TSET specifies timing characteristics for some or all of the timing generators. For a given test pattern, the test engineer selects and uses a particular TSET based on the board function being performed by the test pattern.

The following small example, corresponding to Figure 6.11, illustrates the type of commands used to program the dynamic tester[5].

```
!        Assign phase and format
SET DIGITAL (PD1)
    PHASE = 1,
    FORMAT = $RONE;

SET DIGITAL (PD2)
    PHASE = 2,
    FORMAT = $NONRET;

SET TSET 1 CLOCK = 500 NSEC

    PHASE 1 ASSERT = 300 NSEC
            RETURN = 400 NSEC

    PHASE 2 ASSERT = 100 NSEC
            RETURN = 220 NSEC

USE TSET 1

HI PD1 PD2
LO PD1 PD2
HI PD2
HI PD1
```

In this example, signals PD_1 and PD_2 are first assigned timing generators and waveform formats. Then TSET 1 is defined. It has a period of 500 ns. One timing generator is asserted from 300 to 400 ns and the other is asserted from 100 to 220 ns. The tester is instructed to use TSET 1 for timing and then the logic signals are specified. In this example, it would be possible to define up to eight separate TSETs, and a different TSET could be invoked for each pattern, if the circuit required such a range of configurations.

The separate creation of timing information and logic information has the advantage that timing information is specified merely by selecting a TSET; it is not necessary to associate timing information with every pattern. If a board design is changed such that one or more parts are replaced with similar parts that have identical logical behavior, but different timing characteristics, then it is possible to update the test simply by changing the timing information.

The flexibility of the timing generators makes it quite easy to generate multiphase clocks. In a static tester it requires two test patterns to generate one complete clock pulse. In the dynamic tester, if the programmed data is held constant at 1, and if the Return to Zero format is used, then the pattern applied to a pin will follow the timing generator waveform, hence a complete clock period can be generated in one test pattern. Two or more of the timing generators can be used in this way to generate two or more clocks for a multiphase system.

It is important to note that simulation, including fault simulation, is significantly affected by the TSETs. The fault simulator can claim detection for a fault

6.6 THE IN-CIRCUIT TESTER

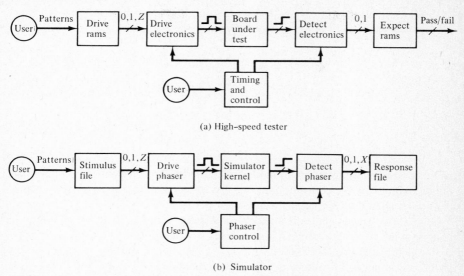

Figure 6.12 High-speed tester and simulator architectures. (© 1983 IEEE.)

only if it differs from the unfaulted machine during the period when an output is being observed and only if the faulted and unfaulted circuits are stable during that period. Therefore, the architecture of the simulator must reflect the architecture of the tester. This is illustrated in Figure 6.12, where the high-speed tester in (a) is contrasted with the simulator in (b). The drive and detect phasers use information in the TSETs to schedule primary input changes at the correct time and check for fault detection on primary outputs at times when the outputs are being monitored[5]. Note that the dynamic tester can be programmed to detect delay faults. If, for example, PW_1 must return to 1 not later than 420 ns into the pattern, then a tester pin can be programmed to strobe the output at 420 ns to determine whether it actually did return to 1.

6.6 THE IN-CIRCUIT TESTER

Despite the increasing power and flexibility of functional board testers, cost of testing continues to escalate dramatically because of the growing number of devices on printed circuit boards and the growing complexity of each of these devices. The cost of initially preparing tests at the board level as well as the cost of debugging the tests, and then, after the tests are certified to be correct, the cost of diagnosing and repairing boards with faults, have all contributed to rising costs. To address these issues, electronics companies are resorting increasingly to the use of in-circuit testers.

The in-circuit testers are able to measure resistances and verify functionality of devices while they are soldered in place on a printed circuit board. The functional test is accomplished by bringing the bed-of-nails fixture in contact with the board

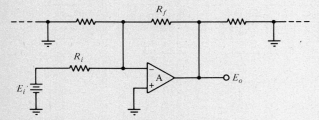

Figure 6.13 The guard circuit.

and then selectively overdriving individual ICs with large currents and monitoring the IC outputs for correct response. The measurement of resistances makes use of the *guard* circuit[6]. This circuit (see Figure 6.13) employs an op-amp. A known voltage E_i is applied through a precision resistor R_i. The op-amp amplifies the voltage at the $(-)$ input terminal and reverses its polarity as it attempts to minimize the voltage difference between its inputs. With a high-gain op-amp the voltage difference is negligible, there is negligible current flow through the op-amp, and the current through R_i is equal to the current through R_f, so we get

$$E_i/R_i = I_x = E_o/R_f$$

Since E_i and R_i are known, we can compute R_f by measuring E_o.

Several advantages are cited for in-circuit test. Among the advantages are:

Programming of tests is simplified.

Common manufacturing errors are rapidly detected and diagnosed.

Most (or all) faults are detected in a single pass through the test station.

Test equipment is cheaper and easier to use.

Test revision due to design changes is usually simpler.

Analog components can be tested.

The process of forcing voltage levels on the inputs of ICs backdrives the outputs of those devices that normally drive the IC. This backdriving operation can damage the devices as it tests them. An investigation into the effects of backdriving reports that failures can be caused by current densities and temperature excursions and can be immediate or cumulative[7]. The high currents used with in-circuit testers can cause failure in poor wire bonds but, interestingly, this may be viewed as a desirable side effect since it may precipitate failure of a potentially unreliable bond. Backdriving is a more serious problem when, after a component is tested, it is then backdriven and damaged while testing another component. The test conditions found acceptable in the report include, for TTL with a 5-V power supply, a force-to-one potential of 3 V and a force-to-zero potential of 0 V. Test lengths of up to 40,000 vectors, of duration up to 100 μs, and 50% duty cycle are tolerable for most devices. Single pulses should be limited to a maximum of 100 ms for standard devices. It is recommended that testing proceed from outputs to inputs in order to

6.6 THE IN-CIRCUIT TESTER

test devices after they are stressed. Furthermore, it is recommended that backdriving of low-output-impedance devices be avoided.

In-circuit testers are provided by their manufacturers with a library of tests for the more commonly available IC types. However, a test from the manufacturer's library may not be usable because of the manner in which a device is used in a circuit. One such example is the use of a D flip-flop as a frequency divider. The output is tied directly back to the input and makes it impossible to drive and monitor the flip-flop simultaneously. In general, any feedback line which may affect an input to the device under test should be examined to verify that it is not going to inhibit testability of that device. Another test problem for in-circuit testers is the practice of tying control lines such as the Set and Reset of a flip-flop or a tri-state output control of an LSI chip directly to ground or to a voltage source. This practice inhibits application of a complete test and requires modification of standard library test programs for the individual devices.

Precautions may have to be taken even in instances where the test can be applied as it exists on the library. Clock lines on flip-flops and complex LSI devices must be protected from transients which can occur when switching large currents[8]. Buses should receive special attention. All devices driving a bus should first be tri-stated to verify that none of the outputs is faulted in such a way as to pull the bus to a low or high value. Then each device can be tested individually while other devices connected to the bus are inhibited. The inhibit technique can be useful for other devices besides those with tri-state outputs. For example, if the output of a device loops back on itself through a NAND gate, then that feedback can be inhibited by forcing another input of the NAND to a 0.

The in-circuit tester requires a large number of connections from the board under test to the tester; it may require several hundreds or even thousands of wires. The number of wires is held down by assigning a single wire to a single net, regardless of how many inputs and outputs are connected to the net. At the tester this circuit is connected to both a driver and a receiver, which are electronically switched depending on whether the wire is presently driving an input or monitoring an output. The use of a single nail or contact probe at any one net has a second advantage in that it increases the probability of detecting an open on a printed circuit board. Consider the net illustrated in Figure 6.14. Suppose terminals 1 and 2 are connected to tri-state outputs and terminals 3, 4, and 5 are connected to IC inputs. If a single nail is used and placed in contact with terminal 1, then an open between terminals 1 and 2 will be detected when terminal 2 is monitored and an

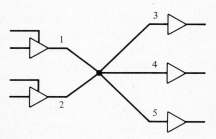

Figure 6.14 Net with multiple connections.

open will be detected between terminal 1 and any of 3, 4, or 5 whenever any of them is to be driven.

The in-circuit tester, despite its attraction, is not a universal panacea for testing problems. It has not, in the past, been able to detect timing problems, although in-circuit testers are appearing which claim to possess this capability[9]. Therefore, a board may pass a test at the in-circuit station and still fail to perform correctly when plugged into a system. Some devices cannot be backdriven. Others, such as complex VLSI devices, require longer backdrive times, and the duration required may exceed safe limits. Failures that occur at a customer's site are frequently more subtle and less likely to be diagnosed by the in-circuit tester. It is possible that a defective device may cause deceptive symptoms; it may pass the in-circuit test but adversely affect another device driving it during actual operation. Shorts between functionally unrelated runs on printed circuit boards may affect operation but go undetected by the in-circuit tester.

The manner in which the circuit board is packaged may prevent it from being tested by the in-circuit tester. A board may contain more nets than the ICT can control. If a board is populated on both sides or if for some other reason nodes are inaccessible, then the in-circuit tester cannot be used. Products that are designed for military use require conformal coating that makes their nodes inaccessible to the in-circuit tester. Some circuits are enclosed within cooling units that make them inaccessible. Dense packaging can make in-circuit test impractical and some circuits are so sensitive that the capacitance of the in-circuit probe will cause the circuit to malfunction[10]. Future packaging practices, such as complete elimination of boards, and three-dimensional wiring, may further restrict the applicability of in-circuit test[11].

6.7 DEVELOPING A BOARD TEST STRATEGY

The test engineer may find it necessary to make a choice between employing a functional board tester or an in-circuit tester, or he may plan to purchase one or more of each and must determine an effective mix of equipment. The strategy chosen will have a significant impact on manufacturing throughput because boards that reach a system with one or more defects will have to be debugged in the system. The system represents revenue; if one or more systems must be available at all times to debug faulty boards, then capital is tied up. The object, then, is to minimize the number of faulty boards that reach the system while also minimizing the cost of equipment and labor.

The in-circuit tester, as pointed out, is quite efficient at finding manufacturing faults; it requires less skill to operate; and test programs are easier to prepare and can be prepared more quickly. In terms of cost of equipment, the in-circuit tester is usually cheaper, but the fixture can prove to be a major expense. On the other hand, the functional board tester provides an environment more closely resembling the environment in which the board will ultimately operate. With a good test program, it will find all of the faults that the in-circuit tester will find as well as performance faults that the in-circuit tester will not find. These additional faults

6.7 DEVELOPING A BOARD TEST STRATEGY

are likely to be those that are most difficult to find when a board is plugged into a system.

The types of testing strategies employed are closely related to the volume of boards manufactured, the number of defects per board, the amount of time required to diagnose and repair the fault, and the cost of labor. A common practice is to send boards through the in-circuit tester and find all of the more obvious problems, and then send the boards through the functional board tester, as illustrated in Figure 6.15.

This strategy uses the in-circuit tester to good advantage to find the more obvious faults at lowest cost; then a functional test is applied prior to the part being tested in a system. If there is a high yield from the in-circuit tester, meaning that most faults are found and removed at the in-circuit tester, then most boards pass at the functional board tester and several in-circuit testers can be used for each functional board tester. If the yield from the in-circuit tester is very high, say in excess of 98%, and the system is relatively inexpensive as compared to a functional board tester, then it may be more economical to omit the functional board test station. The faulty boards that escape detection at the in-circuit tester may be debugged directly in the system. Factored into this approach, of course, must be the cost of more highly trained technicians to debug the boards in a system.

Variations on the approach are commonly employed. If yield of nonfaulted boards from manufacturing is high, then it may be more economical to test directly at the functional board tester and send failing boards to the in-circuit tester. After a board has visited the in-circuit tester, if it still fails at the functional board tester, then it is debugged at the functional board tester.

If it is decided that only one of the two tester types is to be employed, then the specific objectives of the manufacturing environment must be considered. It is generally accepted that the in-circuit tester can be brought on-line more quickly. If faulty boards coming from the tester are not a problem, either because they can be tested in the system or because they can be discarded if the problem is not quickly isolated, then the in-circuit tester is probably a good approach. However, if small numbers of many different types of boards are to be tested, fixture costs may become prohibitive. With a large number of boards easily programmed, the functional board tester is probably the right choice.

Regardless of the strategy chosen, the ultimate goal is to limit the number of units which reach system test with defects. In Chap. 9 we will discuss system test. For now we will simply note that the diagnosis of faults in complex systems is extremely difficult, hence costly, and there is great economic incentive to limit the number of faulty units that reach system test.

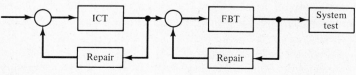

Figure 6.15 Test strategy.

6.8 SUMMARY

The evolution of digital logic from transistors and other discrete components to VLSI has forced a similar evolution in the development of test equipment and test strategy. The test strategies are guided by one dominant theme, to catch a defect as early as possible in the manufacturing process. This has led to improvements in functional test equipment and development of in-circuit test equipment. As we shall see in the next chapter, it has also led to increased emphasis on designing circuits that are more easily testable. With cost of testing assuming a greater proportion of product development cost, improvements in test strategy are, for many companies, vital to survival.

In this chapter we have looked only briefly at some of the test equipment and test strategies presently in use. It is obviously impossible in one short chapter to do justice to such a diverse field, one which is growing in both importance and complexity. The main objective here was to show the architectural features of automatic test equipment in order to see the relationship between the ATE and the test program generation tools which are used to create the test plans. The growing use of dynamic functional testers imposes a requirement that simulators possess corresponding capabilities for fault detection during specific windows in each clock cycle.

An issue which we did not touch upon, but which test managers must consider when planning the purchase of test equipment, is cost of ownership. The initial purchase price is often only a small part of the total cost over the lifetime of the equipment. The cost must also include cost of maintenance, cost of operating the equipment, cost of programming the equipment, and costs associated with not finding faults until later in the manufacturing cycle.

Attempts to peer into the future to see what kinds of test equipment will emerge are complicated by the fact that the type of equipment in use is determined by both product design and product packaging considerations. It does seem safe to predict that test equipment will continue to grow more flexible, thus more accurately emulating the environment in which the circuits are designed to operate. This may require a tighter coupling between the simulators and the test equipment. In the following chapters we will examine some of the other issues that affect the future of test equipment, including design-for-test and the concept of self-test or built-in-test.

Test languages are receiving growing interest. The U.S. Department of Defense has established ATLAS (Abbreviated Test Language for All Systems) as an interim standard test language[12]. ATLAS defines a test in terms of the product to be tested without regard to the tester to be used. It, in effect, defines the test for a virtual machine. Hence, in theory, if a particular tester has a compiler for ATLAS, it can run the test written in ATLAS language. It may ultimately become the means by which test requirements are specified for all government products. The ATLAS language, like the language described here for the L200 series, is characterized by a preamble which defines the test environment, followed by a procedural section which specifies application of stimuli and measurement of response. It permits testing of digital and analog devices and contains numerous constructs for looping and program control. It even has a specific command to leave the ATLAS language so that the user can use non-ATLAS commands to support capabilities which cannot be supported in the ATLAS language.

REFERENCES

1. Durniak, A., "IBM Has a Message: The 4300," *Electronics*, vol. 52, no. 4, Feb. 15, 1979, pp. 85–86.
2. Shapiro, D., "Universal-Grid Bareboard Testers Offer Users Many Benefits," *Electron. Test*, July 1984, pp. 88–94.
3. Sulman, D. L., "Clock-Rate Testing of LSI Circuit Boards," *Proc. 1978 IEEE Test Conf.*, pp. 66–70.
4. Bierman, H., "VLSI Test Gear Keeps Pace with Chip Advances," *Electronics*, April 19, 1984, pp. 125–128.
5. Levin, H., et al., "Design of a New Test Generation System for Performance Testing of LSI Digital Printed Circuit Boards, "*Proc. 1982 Int. Test Conf.*, Oct. 1982, pp. 541–547.
6. Schwedner, F. A., and S. E. Grossman, "In-Circuit Testing Pins Down Defects in PC Boards Early," *Electronics*, Sept. 4, 1975, pp. 98–102.
7. Sobotka, L. J., "The Effects of Backdriving Digital Integrated Circuits during In-Circuit Testing," *Proc. Int. Test Conf.*, Nov. 1982, pp. 269–286.
8. Mastrocola, Aldo, "In-Circuit Test Techniques Applied to Complex Digital Assemblies," *Proc. Int. Test Conf.*, 1981, pp. 124–131.
9. Editorial Staff, "New In-Circuit Tester Catches Digital Device Timing Faults," *Test Measurement World*, Sept. 1984, pp. 146–147.
10. Miklosz, J., "ATE: In-Circuit and Functional," *Electron. Eng. Times*, Jan. 3, 1983, pp. 25–29.
11. Tulloss, R. E., "Automatic Board Testing: Coping with Complex Circuits," *IEEE Spectrum*, vol. 20, no. 7, July 1983, pp. 38–43.
12. *IEEE Standard ATLAS Test Language*, IEEE, New York, 1981.

Chapter 7

Design-for-Test

7.1 INTRODUCTION

Automatic test pattern generators (ATPGs) have been used with varying degrees of success over the years to create test patterns for digital logic. It was pointed out in Chap. 2 that the D-algorithm can, in theory at least, create a test for any fault in combinational logic for which a test exists. In practice, even when a test exists, the ATPG may fail to generate a test for such a fault in a large circuit because of the sheer volume of data that must be maintained and manipulated. However, the real stumbling block for ATPGs has been sequential logic. Because of this failure of ATPGs to adequately deal with sequential logic, many large digital systems are being designed in compliance with design-for-testability rules which attempt to reduce the complexity of the test problem. The growing interest in testable circuit design is evidenced by the growing literature on the subject[1-3].

An inherently untestable circuit clearly cannot be tested. The object of design-for-test is to provide guidelines which insure the creation of testable designs. As an additional benefit, testable designs are frequently easier to design and debug; the same factors that make it difficult to test circuits often make it difficult to pin down design errors. In this chapter we will look at the problems that hinder testability, determine why they are problems, and look at some of the proposed solutions. In Chap. 9 we will look at built-in-test, where test circuits are designed to be an integral part of the operational circuits.

7.2 TEST PROBLEMS IN DIGITAL CIRCUITS

When SSI and MSI were the dominant levels of component integration, large systems were usually partitioned so that data flow paths and control circuits were

7.2 TEST PROBLEMS IN DIGITAL CIRCUITS

on separate boards. Most boards in any given design contained data flow circuits which were not difficult to test with ATPGs. At most a third of the boards contained the more complex control logic and timing signals. Tests for these boards were created by requiring the designer or a test engineer to write test vectors, which were then simulated with a fault simulator to determine their fault coverage. Since the difficult boards constituted a smaller percentage of the boards, the creation of tests was not excessively labor-intensive. Manually written tests for these boards were further simplified by virtue of the fact that transitions through sequential states in the control logic could be observed directly at I/O pins rather than indirectly through observation of their effects on data flow sections.

More recently, the test problem has been complicated because increasing amounts of logic are being placed inside individual chips. The circuits that were easy to test are now on the same board or inside the same chip as the more complex control circuitry. In addition, the increased amount of logic is accompanied by an increasing gate-to-pin ratio, so that there are fewer I/O pins with which to gain access to the logic that must be tested.

The degree of difficulty in testing increases when the circuit is asynchronous. The synchronous circuit is constrained to operate in synchronization with a master clock(s), hence it operates in the discrete time domain. When the clock is external to the circuit under test, the circuit can be controlled by the test equipment and can be held in a particular state until output responses are measured. In an asynchronous circuit the tester may have, at best, only partial control over circuit timing and may not be capable of detecting valid but short-lived pulses on the outputs of the device under test (DUT).

The asynchronous circuit, operating in the continuous time domain, requires that signals change in the right order. Therefore, creation of effective test stimuli requires computing both the correct state transitions and the order in which signal changes must occur to effect these transitions. Circuits which can cause particularly difficult test problems include self-resetting circuits as well as self-timing circuits, in which the completion of an event generates a signal used to trigger the next event. Even circuits which are themselves synchronous, such as circuits with on-board clock sources, can be troublesome if that clock is asynchronous relative to test equipment which must apply stimuli and observe responses at a controlled rate. Furthermore, even synchronous designs usually have some asynchronous features such as Set and Clear lines on flip-flops.

The growing use of logic devices for which there is no available knowledge of internal structure presents another problem for test engineers. Increasing numbers of microprocessors, peripheral chips, and memories are being used on logic boards as components. The test engineer must test for defects in the structure of these devices without knowing their internal structure! If the devices cannot be tested, they must at least somehow be neutralized in order to permit other devices on the board to be tested. All of these factors help to explain why testing costs have risen, not linearly, but exponentially as the number of gates on a chip increases.

Finally, before entering into a discussion of specific problems and solutions, we take note of a unique difference between the design environment and the test environment. Designers are accustomed to thinking about a design in terms of its functionality. An arithmetic logic unit (ALU) may have several functions which are not used in a particular design, or a state machine may have states which cannot be reached, or a counter may be able, as configured, to go through only part of its range. Why bother to test these functional capabilities which cannot be used?

The motive for wanting access to capabilities not present in a functional design stems from the fact that test equipment and test algorithms are often configured for the most general case. The in-circuit tester has a library of tests for standard parts. It assumes that the device is fully controllable. When a particular control or data line is hardwired to ground or V_{CC}, the test must be manually tailored in order to fit that special case. The ATPG frequently encounters problems because it will go up "blind alleys" trying to create a test using circuitry on an LSI part for which some functionality has been inhibited. Consequently, the test engineer may be forced into expending considerable effort to work around a test problem which the designer could sometimes solve quite easily if only he were aware that it was a problem.

7.3 AD HOC DESIGN-FOR-TESTABILITY RULES

Ad hoc techniques are directed toward correcting specific design practices which make it difficult or impossible to create effective tests for logic assemblies. The adverse effects of these practices may be localized or they may be global, such that a single problem may cause an entire logic board to become untestable. Global effects most frequently originate in control logic, where a design practice can make it difficult to put a circuit into a known state. As a result, control signals which pervade the entire board become uncontrollable, and they in turn cause data path functions to become uncontrollable. The solution to such a problem is often quite simple and straightforward, the most difficult part of the problem being the recognition that it is a problem. If corrections are not made, the alternative is to test the board in its operational environment, which usually results in reduced detection and diagnostic capabilities.

Testability problems for digital circuits can be classified as controllability or observability problems (or both). *Controllability* is a measure of the ease or difficulty of driving a net into a known state. *Observability* is a measure of the ability to determine the logic value present on a net. Expressed in terms of controllability and observability, the object of design-for-test is to cause circuit behavior to become easier to control and observe.

We look first at some techniques used to improve controllability and observability of logic circuits, and then look at some design practices which cause problems and the methods that can be used to correct them. The solutions are rather straightforward and there is frequently more than a single solution. The solution chosen will depend on the resources available, such as the amount of board space and/or number of edge pins.

7.3.1 The Solutions

A straightforward approach to solving the observability problem is to connect a tester directly to the output of a gate that causes test problems, if physically pos-

7.3 AD HOC DESIGN-FOR-TESTABILITY RULES

Figure 7.1 Observability enhancement.

sible, or to connect the output of the gate to an I/O pin, as in Figure 7.1. Either approach permits direct observation of a signal; symptoms do not become blocked or transformed by other logic. This simple "fix" can make a great deal of logic testable that might otherwise be untestable because of the inability to propagate tests to an output through complex logic.

The controllability problem is readily corrected by adding an OR gate or an AND gate to the circuit, together with additional I/O pins. The choice depends on whether the difficulty lies in obtaining the logic 0 or logic 1 state. The logic designer may be aware, from his knowledge of the circuit or from some testability analysis tool, that the 0 state is easily obtained but that setting up the 1 state requires an elaborate sequence of events occurring in strict chronological order. In that case a two-input OR gate is used. One input comes from the net which is difficult to control and the other input is tied to an edge pin. In normal use the primary input is grounded through a pull-down resistor; during testing the input is pulled up to the logic 1 state when that value is needed.

If the test environment, including the technology and packaging, permit direct access to the IC pins, then the edge pin connection can be eliminated. The IC pin is tied only to pull-up or pull-down resistors, as in Figure 7.2, and the tester is placed directly in contact with the IC pin by some means.

If both logic values must be controlled, then two gates are used, as in Figure 7.3(a). The first gate inhibits the normal signal when its test input is brought low and the second gate is used to insert the desired test signal. This configuration gives complete control of the signal appearing on the net for both the 0 and 1 states at a cost of two I/O pins and two gates. The inhibit signal for several such circuits can be connected to a single I/O pin to reduce the number of edge pins required. This configuration can be implemented without I/O pins if the tester can be connected directly to the IC pins. If switches are allowed on the board, then controllability of the net can be achieved as shown in Figure 7.3(b).

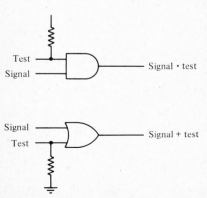

Figure 7.2 Controllability for 1 or 0 state.

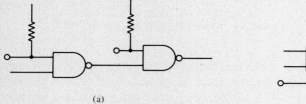

(a)   (b)

Figure 7.3 Total controllability.

Total controllability and observability at a troublesome net is had by bringing the net to a pair of edge pins, shown in Figure 7.4(a). These pins are reconnected at the card slot. This solution may, of course, create its own problems if the extra wire length picks up noise or adds excessive delay time to the signal path. An alternative circuit, shown in Figure 7.4(b), uses a tri-state gate. In normal operation the tri-state control is held at its active state and the bidirectional I/O pin is unused. During test, the bidirectional pin is used to observe logic values when the tri-state control is active or to inject signals when the tri-state control disables the output of the preceding gate. A single tri-state control can disable several gates to minimize the number of I/O pins required.

In-circuit testers are adversely affected by the practice of tying control or enable lines directly to ground or V_{CC}. This is frequently done when a tri-state output only drives a single device; the enable line is tied directly to the signal level which enables the output. In-circuit testers attempt to apply standard tests to ICs and these tests assume that all I/O pins can be controlled. If the test cannot be modified or a new test obtained, then the device cannot be tested. The solution is to tie unused pins to high or low signals through pull-up or pull-down resistors (Figure 7.5). The in-circuit tester can then pull the pin high or low during test.

Some additional solutions to testability problems include[4]:

Use sockets for difficult to test devices such as microprocessors and related peripheral circuits.

Make memory read/write lines accessible at a board edge pin.

Buffer the primary inputs to a circuit.

Put analog devices on separate boards.

Use removable jumper wires.

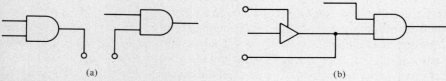

(a)   (b)

Figure 7.4 Total controllability and observability.

7.3 AD HOC DESIGN-FOR-TESTABILITY RULES

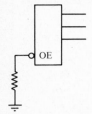

Figure 7.5 Control of unused I/O pin.

Employ standard packaging.

Provide good documentation.

The use of sockets for microprocessors and related peripheral chips is an untidy solution. The board is easier to test but it is also more susceptible to failure. As a result, it bears looking into very closely before a decision is made. The enhanced testability, in some situations, may not justify the increased failure rate due to the presence of sockets.

As explained in Chap. 6, automatic test equipment usually has different drive characteristics from the device that will drive a primary input pin in normal operation. If two or more devices are connected directly to a primary input pin without buffering, critical timing relationships between the signals may not be maintained by the ATE.

Analog devices, such as analog-to-digital and digital-to-analog converters, usually must be tested functionally over their entire range. This becomes exceedingly difficult when they are on the same board with digital logic. Voltage regulators placed on a board with digital logic can, if performing marginally, produce many seemingly different and unrelated symptoms within the digital logic, thus making diagnosis more difficult.

Finally, some practical considerations to aid in diagnosis of faults can provide a substantial return on investment. Removable jumper wires can significantly reduce the amount of time required to diagnose failures. Standard packaging, common orientation, spacing, and numbering can reduce error and confusion during troubleshooting. Good documentation can be invaluable when trying to diagnose the cause of a failure.

7.3.2 The Problems

We now look at some of the design practices which adversely affect controllability and observability. We will relate these practices to the difficulties that they cause for simulation and ATPG software, since an objective of design-for-test is to facilitate creation of test stimuli with a minimum of engineering and computer resources. It is clearly impossible to list all design practices which cause testing difficulties. The emphasis will be on providing an understanding of why certain practices create untestable designs so that the designer can exercise judgment when unsure of whether a particular design practice will be troublesome.

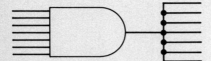

Figure 7.6 A critical node.

In combinational logic, when many signals converge at a single node, such as when an AND gate or an OR gate has many inputs, then observability of fault symptoms along any individual path converging on that gate requires setting all other inputs to 1 for an AND gate or to 0 for an OR gate. If this node in turn fans out to several other gates, then controllability of those gates is diminished in proportion to the difficulty in setting the convergent node to a 0 or 1. For the AND gate of Figure 7.6, a 0 is usually easy to obtain on the output but, if even a single input is difficult to set to 1, then that input could block a test path for all other inputs. Because that node fans out to other logic, it affects observability of logic up to that point and it affects controllability of logic following that node.

The parity checker illustrates another test problem which can occur in combinational logic. The 8-bit bus, shown in Figure 7.7, may carry a 7-bit ASCII code and a parity bit intended to produce even parity. The parity checker output should normally be low unless some fault causes odd parity to occur on the bus. But some faults in the parity checker may inhibit it from going high. To detect these faults it must be possible to get odd parity on the bus.

Most problems which adversely affect testability occur in sequential logic, and some are quite easy to correct. A common practice in sequential circuit design is to tie unused inputs of delay and JK flip-flops directly to ground or to a voltage source. The Set and Clear lines are especially susceptible to this practice in applications where the flip-flops are not required to be set or cleared at system start-up time. This practice, as previously pointed out, hinders the in-circuit tester. Functional test is also adversely affected since an easy means for getting at least one known value on the output of the flip-flop has been lost.

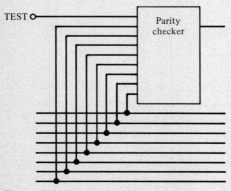

Figure 7.7 Bus parity checker.

7.3 AD HOC DESIGN-FOR-TESTABILITY RULES 261

Disabled Set and Clear lines cause further problems when the flip-flop is used as a frequency divider. In this configuration, shown in Figure 7.8, the $\overline{Q}$ output of a flip-flop is tied back onto its own data input, so that two changes at the clock input cause one change at the Q output. Two or more of these frequency dividers may be tied in series to further divide the main clock frequency. The value at the output of the divider circuit is not known at any given time, nor does it need to be known for correct operation of the circuit, since other control signals are used to synchronize the exchange of data between two devices clocked at different frequencies. What is known is that the output will switch at some fraction of the main clock frequency, and therefore some device(s) will be clocked at the lower rate.

For test and simulation the frequency divider produces the usual problems associated with indeterminate states. However, even when the correct state can be determined, if several of these frequency dividers are connected in series, then a large number of input patterns must be applied to cause a single change at the output of the frequency divider. These patterns could require exorbitant amounts of CPU time to simulate.

The input signal to a frequency divider usually comes from an on-board oscillator which generates clock pulses for the board. The oscillator executes a predictable pattern of output pulses but it is not known when the signal rises and falls. Hence we are again faced with the problem of indeterminate logic values. The use of a pair of gates at the output of either the oscillator or the string of frequency divider flip-flops, as shown in Figure 7.3, permits direct substitution of a clock signal from the tester.

Several techniques exist for creating pulses in sequential circuits and virtually all of them cause test problems. The methods include use of single-shots, self-resetting flip-flops, and circuits which gate a signal with a delayed version of the same signal. The latter method, shown in Figure 7.9, is a problem for the ATPG because the ATPG usually operates only in the logic domain; hence it will see only the logic operation, which is the AND of a signal and its complement, and will therefore always compute a 0 on the output of the AND gate. The addition of the OR gate permits the ATPG to substitute a clock signal directly.

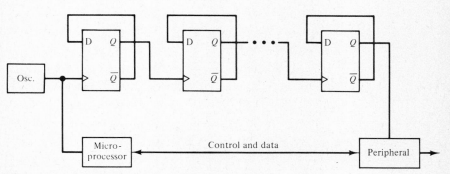

Figure 7.8 Peripheral clocked by frequency divider.

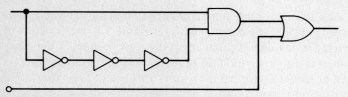

Figure 7.9 Gates used to create a pulse.

The self-resetting flip-flop is shown in Figure 7.10. When clocked, the Q output goes high and the $\overline{Q}$ output goes low. After some elapsed time, determined by the delay element and other circuit delays, the signal at $\overline{Q}$ appears on the Reset line and resets the flip-flop. This configuration causes difficulties for the in-circuit tester because the circuit will usually reset before the output is measured. A controllable buffer gate should be inserted between the delay and the reset line to prevent the device from oscillating while the in-circuit tester is attempting to measure its response. For the ATPG this configuration presents problems which, at first glance, seem similar to those of the frequency divider. Initially, when first powered up, the flip-flop is assumed to be in the x state. If an attempt is made to clock a logic 1 through the flip-flop while an x is present on the $\overline{\text{Reset}}$ input, the flip-flop remains in the x state because it cannot be determined whether or not a reset will actually occur.

An important distinction between this circuit and the frequency divider is the fact that we know how the flip-flop behaves when power is applied. If it comes up with $Q = 0$, then it is in a stable state. If Q is initially a 1 following application of power, then the 0 on $\overline{Q}$ causes it to reset. Therefore, regardless of the initial state, it is predictably in a 0 state within a few nanoseconds after power is applied.

When the state of a device can be determined, the ATPG or simulator can be given an assist. In this instance, any of the following three methods can be used:

1. Represent the circuit as a primitive (a monostable).
2. Specify an initial state for the circuit.
3. Use a dummy reset.

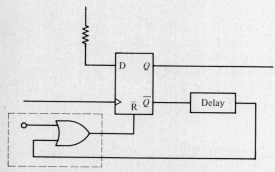

Figure 7.10 Self-resetting flip-flop with dummy reset.

7.3 AD HOC DESIGN-FOR-TESTABILITY RULES

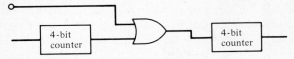

Figure 7.11 Controlling a counter.

If the circuit is modeled as a primitive, then a pulse on the clock input to this primitive causes an output pulse of some duration determined by the delay. Allowing the user to specify an initial state will solve the problem, since either value causes it to achieve a stable state. However, if an indeterminate logic value should reach the clock line at a later point in time, it could cause the circuit to revert to the indeterminate state. The dummy reset, shown in Figure 7.10, could be used in cases where the ATPG or simulator has no facilities for initializing internal nets. It is connected to a fictitious I/O pin or one which exists but which is not used in the design.

Counters, like frequency dividers, cause problems because a counter of several stages may require many thousands of test patterns to drive it to a particular state. The counter can be broken up by means of OR gates as shown in Figure 7.11. If an OR gate is placed between each section of a counter that is made up of 4-bit elements, then any state can be reached with no more than 16 test patterns.

Counters are sometimes used over only part of their counting range. This may happen when a four-state binary counter is used to divide the frequency of an oscillator by a value other than 16, or when two or more clock frequencies are desired from an oscillator. As with other instances of unused functionality, testing can be enhanced if unused pins are tied to ground or V_{CC} through resistors or if some means is provided to reach unused states. If a parallel load counter is loaded with an initial value other than 0, by connecting its inputs to ground or V_{CC}, these connections should be made through resistors.

Some other design practices which cause problems include the following:

Connecting drivers in parallel to get more drive capability

Gating clock lines with data signals

Randomly assigning unused states in state machines

The parallel drivers are shown in Figure 7.12. If one of these drivers should fail, the result may be an intermittent error whose occurrence depends on various environmental factors and internal operating conditions.

Unused states in a state machine are often assigned so as to minimize logic. As a result, an erroneous transition into an unassigned state, followed by a transition

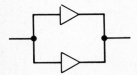

Figure 7.12 Parallel drivers.

to a valid state, may go undetected but may cause data corruption. The severity of the problem depends on the application. To err on the side of safety, a transition into an illegal state should normally cause some noticeable symptom such as an error signal or, at the very least, continued transitions into the same illegal state, that is, a "hang-up," so that an operator can detect the presence of the malfunction.

Transitions into incorrect states can occur when hazards cause unintended pulses on clock lines of flip-flops. One way to avoid this is to avoid gating clock signals with data signals whenever possible. This can be done by using the data signal that would normally be used to gate the clock to control a multiplexer instead, as shown in Figure 7.13. The signal that the designer would normally use to gate the clock is used either to select new data for input to the flip-flop or to reselect the present state of the flip-flop.

7.4 CONTROLLABILITY/OBSERVABILITY ANALYSIS

In the previous section we described some techniques for solving particular testability problems. Some of the configurations virtually always create test problems. Other circuit configurations are not problems in and of themselves but can become problems when they appear in excessive numbers. A small number of flip-flops, connected in a straightforward manner without feedback, apart from that which exists inside the flip-flops, and without critical timing dependencies, can be relatively easy to test. The testability problems occur when large numbers of flip-flops are connected in serial strings such that control of each flip-flop depends on first controlling its predecessors in the chain. Examples that we have seen include the counter and the frequency divider.

Fortunately, the counter and frequency divider are reasonably easy to recognize. In many circuits the nodes that are difficult to test are not so easy to identify. Consider the combinational circuit described in the previous section, which has a node where many signals converged. That node, in turn, controlled several other logic gates. The node may be a problem or it may, in fact, be rather easy to test. Testability measures have been developed which help to determine which of these nodes are most likely to be problems.

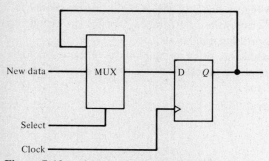

Figure 7.13 Alternate for clock gating circuit.

7.4.1 SCOAP

A testability analysis method called SCOAP (Sandia Controllability Observability Analysis Program)[5] assigns numbers to nodes in a circuit. The numbers indicate the testability of the circuit in terms of the ability to control and observe signals at internal nodes. They are computed in such a way that higher numbers are assigned to nodes which are more difficult to control or observe. The program computes both combinational and sequential controllability and observability numbers for each node and, furthermore, the controllability is broken down into 0-controllability and 1-controllability, acknowledging the fact that it may be relatively easy to produce one of the states at the output of a logic gate while extremely difficult to produce the other state. For example, to get a 0 on the output of an AND gate requires a 0 on any single input. However, to get a 1 on the output requires that 1s be applied to all inputs. That, in general, will be more difficult for gates with larger numbers of inputs.

The Controllability Equations The e-controllability, $e \in \{0, 1\}$, of a node depends both on the logic elements which drive the node and the controllability of the inputs to that element. If the inputs are difficult to control, the output is obviously going to be difficult to control. In similar fashion, the observability of a node is dependent on the elements through which it must be driven to reach an output. Its observability can be no better than the observability of the elements through which it must be driven. Therefore, before applying the SCOAP algorithm, it is necessary to have, for each primitive used in a circuit, equations expressing the 0- and 1-controllability of its output in terms of the controllability of its inputs, and it is necessary to have equations which express the observability of an element input in terms of both the observability of the output of that element and the controllability of some or all of its other inputs.

Consider the three-input AND gate. To get a 1 on the output we require that all inputs be set to 1. Hence, controllability of the output to a 1 state is a function of the controllability of all three inputs. To produce a 0 on the output requires only that a single input be at 0; thus there are three choices and, if there exists some quantitative measure indicating the relative ease or difficulty of controlling each of these three inputs, then it is reasonable to select that input which is easiest to control in order to establish a 0 on the output. Therefore, we can define the combinational 1- and 0-controllabilities, $CC^1(Y)$ and $CC^0(Y)$, of a three-input AND gate with inputs X_1, X_2 and X_3 and output Y as

$$CC^1(Y) = CC^1(X_1) + CC^1(X_2) + CC^1(X_3) + 1$$
and
$$CC^0(Y) = \text{Min}\{CC^0(X_1), CC^0(X_2), CC^0(X_3)\} + 1$$

Controllability to a 1 is additive over all inputs and to a 0 it is merely the minimum over all inputs. In either case the result is incremented by 1 so that, for intermediate nodes, the number reflects, at least in part, distance (measured in numbers of gates) to primary inputs and outputs. The controllability equations for any combinational function can be determined from either its truth table or its cover. If two or more inputs must be controlled to 0 or 1 in order to produce the value e, $e \in \{0, 1\}$, then the controllabilities of these inputs are summed and the result is incremented by 1. If more than one input combination produces the value e, then the controllability number is the minimum over all such combinations.

EXAMPLE ■

Consider the two-input exclusive-OR. The truth table is:

X_1	X_2	Y
0	0	0
0	1	1
1	0	1
1	1	0

The combinational controllability equations are:

$$CC^0(Y) = \text{Min}\{CC^0(X_1) + CC^0(X_2), CC^1(X_1) + CC^1(X_2)\} + 1$$
$$CC^1(Y) = \text{Min}\{CC^0(X_1) + CC^1(X_2), CC^1(X_1) + CC^0(X_2)\} + 1$$ ■ ■

The sequential 0- and 1-controllabilities for combinational circuits, denoted SC^0 and SC^1, are computed using similar equations.

EXAMPLE ■

For the two-input exclusive-OR, the sequential controllabilities are:

$$SC^0(Y) = \text{Min}\{SC^0(X_1) + SC^0(X_2), SC^1(X_1) + SC^1(X_2)\}$$
$$SC^1(Y) = \text{Min}\{SC^0(X_1) + SC^1(X_2), SC^1(X_1) + SC^0(X_2)\}$$ ■ ■

When computing sequential controllabilities through combinational logic, the value is not incremented. The intent of the sequential controllability number is to provide an estimate of the number of time frames needed to provide a 0 or 1 at a particular node. Propagation through combinational logic does not necessitate the use of additional time frames.

When deriving equations for sequential circuits, both combinational and sequential controllabilities are computed, but the roles are reversed. The sequential controllability is incremented by 1, but an increment is not included in the combinational controllability equation. The creation of equations for a sequential circuit will be illustrated by means of an example.

EXAMPLE ■

Consider a positive edge-triggered flip-flop with an active low reset but without a set capability. Then, 0-controllability is computed with

$$CC^0(Q) = \text{Min}\{CC^0(R), CC^1(R) + CC^0(D) + CC^0(C) + CC^1(C)\}$$
$$SC^0(Q) = \text{Min}\{SC^0(R), SC^1(R) + SC^0(D) + SC^0(C) + SC^1(C)\} + 1$$

and 1-controllability is computed with

$$CC^1(Q) = CC^1(R) + CC^1(D) + CC^0(C) + CC^1(C)$$
$$SC^1(Q) = SC^1(R) + SC^1(D) + SC^0(C) + SC^1(C) + 1$$ ■ ■

7.4 CONTROLLABILITY/OBSERVABILITY ANALYSIS

The first two equations state that a 0 can be obtained on the output of the delay flip-flop in either of two ways. It can be obtained by setting the reset line to 0 or it can be obtained by setting the reset line to 1, setting the data line to 0, and then creating a rising edge on the clock line. Since four events must occur in the second choice, the controllability figure is the sum of the controllabilities of the four events. The sequential equation is incremented by 1 to reflect the fact that an additional time image is required to propagate a signal through the flip-flop. (This is not strictly true since a reset will produce a 0 at the Q output in the same time frame.) A logic 1 can be achieved only by clocking a 1 through the data line and that also requires holding the reset line at a 1.

The Observability Equations The observability of a node is a function of both the observability and the controllability of other nodes. This is illustrated in the circuit of Figure 7.14.

In order to observe the value at node P, we must first be able to observe the value on node N. If it is not possible to observe the value on node N, then clearly node P cannot be observed. However, it is also necessary to place nodes Q and R into the 1 state. Therefore, a measure of the difficulty of observing node P can be computed with the following equation:

$$CO(P) = CO(N) + CC^1(Q) + CC^1(R) + 1$$

In general, the combinational observability of the output of a logic gate that is connected to the input of an AND gate is equal to the observability of the AND gate input which, in turn, is equal to the sum of the observability of its output plus the 1-controllabilities of its other inputs, incremented by 1.

For a more general primitive combinational function, the observability of a given input can be computed from its propagation D-cubes (see Sec. 2.6.3). The process is as follows:

> Select a D-cube which has a D or $\overline{D}$ only on the input in question and 0, 1 or x on all other inputs.
>
> Add the 0- and 1-controllabilities corresponding to each input in the cube which has a 0 or 1 assigned.
>
> Select the minimum controllability number so computed and add to it the observability of the output.

EXAMPLE ■

The propagation D-cubes for input 1 of the AND-OR-Invert of Figure 2.9 are $(D, 1, 0, x)$ and $(D, 1, x, 0)$. The combinational observability for input 1 is equal to

$$CO(1) = \text{Min}\{CO(\text{OUT}) + CC^1(2) + CC^0(3), CO(\text{OUT}) \\ + CC^1(2) + CC^0(4)\} + 1$$ ■■

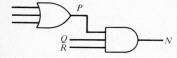

Figure 7.14 Node observability.

The sequential observability for node P is computed using a similar equation:

$$SO(P) = SO(N) + SC^1(Q) + SC^1(R)$$

The sequential observability equations, like the sequential controllability equations, are not incremented by 1 when being computed through a combinational circuit. In general, the sequential controllability/observability equations are incremented by 1 when computed through a sequential circuit and the corresponding combinational equations are not incremented.

EXAMPLE ■

We develop the observability equations for the Reset and Clock lines of the delay flip-flop considered earlier. First consider the Reset line. Its observability can be computed with the following equations:

$$CO(R) = CO(Q) + CC^1(Q) + CC^0(R)$$
$$SO(R) = SO(Q) + SC^1(Q) + SC^0(R) + 1$$

For the clock, we get:

$$CO(C) = \text{Min}\{CO(Q) + CC^1(Q) + CC^1(R) + CC^0(D) + CC^0(C) + CC^1(C),$$
$$CO(Q) + CC^0(Q) + CC^1(R) + CC^1(D) + CC^0(C) + CC^1(C)\}$$
$$SO(C) = \text{Min}\{SO(Q) + SC^1(Q) + SC^1(R) + SC^0(D) + SC^0(C) + SC^1(C),$$
$$SO(Q) + SC^0(Q) + SC^1(R) + SC^1(D)$$
$$+ SC^0(C) + SC^1(C))\} + 1 \quad\blacksquare\blacksquare$$

The equations for the Reset line of the flip-flop assert that observability is equal to the sum of the observability of the Q output, plus the controllability of the flip-flop to a 1, plus the controllability of the Reset line to a 0. Expressed another way, the ability to observe a value on the Reset line depends on the ability to observe the output of the flip-flop, plus the ability to drive the flip-flop into the 1 state and then reset it. Observability of the clock line is described similarly.

The Algorithm Since the equations for the observability of an input to a logic gate or function are dependent on the controllabilities of the other inputs, it is necessary to first compute the controllabilities. We start by assigning initial values to all primary inputs, I, and internal nodes, N:

$$CC^0(I) = CC^1(I) = 1$$
$$CC^0(N) = CC^1(N) = \infty$$
$$SC^0(I) = SC^1(I) = 0$$
$$SC^0(N) = SC^1(N) = \infty$$

Having established these initial values, we then select each internal node in turn and compute the controllability numbers for that node, working from primary inputs to primary outputs, and using the controllability equations developed for the primitives. We then repeat the process until, finally, the calculations stabilize. The nodal values must eventually converge since the controllability numbers are monotonically nonincreasing integers.

7.4 CONTROLLABILITY/OBSERVABILITY ANALYSIS

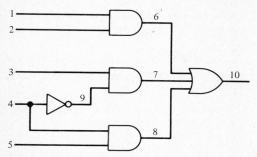

Figure 7.15 Controllability computations.

EXAMPLE ∎

The controllability numbers will be computed for the circuit of Figure 7.15. We initially assign a controllability of 1 to all inputs and ∞ to all internal nodes. After the first iteration the 0- and 1-controllabilities of the internal nodes, in tabular form, are:

N	$CC^0(N)$	$CC^1(N)$	$SC^0(N)$	$SC^1(N)$
6	2	3	0	0
7	2	∞	0	∞
8	2	3	0	0
9	2	2	0	0
10	7	4	0	0

After a second iteration the combinational 1-controllability of node 7 goes to 4 and the sequential controllability goes to 0. If we had numbered the nodes according to the rule that no node is numbered until all its inputs are numbered, the second iteration would have been unnecessary. ∎∎

With the controllability numbers established, it is now possible to compute the observability numbers. The first step in the algorithm is to initialize the primary outputs, Y, and internal nodes, N, with:

$$CO(Y) = 0$$
$$SO(Y) = 0$$
$$CO(N) = \infty$$
$$SO(N) = \infty$$

Then select each node in turn and compute the observability of that node. Continue until the numbers converge to stable values. As with the controllability numbers, the observability numbers must eventually converge. They will usually converge much more quickly, with the fewest number of iterations, if the nodes closest to outputs are selected first and those closest to the inputs selected last.

EXAMPLE ■

We compute the observability numbers for the circuit of Figure 7.15. After the first iteration we get the following table:

N	CO(N)	SO(N)
9	∞	∞
8	5	0
7	5	0
6	5	0
5	7	0
4	7	0
3	8	0
2	7	0
1	7	0

On the second iteration the combinational and sequential observabilities of node 9 settle at 7 and 0, respectively. ■■

7.4.2 Other Testability Measures

Other algorithms exist, similar to SCOAP, which place different emphasis on circuit parameters. COP (controllability and observability program)[6] computes controllability based on the number of primary inputs that must be controlled in order to establish a value at a node. The numbers therefore do not reflect the number of levels of logic between the node being processed and the primary inputs. The SCOAP numbers, which encompass both the number of levels of logic and the number of primary inputs affecting the C/O numbers for a node, give a more accurate estimate of the amount of work that must be performed by an ATPG. However, the number of primary inputs affecting C/O numbers probably reflects more accurately the probability that a node will be switched to some value randomly, hence it may be that it more closely correlates with the probability of random fault coverage when simulating test vectors.

An extension of testability analysis to functional level primitives has been accomplished in FUNTAP (functional testability analysis program)[7]. In this implementation, advantage is taken of structures such as n-wide data paths. Whereas the single net may have binary values 0 and 1, and these values can have different C/O numbers, the n-wide data path made up of binary signals may have a value ranging from 0 to $2^n - 1$. In FUNTAP no significance is attached to these values; it is assumed that the data path can be set to any value i, $0 \leq i \leq 2^n - 1$, with equal ease or difficulty. Therefore, a single controllability number and a single observability number are assigned to all nets in a data path, independent of the logic values assigned to individual nets which make up the data path.

The ITTAP program[8] computes controllability and observability numbers but, in addition, it computes parameters TL0, TL1, and TLOBS, which measure the length of the sequence needed in sequential logic to set a net to 0 or 1 or to observe the value on that node. For example, if a delay flip-flop has a reset which can be used to reset the flip-flop to 0, but can only get a 1 by clocking it in from the Data input, then TL0 = 1 and TL1 = 2.

7.4 CONTROLLABILITY/OBSERVABILITY ANALYSIS

A more significant feature of ITTAP is its selective trace capability. This feature is based on two observations. First, controllabilities must be computed before observabilities, and second, if the numbers were once computed, and if a change is made to enhance testability, numbers need only be recomputed for those nodes where the numbers can change. The selection of elements for recomputation is similar to event-driven simulation. If the controllability of a node changes because of the addition of a test point, then elements driven by that element must have their controllabilities recomputed. This continues until primary outputs are reached or elements are reached where the controllability numbers at the outputs are unaffected by changing numbers at the inputs. At that point, the observabilities are computed back toward the inputs for those elements with changed controllability numbers on their inputs.

The use of selective trace provides a savings in CPU time of 90 to 98% compared to the time for recomputing all numbers for a given circuit. This makes it ideal for use in an interactive environment. The designer can visually inspect either a circuit or a list of nodes at a video display terminal; he can assign a test point and then immediately view the results. Because of the quick response, he can then shift the test point to some other node and recompute the numbers. After several such iterations, he can settle on the node which best improves the C/O numbers.

The interactive strategy has pedagogical value. Placing a test point at the node with the worst C/O numbers is not always the best solution. It may be more effective to place a test point at some node which controls the node in question, since this may improve controllability of several nodes. Also, since observability is a function of controllability, testability may sometimes be best improved by assigning a test point as an input to a gate rather than as an output even though the analysis program indicates that the observability is poor. The engineer who uses the interactive tool, particularly recent graduates who may not have given much thought to testability issues, may learn with such an interactive tool how best to design for testability.

7.4.3 Test Measure Effectiveness

There has been some research into the effectiveness of testability analysis. In one such investigation, controllability of a fault is defined as the fraction of input vectors that will set a faulty net to a value opposite its stuck-at value, and observability is defined as the fraction of input vectors that will propagate the fault effect to an output[9]. Testability is then defined as the fraction of input vectors that detect the fault. Obviously, to test a fault, it is necessary to both control and observe the fault symptom, hence testability can be construed to be the intersection of the sets of inputs for controllability and observability. But there may be two reasonably large sets whose intersection is empty. A simple example is shown in Figure 7.16. The controllability for the bottom input of the gate numbered 1 is $\frac{1}{2}$. The observability is $\frac{1}{4}$. Yet, the input cannot be tested for an S-A-1 fault because it is redundant.

In another investigation of testability measures, the authors attempted to determine a relationship between testability figures and the detectability of a fault[10].

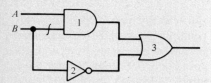

Figure 7.16 An untestable fault.

They partitioned faults into classes based on testability figures for the faults and then plotted curves of fault coverage versus vector number for each of these classes. The curves were reasonably well behaved, the fault coverage curves rising more slowly, in general, for the more difficult to test fault classes, although occasionally a curve for some particular class would rise more rapidly than the curve for a supposedly easier to test class of faults. They concluded that testability data had little chance of predicting detection of individual faults but that general information at the circuit level was available and useful. Further, if some percentage, say 70%, of a class of difficult to test faults are tested, then any fixes made to the circuit for testability purposes have only a 30% chance of being effective.

7.4.4 Using the Test Pattern Generator

If tests for a circuit are to be generated by an ATPG, then the most direct way in which to determine its testability is to simply run the ATPG on the circuit. The ability (or inability) of an ATPG to generate tests for all or part of a design is the best criterion for testability. Furthermore, it is a good practice to run test pattern generation on a design before the circuit has been fabricated. After a board or IC has been fabricated, the cost of incorporating changes to make the design more testable increases dramatically.

A technique employed by at least one ATPG[11] incorporates a preprocess mode in which it attempts to set latches and flip-flops in a circuit to both the 0 and 1 states before it attempts to create tests for specific faults. The objective is to find troublesome circuits before going into automatic test pattern generation. The ATPG compiles a list of those flip-flops for which it could not establish either the 0 or 1 state, or both. Whenever possible, it indicates the reason for the failure to establish desired value(s). The failure may result from such things as races in which relative timing of the signals is too close to call with confidence, or it could be caused by bus conflicts resulting from inability to set one or more tri-state control lines to a desired value. It also has criteria for determining whether the establishment of a 0 or 1 state took an excessive amount of time. At the end of the preprocess mode it provides a report to the user listing those nodes which either could not be set to one or both logic states or required excessive time to reach one or both states.

Analysis of information in this preprocess mode of operation may reveal clusters of nodes that are all affected by a single uncontrollable node. It is also important to bear in mind that nodes which require a great deal of time to initialize can be as detrimental to testability as nodes which cannot be initialized. An ATPG may set arbitrary limits on the amount of time to be expended in trying to set up a test for a particular fault. When that threshold is exceeded, the ATPG will give up on the fault even though a test may exist.

7.5 THE SCAN PATH

A variation of testability analysis is the use of C/O numbers in the ATPG. Since the ATPG processes the same circuit elements many times during a typical run, and is constantly faced with making choices, the C/O numbers may significantly reduce the amount of time required to create test patterns by virtue of their ability to influence choices made by the ATPG during propagation and justification of signal paths (see Sec. 2.5.2).

7.5 THE SCAN PATH

A formal technique called scan path makes it possible to achieve total or near total controllability and observability in sequential circuits. In this approach the flip-flops and/or latches are designed to be able to operate in either parallel load or serial shift mode. In the normal mode of operation flip-flops and latches are configured for parallel load. For test purposes the flip-flops are switched to serial shift mode. In serial mode, any needed test values can be loaded by serially clocking in the desired values. In similar fashion, any value present in the flip-flops can be observed by clocking out their contents while in the serial shift mode.

A simple means for creating the scan path consists of placing a multiplexer just ahead of each flip-flop, as illustrated in Figure 7.17. One input to this 2-to-1 multiplexer is sourced by normal operational data and the other input, with one exception, is sourced by the output of another flip-flop. On one of the multiplexers, the serial input is connected to a primary input pin. Likewise, one of the flip-flop outputs is connected to a primary output pin. The multiplexer control line, also connected to a primary input pin, is now a mode control; it can permit parallel load for normal operation or it can select serial shift in order to go into the test mode. When the serial mode is selected, there is a complete serial shift path from an input pin to an output pin.

Since we can load arbitrary values into, and read the contents out of, the flip-flops by means of a shift path, it becomes possible to enormously simplify the ATPG used to create tests for digital circuits. A preprocessor modifies the circuit model by deleting the flip-flops and substituting pseudo-inputs and pseudo-outputs. The circuit has, in effect, been converted into a combinational circuit for test

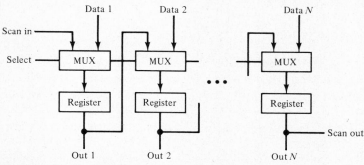

Figure 7.17 The scan path.

purposes. The payoff is that the complexity of testing has been significantly reduced because it is no longer necessary to propagate tests through the time dimension represented by sequential circuits. The scan path can be separately tested from the sequential logic by shifting alternating 1s and 0s through the path. Therefore, faults inside the flip-flops can be ignored by the test generator, as well as the stuck-at fault on the clock line itself, which is so often difficult to classify as detected or not detected due to the problems caused by inability to initialize the flip-flop.

When the ATPG is propagating a sensitized path, it stops at a flip-flop input just as it would stop at a primary output. When justifying logic assignments, the ATPG stops at the output of flip-flops just as it would stop at primary inputs. In effect, flip-flop outputs are treated as though they were primary inputs and their inputs are treated as though they were primary outputs. The only difference between actual I/O pins and the flip-flop "I/O pins" is the fact that values assigned to the flip-flops must be serially shifted in when used as inputs and serially shifted out when used as outputs.

When using scan path elements as I/O pins, flip-flops in the scan path must be identified as well as the order in which they are interconnected. The ATPG does not have to distinguish between real and pseudo-I/O pins for the purpose of creating tests, but the test equipment applies stimuli directly to physical I/O pins, while stimuli for pseudo-I/O pins must be serially shifted in via the scan path, and those stimuli must be in the correct order. In addition, the tester must also know which physical I/O pins are used to implement the scan path: which pin serves as the scan-in, which serves as the scan-out, and which pin is used for the mode control.

When the circuit with a scan path is being used in its normal operational environment, the mode control is set for parallel load. The multiplexer selects normal operational data and, except for the delay time through the multiplexer, the scan circuitry is transparent. When the device is being tested, the mode control alternates between parallel load and serial shift. To apply a test to a circuit with a scan path, the tester must:

> Switch the mode select line to serial shift.
>
> Serially clock test data into the flip-flops using the scan input.
>
> Switch the mode select line to operational mode.
>
> Apply test data to the I/O pins.
>
> Exercise the clock.
>
> Evaluate response at the output pins.
>
> Switch the mode line to serial shift mode.
>
> Serially shift out the updated contents of the flip-flops via the scan output pin.

Note that new test input data can be serially shifted into the scan path while test results are being clocked out.

The effectiveness of scan path can be demonstrated with the circuit of Figure 7.18. It is an eight-state sequential machine which can be implemented with about

7.5 THE SCAN PATH

40 logic gates. It has three inputs — a clock, a reset (not shown) which can set $Q_2, Q_1, Q_0 = 0, 0, 0$, a single data input—and it has a single output. With the correct sequence of inputs, it can be fully tested for all single stuck-at faults. However, the reader who attempts to create tests for specific faults, for instance, input 3 of gate 23, will appreciate the problems facing an ATPG, which must work with a gate-level description of the circuit. In trying to create a test for the fault in question, it is not difficult to figure out that values $Q_2, Q_1, Q_0 = 1, 0, 0$ must be contained in the flip-flops. The difficult task is to determine, given that the machine is presently in some other state, and using only a gate-level description, how to obtain those values in the flip-flops. If the flip-flops are connected into a serial scan path, then the circuit is placed in the serial shift mode and the required values are shifted into Q_2, Q_1, and Q_0.

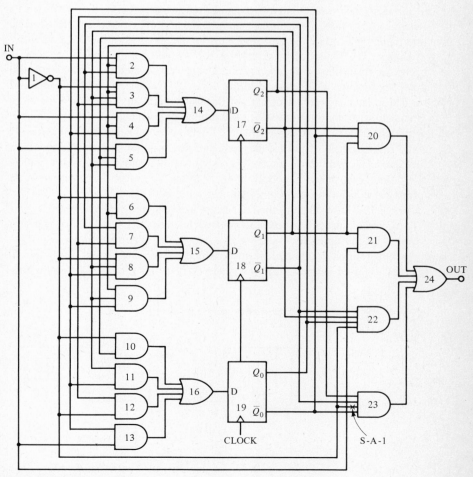

Figure 7.18 State machine.

7.5.1 Level-Sensitive Scan Design

The intent of scan is to make a circuit testable by making it appear to be strictly combinational logic to an ATPG. However, not all circuits can be directly transformed into combinational circuits simply by adding a scan path. Consider the self-resetting flip-flop discussed earlier. We could put it into a serial scan path as illustrated in Figure 7.19. However, any attempt to serially shift data will be defeated by the self-resetting capability of the flip-flop. A number of other circuit configurations create similar complications. In particular, configurations in which flip-flops can be asynchronously set or reset or in which the clock input is data-dependent pose problems for the scan technique.

Level-sensitive scan design (LSSD) is a design methodology which addresses the problems associated with implementing the scan path. It augments the scan path concept with a number of additional rules which cause a design to become level-sensitive. A *level-sensitive* system is one in which the steady-state response to any allowed input state change is independent of circuit and wire delays within the system. In addition, if an input state change involves the changing of more than one input signal, then the response must be independent of the order in which they change[12]. The object of these rules is to preclude the creation of designs in which correct operation is dependent on critical timing factors. To achieve this objective, the memory devices used in the design of the system are level-sensitive latches. These latches permit a change of internal state at any time when the clock is in one state, usually the high state, and inhibit state changes when the clock is in the opposite state. Unlike edge-sensitive flip-flops, the latches are insensitive to rising and falling edges of pulses and therefore the designer cannot create circuits in which correct operation is dependent on pulses that are themselves critically dependent on circuit delay. The only timing that must be taken into account is the total propagation time through combinational logic between the latches.

In the LSSD environment, the latches are used in pairs as illustrated in Figure 7.20. These latch pairs are called shift register latches (SRLs) and their operation is controlled by multiple clocks, denoted A, B, and C. The Data input is used in the normal operational mode and Scan-in, which is connected to the L2 output of another SRL, is used in the serial scan mode. When functioning in its operational

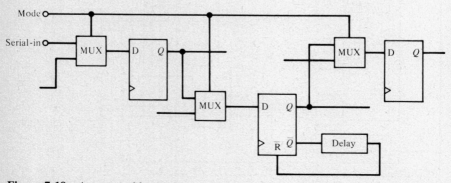

Figure 7.19 A reset problem.

7.5 THE SCAN PATH

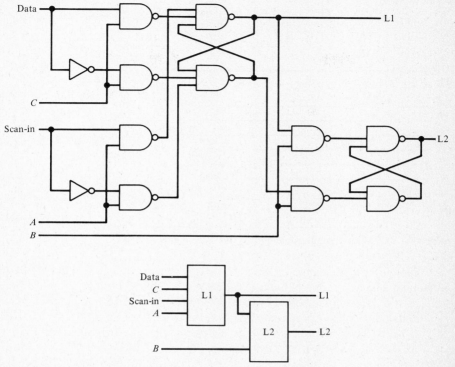

Figure 7.20 The shift register latch.

mode the *A* clock is inactive. The *C* clock is used to clock data into L1 from the Data input, and output can be taken from either L1 or L2. If it is taken from L2 then two clock signals are required. The second signal, called the *B* clock, clocks data into L2 from the L1 latch. This configuration is sometimes referred to as a "double latch" design.

When the scan path is used for testing purposes the *A* clock is used in conjunction with the *B* clock. Since the *A* clock causes data at the Scan-in input to be latched into L1, and the Scan-in signal comes from the L2 output of another SRL (or a primary input pin), alternately switching the *A* and *B* clocks serially shifts data through the scan path from the Scan-in terminal to the Scan-out terminal. The operation, in general, is similar to that described in the previous section for serial scan, except that the mode switch is not necessary with LSSD.

So far, the LSSD concept appears quite similar to the serial scan concept, which uses a multiplexer as described in the previous section, except for the employment of latches in place of the edge-triggered flip-flops and the use of multiple clocks in place of the mode switch. However, there is more to the LSSD concept, namely a set of rules governing the manner in which logic is clocked.

Consider the circuit of Figure 7.21. If S1, S2, and S3 are L1 latches, the correct operation of the circuit depends on relative timing between the clock and data signals. When the clock is high there is a direct combinational logic path from

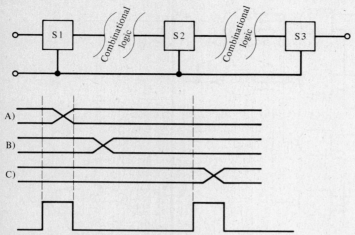

Figure 7.21 Some timing problems.

the input of S1 to the output of S3. Since the clock signal must stay high for some minimum amount of time in order to latch the data, this direct combinational path will exist for some finite period of time. In addition, the signal from S1 to S2 may go through a very short propagation path. If the clock does not drop in time, input data to the S1 latch may not only be latched in S1 but may reach S2 and be latched into S2 a clock period earlier than intended. Hence, as illustrated in waveform A the short propagation path can cause unpredictable results. Waveform C illustrates the opposite problem. The next clock pulse appears before new data reaches S2. For optimum performance it is necessary that the clock cycle be as short as possible, but it must not be shorter than the propagation time through combinational logic.

The use of the double latch design can eliminate the situation in waveform A. To resolve this problem, LSSD imposes restrictions on the clocking of latches. We first list the rules and then discuss their effect on the circuit of Figure 7.21.

> Latches are controlled by two or more nonoverlapping clocks such that a latch X may feed the data port of another latch Y if and only if the clock that sets the data into latch Y does not clock latch X.
>
> A latch X may gate a clock C_1 to produce a gated clock C_2 which drives another latch Y if and only if clock C_3 does not clock latch X, where C_3 is any clock produced from C_1.
>
> It must be possible to identify a set of clock primary inputs from which the clock inputs to SRLs are controlled either through simple powering trees or through logic that is gated by SRLs and/or nonclock primary inputs.
>
> All clock inputs to all SRLs must be at their off states when all clock primary inputs are held to their off states.
>
> The clock signal that appears at any clock input of an SRL must be controlled from one or more clock primary inputs such that it is possible

7.5 THE SCAN PATH

to set the clock input of the SRL to an on state by turning any one of the corresponding primary inputs to its on state and also setting the required gating condition from SRLs and/or nonclock primary inputs.

No clock can be ANDed with either the true value or the complement value of another clock.

Clock primary inputs may not feed the data inputs to latches, either directly or through combinational logic, but may only feed the clock input to the latches or the primary outputs.

The first LSSD design rule forbids the configuration shown in Figure 7.21. A simple way to comply with the rules is to use both the L1 and L2 latches and control them with nonoverlapping clocks as shown in Figure 7.22. Then the situation illustrated in waveform A will not occur. The contents of the L2 latch cannot change in response to new data at its input as long as the B clock remains low. Therefore, the new data entering the L1 latch of SRL S1, as a result of clock C being high, cannot get through its L2 latch, because the B clock is low, and hence cannot reach the input of SRL S2. The input to S2 remains stable and is latched by the C clock.

The use of nonoverlapping clocks will protect a design from problems caused by short propagation paths. However, the time between the fall of clock C and the rise of clock B is "dead time"; that is, we want the clocks to be nonoverlapping but once the data is latched into L1, we want to get it into L2 as quickly as possible in order to realize maximum performance. Hence, we want the interval between the fall of C and the rise of B to be as brief as possible without, however, making the time too short. In a chip with a great many wire paths, the two clocks may be nonoverlapping at the I/O pins and yet may overlap at one or more SRLs inside the chip due to signal path delays. This condition is referred to as "clock skew." When debugging a design, experimentation with clock edge separation can help to determine whether clock skew is causing problems. If clock skew problems exist, then it may be necessary to change the layout of a chip or board, or it may require permanently separating clock edges by a greater amount to resolve the problem.

The designer must still be concerned with the configuration in waveform C; that is, the clock cycle must exceed the propagation delay in the longest propagation path. However, it is a relatively straightforward task to compute propagation delays along combinational logic paths. Timing verification, as described in Chap. 4, can

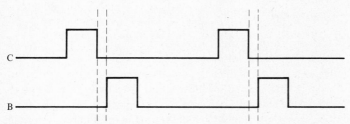

Figure 7.22 The two-clock signal.

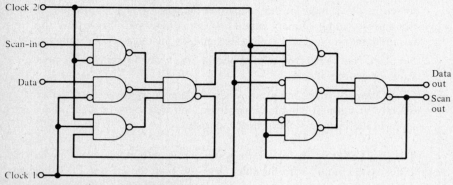

Figure 7.23 Flip-flop with dual clock.

be employed to compute the delay along each path and print out all critical paths which exceed some specified threshold. The design team can elect to redesign the critical paths or increase the clock cycle.

Test program development using the LSSD scan path closely parallels the technique used with other scan paths. One interesting variant when testing is the fact that the scan path itself can be checked with a so-called flush test[13]. In a flush test the *A* and *B* clocks are both set high. This creates a direct combinational path from the scan-in to the scan-out. It is then possible to apply a logic 1 and 0 to the scan-in and observe them directly without further exercising the clocks. This flush test exercises a significant portion of the scan path. This is followed by clocking 1s and 0s through the scan path to insure that the clock lines are fault-free.

Another significant feature of LSSD, as implemented, is the fact that it is an integral part of a design automation system[14]. As a part of the design automation system, it affords an opportunity to check the design for compliance with design rules. Violations detected by the checking programs can be corrected before the design is fabricated, thus insuring that design violations will not compromise the testability goals that were the object of the LSSD rules.

7.5.2 Other Scan Methods

In its basic form the serial scan path connects all operational flip-flops into a serial/parallel shift register. There are, in addition to LSSD, other variations of the basic serial scan path concept and there are random scan methods which scan data into and out of flip-flops without employing serial shift capability.

The Dual Clock Serial Scan Another implementation of the serial scan with dual clocks is shown in Figure 7.23[15]. In this implementation clock 1 is used in the normal operational mode and clock 2 is the test or scan clock. In normal operation, when clock 1 is low, system data can be loaded into the first latch. When it goes high, data is transferred to the second latch and the first latch is inhibited from receiving data. In the test mode, the operational data input is disabled by virtue of clock 1 being held high. Clock 2 then clocks data in at the scan-in input. A low

7.5 THE SCAN PATH

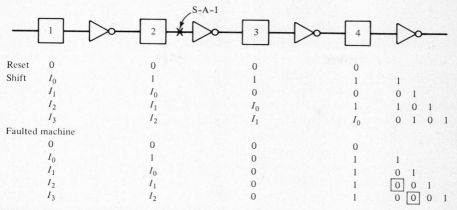

Figure 7.24 Diagnosing scan path fault.

on the clock 2 input causes test data to be latched into latch 1 and a high clock signal then latches data into latch 2 and disables latch 1.

By inserting inverters in the scan path, in conjunction with a reset line, it is possible to improve diagnostic capability[16]. This is illustrated in Figure 7.24. When the reset is applied, all flip-flops contain the logic value 0. After shifting, the output for a fault-free shift path is a sequence of 1s and 0s. If a fault occurs in the shift path, as shown in the figure, where we assume the output of flip-flop number 2 is S-A-1, the alternating string of 1s and 0s is disrupted.

Addressable Registers Improved controllability and observability of sequential elements can be obtained through the use of addressable registers[17]. Here, again, the intent is to gain access and control of sequential memory elements in a circuit. This approach uses x and y "address lines" (see Figure 7.25). Each latch has an x address and a y address. The latches also have clear and preset as well as the

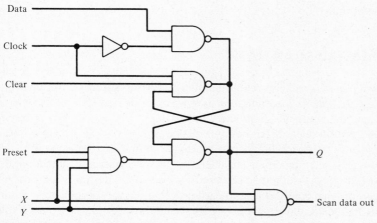

Figure 7.25 Addressable flip-flop.

usual clock and data lines. A scan address goes to x and y decoders for the purpose of generating the x and y signals which select a latch to be loaded. The latch is then preset by setting the address lines and then pulsing the preset line.

Readout of data is also accomplished by means of the x and y addresses. The selected latch is gated to the SDO (scan data out), from which it can be observed. If there are more address lines decoded than are necessary to observe flip-flops, then x and y addresses can be used to observe nodes in combinational logic. The node to be observed is input to a NAND gate along with x and y signals, just as a latch would be, and when selected its value appears at SDO.

The addressable latch technique requires just a few gates for each storage element. During normal operation its effect on operation is almost negligible, due mainly to loading caused by the NAND gate attached to the Q output. The scan address could require several I/O pins, but it could also be generated internally by a counter which is reset initially and then clocked through consecutive addresses to permit loading or reading of the flip-flops.

The random access scan is attactive because of its negligible effect on IC performance and "real estate." However, with shrinking component size the amount of area taken by interconnections inside an IC grows more significant; it represents a larger percentage of total chip area. The addressable latches require that several signal lines be routed to each addressable latch and the chip area occupied by these signal lines may become a relevant factor when assessing the various methods.

A significant difference between LSSD and other scan methods is that the LSSD concept makes it almost impossible to design untestable configurations, and is backed up by design rule enforcement incorporated into a design automation system. The addressable latches and non-LSSD techniques do not, in and of themselves, inhibit some design practices which traditionally have caused problems for ATPGs. They require design discipline imposed either by the designers themselves or by some designated testability supervisor. By requiring that designs be entered into a design data base via design automation programs that can check for rule violations, as is the case with LSSD, it becomes difficult to incorporate design violations without concurrence of the very people who are ultimately responsible for testing the design.

7.6 THE PARTIAL SCAN PATH

The use of scan provides excellent controllability and observability. Unfortunately, it is not always feasible to incorporate it into a design. It requires that designers have control over the entire design process, including design and fabrication of the ICs used in the end product. Then they can dictate the use of scan inside the ICs. However, standard, off-the-shelf LSI parts generally do not incorporate scan. This becomes an increasingly important consideration as more standard functions are absorbed into LSI chips, which then become available to the design community at a small fraction of what it would cost to design the equivalent function. Economic considerations frequently dictate the use of these standard, off-the-shelf parts, and this in turn negates the possibility of using the scan techniques. Memory chips pose

7.6 THE PARTIAL SCAN PATH

similar problems. Whereas designers would once dedicate entire boards to memory, now in many applications it is possible to get all of the needed memory onto part of a board. The unused board space is available for use by other logic. Testing strategies for memories and random logic must now somehow coexist.

When testing printed circuit boards that use off-the-shelf components, a partial scan path may help test the circuitry surrounding the LSI parts. It can range from use of scan for everything except a few LSI parts to use of scan for just a few of the more troublesome circuits, such as status registers, counters, and configurations of flip-flops which require long, tedious sequences to test in the operational mode. Testability analysis tools such as SCOAP can help to determine where the partial scan would be most effective.

One example of partial scan is called Scan/Set[18]. It is a less formal technique than the other methods previously described. Scan/Set provides parallel/serial flip-flops which can be loaded and read out via a scan path, but the registers are separate from the functional logic. They therefore have somewhat less effect on the performance of the functional logic. The Set feature, which loads operational flip-flops from the Scan/Set flip-flops, is used only for flip-flops judged to be difficult to control. Multiplexers are also used to route signals to output pins. Several internal points can be observed by the multiplexer by suitably controlling the multiplexer control lines. Ad-hoc design rules exist as part of the system. These rules both prohibit certain design practices and help to select nodes to be scanned or set.

Some vendor parts are appearing with provisions for test. One such example is the diagnostic pipeline register (DPR)[19]. This circuit, shown in Figure 7.26, contains an 8-bit *shadow register* which operates in parallel with the 8-bit pipeline register used for normal operations. The shadow register can be serially loaded from the SDI input. Then the pipeline register is parallel-loaded from the shadow register. The contents of the pipeline register can be observed by first parallel-loading them into the shadow register, then serially shifting out through the SDO pin. The diagnostic clock (DCLK) is used to serially shift data through the shadow register. The SDI and SDO pins can be connected together to allow n of these devices to provide a single signal path of width $8 \cdot n$. When $\overline{OEY}$ is high, the shadow register can be loaded with information from a bus and thus allow system information to be captured and scanned out for analysis. This can be useful in isolating problems after an error has been detected on the bus. Another feature of the DPR is the ability to write data back into a writable control store. The Data inputs (D_7–D_0) of the DPR are normally driven by control store in a microprogrammable architecture. In a writable control store the flip-flop can enable the shadow register contents onto the Data inputs so that diagnostic data or programs can be loaded into control store.

When partial scan is used, questions must still be answered. What is to be done with LSI devices that are not included in the scan path? Can they be fenced off, that is, made transparent to the ATPG? Do we create simulation and test models that an ATPG can process? The cost of creating gate-level models for complex LSI devices can be an unacceptable financial drain. Can partial models be made which emulate the real device to some extent, and thus verify at least part of the functionality of the chip and, hopefully, detect faults that are most likely to occur?

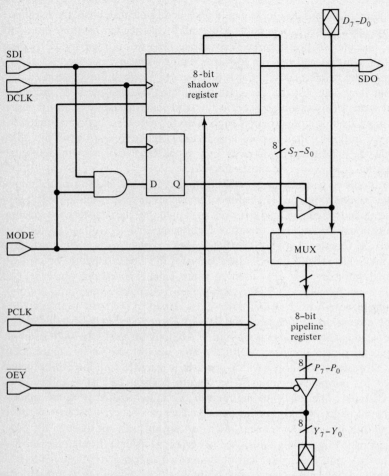

Figure 7.26 Diagnostic pipeline register. (Copyright © 1983 Advanced Micro Devices, Inc. Reprinted with permission of copyright owner. All rights reserved.)

Answers to these questions carry significant implications. For example, a complete scan path obviates the need for an ATPG with sequential test pattern generation capability. A partial scan path which excludes some sequential elements but leaves others in a circuit that is processed by an ATPG implies that the ATPG must still have processing capability for sequential circuits; hence part of the benefit of scan is lost.

If there are just a few LSI devices that cannot be processed, the best approach may be to fence off the devices. They must be designed into the circuit in such a way that they can be controlled; their behavior must be inhibited from disrupting other tests. If an LSI device which connects to a bus is not modeled, then it must be possible to force it to the high-impedance state and to hold it there for the duration of the test. If a microprocessor is the only device that can generate addresses for a memory device on the board, there must be some alternative method for generating addresses when the microprocessor is disabled. If the microprocessor address lines go to a memory address register, that register can be connected into a scan path. Control signals from the microprocessor may be important for exercising other logic on a board. A microprocessor may provide a "valid memory address" without which the memory will not function. That signal and other such signals must be replaced with some suitable substitute signal.

If an LSI device is excluded from the circuit model, a dummy model of the LSI device may be used in its place. This was shown in Figure 5.19, where signals from the microprocessor were replaced by signals from primary inputs used strictly for test purposes. The primary inputs can be replaced with a scan path so that any number of signals can be substituted at the cost of only three I/O pins.

When partial scan is used, careful attention must be given to clock controls. It must be possible to shift data into and out of the scan path without clocking those sequential devices that were excluded from the scan path. If the nonscannable devices are clocked while data is being scanned in, their contents become unpredictable.

7.7 SUMMARY

Logic designers continue to make circuits smaller and faster, sometimes with the aid of design practices that produce untestable circuits. For the large mainframes of a former era it was, apparently, more cost-effective to design for maximum performance and keep a field engineer nearby. However, cost of hardware continues to decrease while cost of the field engineer increases. In little more than two decades we have progressed from huge systems that occupy large rooms to the point where we now have "throw-away" computers that sit on top of a desk. These desktop computers present problems quite different from the large mainframe. It is not practical to send a field engineer to the site of a failed desktop computer. Furthermore, it is a very competitive industry. It is necessary to detect, diagnose, and correct problems quickly, accurately, and economically in a mass production environment. Even with throw-away computers we only want to throw away the bad ones.

The problems of designing testable logic have their parallel in software development, where it was recognized years earlier that complex systems, developed by people with a diverse range of skills and styles, will result in chaos if maintainability is ignored until after the product is designed and developed. In either case, software or hardware design, it is becoming more widely recognized and accepted that the designer must ask, before pencil is put to paper on the first design document, "How am I going to diagnose the problems *when* this thing fails?" For the software engineer the answer is structured design. For the logic designer the answer must be design-for-testability. This requires that the designer understand testability issues and be able to anticipate testability problems in the design.

With a growing trend toward gate arrays and semicustom ICs, the need for testability grows. For custom ICs with sales volume in the hundreds of thousands (or millions), expensive test programs can be amortized at negligible unit cost. However, the gate arrays and semicustom ICs tend to be low-volume devices, hence cost becomes a greater percentage of product cost.

Many of the concepts discussed in this chapter are based on common sense. A testable design is one in which individual circuits can be exercised and observed without requiring elaborate sequences of inputs. Testability analysis algorithms can provide an indication of the degree of difficulty in testing a circuit but, while experience indicates that they can be effective, they can also be deceptive and should be interpreted accordingly. Numbers of occurrences of particular circuit components can cause a circuit to be rated difficult to test when, in fact, the circuit may be easily tested. Another circuit with only a single problem configuration, and a testability score indicating that it is easy to test, may be impossible to test because the effects of that single untestable configuration are so pervasive as to adversely affect testability of the entire circuit.

Design-for-testability techniques, such as the use of scan path, were mentioned as early as December 1963[20] and again in April 1964[21]. A scan path and its proposed use for testability and operational modes are described in a patent filed in 1968[22]. Discussion of scan path and derivation of a formal cost model were published in 1973[23]. What is new about the DFT techniques is the attention they are getting.

The reluctance to make concessions to testability can be attributed, at least in part, to an emphasis on designing for maximum performance and minimum feature size, but it also resulted, in no small part, from customer tolerance of extended downtime and a lack of understanding on the part of the designer as to what practices cause designs to be untestable.

These factors are changing. Customers are less tolerant of extended mean time to repair and project managers are beginning to realize that the true cost of a digital product is expressed as:

$$\text{Cost(Design + Test)} \leq \text{Cost(Design)} + \text{Cost(Test)}$$

This equation simply says that product cost is best minimized by regarding the design and test of a product as one integral activity rather than two disjoint, unrelated activities. When they are treated as separate issues, relationships become obscured. Decisions are made on the basis of their impact on number of I/O pins, amount of board real estate taken up, and number of nanoseconds impact on performance, without considering the impact on cost of test development, cost of test application, mean time to repair, scrapped units, rework, retest, and loss of customer goodwill.

PROBLEMS

7.1 Given a four-input AND gate with CC^0 and CC^1 numbers $\{1/1, 1/1, 3/8, \infty/\infty\}$ on its inputs, compute the controllability and observability numbers at its output.

7.2 Derive the controllability/observability equations for the two-input NAND gate.

REFERENCES

7.3 Compute the controllability/observability numbers for the exclusive-OR, using the NAND gate version of the EXOR shown in Figure 2.16(d).

7.4 Derive combinational controllability/observability equations for the circuit whose truth table is given below. Use the equations to compute the controllability/observability numbers.

X_1	X_2	X_3	F
0	0	0	0
0	0	1	0
0	1	0	0
0	1	1	1
1	0	0	1
1	0	1	1
1	1	0	0
1	1	1	0

7.5 For the delay flip-flop discussed in the text, derive the observability equations for the Data input.

7.6 Given a set of two-input AND gates connected as a binary tree with output F. For a tree of depth k in which the inputs are equidistant from the output (same number of nodes between each input and the output), show that:

$$CC^1(F) = 2^{k+1} - 1$$
$$CC^0(F) = k + 1$$

7.7 For the circuit in Figure 7.18, find a test for the middle input of gate 5 S-A-1. Assume the existence of a master reset which initially resets all DFFs to 0 (S0).

7.8 For the scan path of Figure 7.27, if the output of the second flip-flop is S-A-1, and the scan path contents are scanned out after a reset is applied, what is the resulting output? State a general rule for diagnosing faults in the scan path.

REFERENCES

1. Williams, T. W., and K. P. Parker, "Design for Testability—A Survey," *IEEE Trans. Comput.*, vol. C-31, no. 1, Jan. 1982, pp. 2–15.
2. Williams, T. W., and L. H. Goldstein, eds., Joint Special Issue on Design for Testability, *IEEE Trans. Comput.*, vol. C-30, no. 11, Nov. 1981, and *IEEE Trans. Circuits Syst.*, vol. CAS-28, no. 11, Nov. 1981.
3. Anderson, K. R., and H. A. Perkins, eds., Special Issue on Testing, *Computer*, vol. 12, no. 10, Oct. 1979.
4. "Designing Digital Circuits for Testability," Hewlett-Packard Application Note 210-4, Hewlett Packard, Loveland, CO 80537.
5. Goldstein, L. H., "Controllability/Observability Analysis of Digital Circuits," *IEEE Trans. Comput.*, vol. CAS-26, no. 9, Sept. 1979, pp. 685–693.
6. Powell, T., "Software Gauges the Testability of Computer-Designed ICs," *Electron. Des.*, Nov. 24, 1983, pp. 149–154.
7. Fong, J. Y. O., "On Functional Controllability and Observability Analysis," *Proc. 1982 Int. Test Conf.*, Nov. 1982, pp. 170–175.

8. Goel, D. K., and R. M. McDermott, "An Interactive Testability Analysis Program—ITTAP," *Proc. 19th Design Automation Conf.*, 1982, pp. 581–586.
9. Savir, J., "Good Controllability and Observability Do Not Guarantee Good Testability," *IEEE Trans. Comput.*, vol. C-32, no. 12, Dec. 1983, pp. 1198–1200.
10. Agrawal, V. D., and M. R. Mercer, "Testability Measures—What Do They Tell Us?", *Proc. 1982 Int. Test Conf.*, Nov. 1982, pp. 391–396.
11. LASAR User's Manual, Teradyne Corp., Boston, MA 02111.
12. Eichelberger, E. B., and T. W. Williams, "A Logic Design Structure for LSI Testability," *Proc. 14th Design Automation Conf.*, June 1977, pp. 462–468.
13. Bottorff, P. S., et al., "Test Generation for Large Logic Networks," *Proc. 14th Design Automation Conf.*, June 1977, pp. 479–485.
14. Godoy, H. C., et al., "Automatic Checking of Logic Design Structures for Compliance with Testability Ground Rules," *Proc. 14th Design Automation Conf.*, June 1977, pp. 469–478.
15. Funatsu, S., et al., "Test Generation Systems in Japan," *Proc. 12th Design Automation Conf.*, June 1975, pp. 114–122.
16. Miller, H. W., " 'Design for Test' via Standardized Design and Display Techniques," *Electron. Test*, vol. 6, no. 10, Oct. 1983, pp. 108–116.
17. Ando, H., "Testing VLSI with Random Access Scan," *Digest CompCon. 1980*, Feb. 1980, pp. 50–52.
18. Stewart, J. H., "Future Testing of Large LSI Circuit Cards," *Proc. 1977 Cherry Hill Test Conf.*, Oct. 1977, pp. 6–17.
19. Lee, F., et al., "On-Chip Circuitry Reveals System's Logic States," *Electron. Des.*, April 14, 1983, pp. 119–123.
20. Maling, K., and E. L. Allen, "A Computer Organization and Programming System for Automated Maintenance," *IEEE Trans. Electron Comput.*, vol. EC-12, Dec. 1963, pp. 887–895.
21. Carter, W. C., et al., "Design of Serviceability Features for the IBM System/360," *IBM J. Res. Dev.*, vol. 8, April 1964, pp. 115–126.
22. Hirtle, A. C., et al., "Data Processing System Having Auxiliary Register Storage," U.S. Patent 3,582,902, filed Dec. 30, 1968.
23. Williams, M. J. Y., and J. B. Angell, "Enhancing Testability of Large-Scale Integrated Circuits via Test Points and Additional Logic," *IEEE Trans. Comput.*, vol. C-22, no. 1, Jan. 1973, pp. 46–60.

Chapter 8

Memory System Design and Test

8.1 INTRODUCTION

Memories possess a high degree of regularity. That makes it easy to devise algorithms to test them. However, because of the growing capacity of memories, some of the tests may run for prohibitively long periods of time. A significant problem then, in testing memories, is to identify the kinds of faults that may occur and apply the most efficient test for the faults of interest.

Because of shrinking cell size there is a greater incidence of soft errors. These are intermittent errors caused by, among other things, alpha particles. The term soft error refers to the fact that the error is not easily repeatable and the conditions leading up to its occurrence cannot be duplicated. Therefore a specific test to detect its presence does not exist, in contrast to permanent or hard errors, for which tests can be created. Soft errors must be dealt with by means of error-correcting codes (ECCs), also called error detection and correction (EDAC) codes. We will look at hard faults, tests devised to detect these faults, and error-correcting codes used to recover from the effects of soft errors.

8.2 SEMICONDUCTOR MEMORY ORGANIZATION

Memories fall into several categories, according to the following properties:

- Volatile or nonvolatile
- Static or dynamic
- Serial or parallel access
- Destructive or nondestructive readout

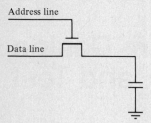

Figure 8.1 Dynamic memory cell.

A *volatile* memory is one which cannot retain information when power is removed. Semiconductor memories fall into this category. Memories which can retain information after power is removed, such as magnetic cores, magnetic tapes, disks, and bubble memories, are termed *nonvolatile*. Some memory devices may lose information despite the presence of a constant power source; these devices are called *dynamic* memories. As illustrated in Figure 8.1, they are basically capacitances, in which the charge can leak away over time. The memory system must employ refresh circuitry which periodically reads the cells and writes back a suitably amplified version of the signal. A memory in which the cells can retain their information without the need for refresh is called a *static* memory. Such cells are typically flip-flops, which, with their two stable states, can remain in a given state indefinitely.

Memories can be classified according to whether they employ *serial* or *random* access. Serial memories are those in which data can be accessed only in a fixed, predetermined sequence. Magnetic tape units are an example of serial access. To read a record it is necessary to read the entire tape up to the point where the desired data exists. A random access memory (RAM) permits reading of data at any specific location without first reading other data. If the contents of a memory device are destroyed by a read operation, then it is classified as a *destructive readout* (DRO); otherwise it is a *nondestructive readout* (NDRO) device.

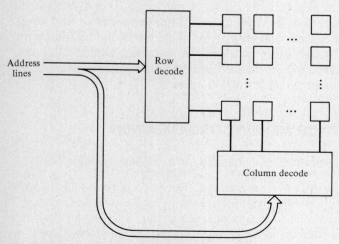

Figure 8.2 Semiconductor memory organization.

Semiconductor memories usually employ an organization called *2-D*. In this organization a $2^m \times 1$ memory with m address lines is actually organized into a matrix with 2^r rows and 2^c columns, where $r + c = m$. The address lines are split up into two groups such that r lines go to a row decoder and c lines go to a column decoder. This is illustrated in Figure 8.2. A specific cell is activated for a read or write operation when both its row select and column select lines have been activated.

8.3 MEMORY FAULTS

A number of different failure types can occur in semiconductor memories, affecting memory cell contents, cell addressing, and the time required to read out data. Some of the more common failures include the following[1]:

Cell opens or shorts

Address nonuniqueness

Cell/column/row disturb sensitivity

Sense amplifier interaction

Slow access time

Slow write recovery

Data sensitivity

Refresh sensitivity

Static data losses

Opens and shorts within semiconductor memory cells may occur because of faulty processing, including misaligned masks or imperfect metallization. These failures are characterized by a general randomness in their nature. Opens and shorts may also occur at the chip connections to a printed circuit board. In a $km \times n$ memory system containing km words of n bits each, and made up of memory chips of size $m \times 1$, a fault that exists in bit position i of m consecutive bits often indicates either a totally failed chip or one in which there is an open or short between the chip and the circuit board on which it is mounted.

Address nonuniqueness results from address decoder failures, which may cause the same memory cell to be accessed by several different addresses or several cells to be accessed by a single address. These failures often cause some cells to be physically inaccessible. An effective test must insure that each read or write operation accesses one, and only one, memory cell.

Disturb sensitivity between adjacent cells or between cells in the same row or column can result from capacitive coupling. Slow access time can be caused by slow decoders, overloaded sense amplifiers, or an excessive capacitive charge on output circuits. Slow write recovery may indicate a saturated sense amplifier which cannot recover from a write operation in time to perform a subsequent read operation.

A memory cell can be affected by the contents of neighboring cells. Worse still, the cell may be affected only by particular combinations on neighboring cells. This problem grows more serious as the distance between neighboring cells diminishes. Refresh sensitivity in dynamic RAMs may be induced by a combination of data sensitivity and temperature or voltage fluctuations. Static RAM cells are normally able to retain their state indefinitely. However, they may lose data due to leakage current or opens in resistors or feedback paths.

The memory defects just described can be grouped into three broad categories: address faults, cell faults, and parametrics. The breakdown of faults in each of these categories is estimated to be[2]:

40%—address fault
40%—cell failure
20%—parametrics

8.4 MEMORY TEST PATTERNS

Memory tests fall into two categories: functional and dynamic. The *functional test* attempts to find defects within a memory cell or failures wherein cell contents are altered by a read or write of another cell. A *dynamic test* attempts to find access time failures. The *all-1s* or *all-0s* tests are examples of functional tests. These tests write 1s or 0s into all memory cells in order to detect individual cell defects including shorts and opens. However, these tests are not effective at finding other failure types.

A memory test pattern that tests for address nonuniqueness and other functional faults in memories, as well as some dynamic faults, is the *Galpat*, sometimes referred to as a ping-pong pattern. This pattern accesses each address repeatedly, using, at some point, every other cell as a previous address. It writes a background of zeros into all memory cells. Then the first cell becomes the test cell. It is complemented and read alternately with every other cell in memory. Each succeeding cell then becomes the test cell in turn and the entire read process is repeated. All data is complemented and the entire test is repeated. If each read and compare is counted as one operation, then Galpat has an execution time proportional to $4N^2$, where N is the number of cells. It is effective for finding cell opens, shorts, address uniqueness faults, and sense amplifier interaction and access time problems. The algorithm is as follows:

```
1    FOR e = 0 TO 1          ;GALPAT FOR N x 1 MEMORY
2      FOR i = 0 TO N-1
3        a(i) = e             ;CREATE BACKGROUND OF ALL e (0 or 1)
4      NEXT i

5      FOR i = 0 TO N-1
6        a(i) = ē
7        FOR j = 0 TO N-1
8          IF j ≠ i THEN READ a(j) ELSE GOTO 10
```

8.4 MEMORY TEST PATTERNS

```
9          If a(j) ≠ e THEN RECORD ERROR
10         READ a(i)
11         If a(i) ≠ ē THEN RECORD ERROR
12       NEXT j
13     NEXT i
14  NEXT e                    ;PERFORM TEST FOR BACKGROUND OF
                              ; -- ALL 0's AND ALL 1's
```

The *Walking Pattern* is similar to the Galpat except that the test cell is read once and then all other cells are read. To create a Walking Pattern test from the Galpat program, steps 10 and 11 are omitted. The Walking Pattern has an execution time proportional to $2N^2$. It checks memory for cell opens and shorts and address uniqueness.

The *March*, like most of the algorithms, begins by writing a background of zeros. Then it reads the data at the first location and writes a 1 to that address. It continues this read/write procedure sequentially with each address in memory. When the end of memory is reached, each cell is read and changed back to zero in reverse order. The test is then repeated using complemented data. Execution time is of order N. It can find cell opens, shorts, address uniqueness, and some cell interactions.

The *Galloping Diagonal* is similar to Galpat in that a 1 is moved through memory. However, it is moved diagonally, checking both row and column decoders simultaneously. It is of order $4N^{3/2}$. Row and column Galpats of order $4N^{3/2}$ also exist.

The *Sliding Diagonal* writes a complete diagonal of 1s against a background of 0s and then, after reading all memory cells, shifts the diagonal horizontally. This continues until the diagonal of 1s has passed through all memory locations. The Diagonal test, of order N, will verify address uniqueness at a significant speed enhancement over the Walk or Galpat.

The *Checkerboard* writes 1s and 0s into alternate memory locations in a checkerboard pattern. After a time delay, which may be several seconds, the pattern is read from memory. This pattern is used to evaluate data retention in static RAMs.

Surround Read Disturb starts by creating a background of all 0s. Then each cell in turn becomes the test cell. The test cell is complemented and the eight physically adjacent cells are repeatedly read. After some number of iterations the

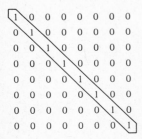

Figure 8.3 The diagonal test.

test cell is read to determine whether it has been affected by the read of its neighbors. Then the operation is repeated for a background of 1s. The intent is to find disturbances caused by adjacent cell operations. Execution time depends on the number of read cycles but is on the order of N. The *Surround Write Disturb* is identical to the Surround Read Disturb except that a write rather than a read is performed.

The *Write Recovery* writes a background of 0s. Then the first cell is established as the test cell. A 1 is written into the second cell and the first (test) cell is read. The second cell is restored to 0 and the test cell is read again. This is repeated for the test cell and every other cell. Every cell then becomes the test cell in turn. The entire process is repeated using complemented data. This is an N^2 test which is directed at write recovery type faults. It also detects faults that are detected by Galpat.

Address Test writes a unique value into each memory location. Typically, this could be the address of that memory cell; that is, the value n is written into memory location n. After writing all memory locations, the data is read back. The purpose of this test is to check for address uniqueness. This algorithm requires that the number of bits in each memory word equals or exceeds the number of address bits.

The *Moving Inversions* test[3] inverts a memory filled with 0s to 1s and conversely. After initially filling the memory with 0s, a word is read. Then a single bit is changed to a 1, and the word is read again. This is repeated until all bits in the word are set to 1 and then repeated for every word in memory. The operation is then reversed, setting bits to 0 and working from high memory to low memory.

For a memory with n address bits the process is repeated n times. However, on each repetition, a different bit of the address is taken as the least significant bit for incrementing through all possible addresses. An overflow generates an end around carry so all addresses are generated but the method increments through addresses by 1s, 2s, 4s, and so on. For example, on the second time through, bit 1 (when regarding bit 0 as least significant bit, LSB) is treated as the LSB so all even addresses are generated out to the end of memory. After incrementing to address 111...110, the next address generated is address 000...001, and then all consecutive odd addresses are generated out to the end of memory. The pattern of memory address generation for the second iteration is as follows:

$$0...0000$$
$$0...0010$$
$$0...0100$$
$$0...0110$$
$$...$$
$$...$$
$$...$$
$$1...1001$$
$$1...1011$$
$$1...1101$$
$$1...1111$$

8.5 REPAIRABLE MEMORIES

The Moving Inversions test pattern has $12BN \log_2 N$ patterns, where B is the number of bits in a memory word. It detects addressing failures and cell opens and shorts. It is also good at checking access times.

8.5 REPAIRABLE MEMORIES

With the growing number of cells in each memory IC, and with dimensions shrinking, yield becomes a more critical problem. Shrinking dimensions permit more ICs on an individual wafer, and with smaller die size and all other factors held constant, the probability of a defect of a given size on a die decreases. However, with smaller feature sizes, incorrect operation can be caused by mask imperfections or pinhole defects that might not have caused errors in a die having larger dimensions. These dice are normally discarded. However, the objective of memory manufacturing is to maximize the number of correctly behaving dice obtained from each wafer, and a significant portion of the faulty dice contain only a few faulty memory cells, so

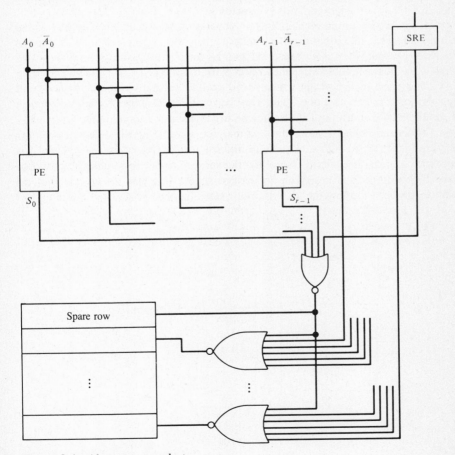

Figure 8.4 Alternate row select.

why not repair them[4]? Because of the regular structure of RAM chips, extra rows and columns can be added to improve yield. During test, if a memory chip is discovered to have just a few bad cells, then a spare row(s) or column(s) is substituted for the faulty row(s) or column(s) containing the bad cell(s) to create a good memory chip.

The substitution of a row or column for another one is achieved by means of fuses. In Figure 8.4 the SRE (spare row enable) signal is normally held high so the output of the spare row NOR decoder is held low. If the spare row is to be used, then an SRE fuse is blown, which enables the spare row NOR. If the spare row NOR is selected, it disables all other NORs. There is a programming element (PE) for each row address line. The PEs, shown in Figure 8.5, determine the row address to be selected. During test, if it is determined that the die can be repaired by substituting a spare row for a row with failed cells, then the SRE signal is activated by blowing a fuse. The address of the failed row is then programmed into the corresponding PE. If the fuse in the PE is blown, the output S_i is connected to $\bar{A}_i$ because T_1 is enabled through transistor D. if the fuse is not blown, then the transistor T_2 is activated and S_i is connected to A_i.

Each programming element has a unique address. If the address fuse in that PE is addressed, its output enables transistor P and a large current flows through the fuse, causing it to open. These PE address lines and V_{DP} are accessible on the die but are not accessible after the chip has been packaged.

The spare row concept can also be applied to spare column replacement. Furthermore, more than one spare row and column can be provided. Practical considerations usually limit the spares to two rows and two columns, since additional rows and columns cause die size to grow, countering the objective of maximizing yield. When a row or column has replaced another row or column, it is necessary to retest the die to insure that the substitute row or column is not defective. In addition, it is necessary to verify that the fuse has blown and that the mechanism used to blow the fuse has not caused damage to surrounding circuitry.

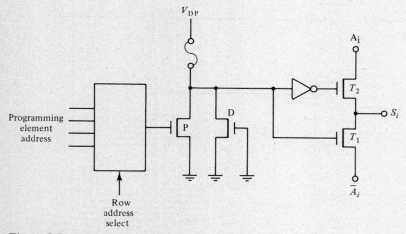

Figure 8.5 Programming element.

8.6 ERROR-CORRECTING CODES

There appears to be negligible effect on memory access time due to rows or columns being substituted. The presence of the additional transistor, T_1 or T_2, causes about 8% increase in access time. There is concern that redundant rows and columns may adversely affect memory tests intended to uncover disturb sensitivities, but comparison of test data between devices with and without redundancies showed no significant differences[4].

8.6 ERROR-CORRECTING CODES

In 1948 Claude Shannon published his classic paper entitled "The Mathematical Theory of Communication"[5]. In that paper he proved the following theorem:

Theorem 8.1. Let a source have entropy H (bits per symbol) and a channel have a capacity C (bits per second). Then it is possible to encode the output of the source in such a way as to transmit at the average rate $(C/H) - e$ symbols per second over the channel, where e is arbitrarily small. It is not possible to transmit at an average rate greater than C/H.

This theorem asserts the existence of codes which permit transmission of data through a noisy medium with arbitrarily small error rate at the receiver. The alternative, when transmitting through a noisy medium, is to increase transmission power to overcome the effects of noise. An important problem in data transmission is that of minimizing the frequency of occurrence of errors at a receiver with the most economical mix of transmitter power and data encoding.

An analogous situation exists with semiconductor memories. They continue to shrink in size; hence error rates increase due to adjacent cell disturbance caused by the close proximity of cells to one another. Errors can also be caused by reduced charge densities[6]. Helium nuclei from impurities found in the semiconductor packaging materials can migrate toward the charge area and neutralize enough of the charge in a memory cell to cause a logic 1 to be changed to a 0. These soft errors, intermittent in nature, are growing more prevalent as chip densities increase. One solution is to employ a parity bit with each memory word to aid in the detection of memory bit errors. A single parity bit can detect any odd number of bit errors. Detection of a parity error, if caused by a soft error, may necessitate reloading of a program and/or data area.

If memory errors entail serious consequences, then the alternatives are to use more reliable memories or, as in communications, employ error-correcting codes or some combination of the two alternatives to reach a desired level of reliability at an acceptable cost. Since Shannon's paper in 1948, many families of error-correcting codes have been discovered. In memory systems the Hamming codes have proved to be popular.

8.6.1 Vector Spaces

An understanding of Hamming codes requires some understanding of vector spaces, so we introduce some definitions. A *vector* is an ordered *n*-tuple containing

n elements called *scalars*. In this discussion, the scalars will be restricted to the values 0 and 1. Addition of two vectors is on an element-by-element basis, for instance

$$v_1 + v_2 = (v_{11}, v_{12}, \ldots, v_{1n}) + (v_{21}, v_{22}, \ldots, v_{2n})$$
$$= (v_{11} + v_{21}, v_{12} + v_{22}, \ldots, v_{1n} + v_{2n})$$

The addition operation, denoted by $+$, will be the mod 2 operation (exclusive-OR) in which carries are ignored.

EXAMPLE. ■

If $v_1 = (0, 1, 1, 0)$ and $v_2 = (1, 1, 0, 0)$,
then $v_1 + v_2 = (0 + 1, 1 + 1, 1 + 0, 0 + 0) = (1, 0, 1, 0)$. ■ ■

Multiplication of a scalar $a \in \{0, 1\}$ and a vector v_1 is defined by

$$av_1 = (av_{11}, av_{12}, \ldots, av_{1n})$$

The *inner product* of two vectors v_1 and v_2 is defined as

$$v_1 \cdot v_2 = (v_{11}, v_{12}, \ldots, v_{1n}) \cdot (v_{21}, v_{22}, \ldots, v_{2n})$$
$$= (v_{11} \cdot v_{21} + v_{12} \cdot v_{22} + \cdots + v_{1n} \cdot v_{2n})$$

If the inner product of two vectors is 0, they are said to be *orthogonal*.

A *vector space* is a set V of vectors which satisfy the property that all linear combinations of vectors contained in V are themselves contained in V, where the linear combination u of the vectors $v_1, v_2, \ldots, v_n$ is defined as

$$u = a_1 v_1 + a_2 v_2 + \cdots + a_n v_n \qquad a_i \in \{0, 1\}$$

The following additional properties must be satisfied by a vector space:

1. If $v_1, v_2 \in V$, then $v_1 + v_2 \in V$
2. $(v_1 + v_2) + v_3 = v_1 + (v_2 + v_3)$
3. $v_1 + e = v_1$ for some $e \in V$
4. For $v_1 \in V$, there exists v_2 such that $v_1 + v_2 = e$
5. The product $a \cdot v_1$ is defined for all $v_1 \in V, a \in \{0, 1\}$
6. $a(v_1 + v_2) = av_1 + av_2$
7. $(a + b)v_1 = av_1 + bv_1$
8. $(ab)v_1 = a(bv_1)$

A set of vectors $v_1, v_2, \ldots, v_n$ is *linearly dependent* if there exist scalars $c_1, c_2, \ldots, c_n$, not all zero, such that

$$c_1 v_1 + c_2 v_2 + \cdots + c_n v_n = 0$$

If the vectors $v_1, v_2, \ldots, v_n$ are not linearly dependent, then they are said to be *linearly independent*.

8.6 ERROR-CORRECTING CODES

Given a set of vectors S contained in V, the set $L(S)$ of all linear combinations of vectors of S is called the *linear span* of S. If the set of vectors S is linearly independent, and if $L(S) = V$, then the set S is a *basis* of V. The number of vectors in S is called the *dimension* of V.

A subset U contained in V is a subspace of V if $u_1, u_2 \in U$ implies that $c_1 v_1 + c_2 v_2 \in U$ for $c_1, c_2 \in \{0, 1\}$.

The following four theorems follow from the above definitions:

Theorem 8.2. The set of all n-tuples orthogonal to a subspace V_1 of n-tuples forms a subspace V_2 of n-tuples. This subspace V_2 is called the *null space* of V_1.

Theorem 8.3. If a vector is orthogonal to every vector of a set which spans V_1, it is in the null space of V_1.

Theorem 8.4. If the dimension of a subspace of n-tuples is k, the dimension of the null space is $n - k$.

Theorem 8.5. If V_2 is a subspace of n-tuples and V_1 is the null space of V_2, then V_2 is the null space of V_1.

EXAMPLE ■

The vectors in the following matrix, called the *generator matrix* of V, are linearly independent. They form a basis for a vector space of 16 elements.

$$G = \begin{bmatrix} 1 & 0 & 0 & 0 & 0 & 1 & 1 \\ 0 & 1 & 0 & 0 & 1 & 0 & 1 \\ 0 & 0 & 1 & 0 & 1 & 1 & 0 \\ 0 & 0 & 0 & 1 & 1 & 1 & 1 \end{bmatrix} \tag{8.1}$$

The dimension of the subspace defined by the vectors is 4. The vectors 0111100, 1011010, 1101001 are orthogonal to all of the vectors in G, hence they are in the null space of G. Furthermore, they are linearly independent, so they define the following generator matrix H for the null space of V:

$$H = \begin{bmatrix} 0 & 1 & 1 & 1 & 1 & 0 & 0 \\ 1 & 0 & 1 & 1 & 0 & 1 & 0 \\ 1 & 1 & 0 & 1 & 0 & 0 & 1 \end{bmatrix} \tag{8.2}$$

■ ■

8.6.2 The Hamming Codes

From Theorem 8.5 we see that a vector space can be defined in terms of its generator matrix G or in terms of the generator matrix H for its null space. Since a vector $v \in V$ is orthogonal to every vector in the null space, it follows that

$$vH^T = 0 \tag{8.3}$$

where H^T is the transpose of H.

We define the *Hamming weight* of a vector v as the number of nonzero components in the vector. The *Hamming distance* between two vectors is the number of positions in which they differ. In the vector space generated by the matrix G in Eq. (8.1), the nonzero vectors all have Hamming weights that equal or exceed three. This follows from Eq. (8.3), where the vector v selects columns of H which sum, mod 2, to the 0 vector. Since no column of H contains all zeros, and no two columns are identical, v must select at least three columns of H in order to sum to the 0 vector.

We now represent a set of binary information bits by the vector $J = (j_1, j_2, \ldots, j_k)$. If G is a $k \times n$ matrix, then the product $J \cdot G$ encodes the information bits by selecting and creating linear combinations of rows of G corresponding to nonzero elements in J. Each information vector is mapped into a unique vector in the space V defined by the generator matrix G. Furthermore, if the columns of the generator matrix H of the null space are all nonzero, and if no two columns of H are identical, then the encoding produces code words with minimum Hamming weight equal to three. Since the sum of any two vectors is also contained in the space, the Hamming distance between any two vectors must be at least three. Therefore, if one or two bits are in error, it is possible to detect the fact that the encoded word has been altered.

If we represent an encoded vector as v and an error vector as e, then

$$(v + e)H^T = vH^T + eH^T = eH^T$$

If e represents a single-bit error, then the product eH^T matches the column of H corresponding to the bit in e which is nonzero.

EXAMPLE ■

Using the matrix G in Eq. (8.1), if $J = (1, 0, 1, 0)$, then $v = J \cdot G = (1, 0, 1, 0, 1, 0, 1)$. If $e = (0, 0, 0, 1, 0, 0, 0)$, then

$$v + e = (1, 0, 1, 1, 1, 0, 1)$$

so,
$$(v + e)H^T = (1, 0, 1, 1, 1, 0, 1) \begin{bmatrix} 0 & 1 & 0 \\ 0 & 1 & 1 \\ 0 & 0 & 1 \\ 1 & 1 & 1 \\ 1 & 0 & 1 \\ 1 & 0 & 0 \\ 1 & 1 & 0 \end{bmatrix} = (1, 1, 1)$$

The product matches the fourth column of H, hence we conclude that the fourth bit of the message vector is in error. Since the information bits are binary, it is a simple matter to invert the fourth bit to get the original vector $(1, 0, 1, 0, 1, 0, 1)$. ■■

Note that in this encoding the first four columns of G form the identity matrix; hence when we multiply J and G, the first four elements of the resulting vector match the original information vector. Such a code is called a *systematic* code. In

8.6 ERROR-CORRECTING CODES

the general case the columns of G could be permuted so that the columns which make up the identity matrix could appear anywhere in the matrix. The systematic code is convenient for use with memories since it permits data to be stored in memory exactly as it exists outside memory. A general form for G and H, as systematic codes, is

$$G = [I_k : P_{kx(n-k)}]$$
$$H = [P^T_{(n-k)xk} : I_{(n-k)}]$$

where I_n is the identity matrix of dimension n, the parameter k represents the number of information bits, n is the number of bits in the encoded vector, and $n - k$ is the number of parity bits. The matrix P is called the *parity* matrix, the generator matrix H is called the *parity check* matrix, and the product $v \cdot H^T$ is called the *syndrome*. When constructing an error-correcting code, parameters n and k must satisfy the expression $2^{n-k} - 1 \geq n$.

Error-correcting codes employ *maximum likelihood* decoding. This simply says that if the syndrome is nonzero, the code vector is mapped into the most likely message vector. In the code described above, if the syndrome is $(1, 1, 1)$, it is assumed that bit 4 of the vector is in error. But notice that the 2-bit error $e = (1, 0, 0, 0, 1, 0, 0)$ could have produced the same syndrome. This can cause a false correction because maximum likelihood decoding assumes that one error is more probable than two errors; that is, if P_i is the probability that the ith bit is received correctly, then $P_i > Q_i = 1 - P_i$, where Q_i is the probability of receiving the incorrect bit.

To avoid the possibility of an incorrect "correction," an additional bit can be added to the code vectors. This bit is an even parity check on all of the preceding bits. The parity matrix P for the preceding example now becomes:

$$P = \begin{bmatrix} 0 & 1 & 1 & 1 \\ 1 & 0 & 1 & 1 \\ 1 & 1 & 0 & 1 \\ 1 & 1 & 1 & 0 \end{bmatrix}$$

Since the information vectors must now be even parity, any odd number of errors can be detected. The decoding rule is:

1. If the syndrome is 0, assume no error has occurred.
2. If the last bit of the syndrome is one, assume a single-bit error has occurred; the remaining bits of the syndrome will match the column vector in H corresponding to the error.
3. If the last bit of the syndrome is zero, but other syndrome bits are one, an uncorrectable error has occurred.

In case 3, an even number of errors has occurred; consequently it is beyond the correcting capability of the code. An error bit may be set when that situation is detected, or, in a computer memory system, an uncorrectable error may trigger an interrupt so that the operating system can take corrective action.

8.6.3 ECC Implementation

An ECC encoder circuit must create parity check bits based on the information bits to be encoded and the generator matrix G to be implemented. Consider the information vector $J = (j_1, j_2, \ldots, j_k)$ and $G = [I_k : P_{k \times r}]$, where $r = n - k$ and

$$P_{k \cdot r} = \begin{bmatrix} p_{11} & p_{12} & \cdots & p_{1r} \\ p_{21} & p_{22} & \cdots & p_{2r} \\ & & \cdots & \\ p_{k1} & p_{k2} & \cdots & p_{kr} \end{bmatrix}$$

In the product $J \cdot G$, the first k bits remain unchanged. However, the $(k + s)$th bit, $1 \leq s \leq r$, becomes

$$g_s = j_1 \cdot p_{1s} + j_2 \cdot p_{2s} + \cdots + j_k \cdot p_{ks}$$

$$= \sum_{m=1}^{k} j_m \cdot p_{ms}$$

Therefore, in an implementation, the $(k + s)$th symbol is a parity check on information bits corresponding to nonzero elements in the sth column of P.

The encoded vector is decoded by multiplying it with the parity generator H to compute the syndrome. This gives:

$$(v + e) \cdot H^T = (v_1, v_2, \ldots, v_n) \cdot \begin{bmatrix} P_{k \cdot r} \\ I_r \end{bmatrix} + e \cdot \begin{bmatrix} P_{k \cdot r} \\ I_r \end{bmatrix}$$

$$= (j_1, j_2, \ldots, j_k, p_1, p_2, \ldots, p_r) \cdot \begin{bmatrix} P_{k \cdot r} \\ I_r \end{bmatrix} + e \cdot \begin{bmatrix} P_{k \cdot r} \\ I_r \end{bmatrix}$$

Therefore, to decode the vector, encode the information bits as before, and then exclusive-OR them with the parity bits to produce a syndrome. Use the syndrome to correct the data bits. If the syndrome is 0, no corrective action is required. If the error is correctable, use the syndrome with a decoder to select the data bit that is to be inverted. The correction circuit is illustrated in Figure 8.6. With suitable control circuitry, the same syndrome generator can be used to generate the syndrome bits.

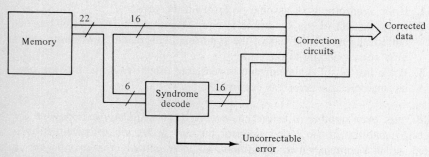

Figure 8.6 Error correction circuit.

8.6 ERROR-CORRECTING CODES

Error-correcting codes have been designed into memory systems with word widths as great as 64 bits[7] and have been designed into 4-bit-wide memories and implemented directly on-chip[8]. Since the number of additional bits in an SEC-DED (single error correcting-double error detecting) Hamming code with a 2^n-bit word is $n + 2$, the additional bits as a percentage of data word width decrease with increasing memory width. For a 4-bit memory, 3 bits are needed for SEC and 4 bits for SEC-DED. For a 64-bit memory, SEC requires 7 bits and SEC-DED requires 8 bits.

8.6.4 Reliability Improvements

The improvement in memory reliability provided by ECCs can be expressed as the ratio of the probability of a single error in a memory system without ECC to the probability of a double error in a memory with ECC[9]. Let $R = e^{-\lambda t}$ be the probability of a single memory device operating correctly, where λ is the failure rate of a single memory device. Then the probability of the device failing is

$$Q = 1 - R = 1 - e^{-\lambda t}$$

Given m devices, the binomial expansion yields

$$(Q + R)^m = R^m + mR^{m-1}Q + \cdots + Q^m$$

Hence, the probability of all devices operating correctly is R^m and the probability of one failure is $P_1 = mR^{m-1}Q$. The probability of two errors in a memory with $m + k$ bits is

$$P_2 = \frac{(m + k)(m + k - 1)R^{(m+k-2)}}{2}(1 - R)^2$$

The improvement ratio is

$$R_i = \frac{p_1}{p_2} = \frac{2m}{(m + k)(m + k - 1)} \times \frac{1}{R^{(k-1)}(1 - R)}$$

EXAMPLE ■

Using an SEC-DED for a memory of 32-bit width requires 7 parity bits. If $\lambda = 0.1\%$ per thousand hours, then after 1000 hours

$$R = 0.9990005$$
$$1 - R = 0.0009995$$
$$R_i = \frac{2 \times 32}{39 \times 38} \times \frac{1}{0.9940 \times 0.0009995} = 43.5 \quad ■■$$

If we compute the reliability at $t = 10{,}000$ hours, we find that $R_i = 3.5$. This is interpreted to mean that the likelihood of a single chip failure increases with time. Therefore the likelihood of a second, uncorrectable, error increases with time. Consequently, maintenance intervals should be scheduled to locate and replace failed devices in order to hold the reliability at an acceptable level. Also note that reliability is inversely proportional to memory word width. As word size increases, the number of parity bits as a percentage of memory decreases, hence reliability also decreases.

The equations for reliability improvement were developed for the case of permanent bit-line failures; that is, the bit position fails for every word of memory where it is assumed that one chip contains bit i for every word of memory. Data on 4K RAMs shows that 75–80% of the RAM failures are single-bit errors[10]. Other errors, such as row or column failure, may also affect only part of a memory chip. In the case of soft errors or partial chip failure, the probability of a second failure in conjunction with the first is more remote. The reliability improvement figures may therefore be regarded as lower bounds on reliability improvement.

When should ECC be employed? The answer to this question depends on the application and the extent to which it can tolerate memory bit failures. ECC requires extra memory bits and logic and introduces extra delay in a memory cycle; furthermore, it is not a cure for all memory problems since it cannot correct address line failures and, in memories where data can be stored as bytes or half-words, use of ECC can complicate data storage circuits. Therefore, it should not be used unless a clear-cut need has been established. To determine the frequency of errors, the mean time between failures (MTBF) can be used. The equation is

$$\text{MTBF} = 1/d\lambda$$

where λ is again the failure rate and d is the number of devices. Reliability numbers for MTBF for a single memory chip depend on the technology and the memory size, but may lie in the range 0.01 to 0.2% per thousand hours. A 64K × 8 memory using eight 64K RAM chips with 0.1% per thousand hours would have an MTBF of 125,000 hours. A much larger memory, such as 1 megaword, 32 bits per word, using the same chips would have an MTBF of 2000 hours, or about 80 days between hard failures. Such failure rates may be acceptable, but the frequency of occurrence of soft errors may still be intolerable.

Other factors may also make ECC attractive. For example, on a board populated with many chips, the probability of an open or short between two IC pins increases. ECC can protect against many of those errors. If memory is on a separate board from the CPU, it may be a good practice to put the ECC circuits on the CPU board so that errors resulting from bus problems, including noise pickup and open or high-resistance contacts, can be corrected. A drawback of this approach is the fact that the bus width must be expanded to accommodate the ECC parity bits.

It is possible to achieve error correction beyond the number of errors predicted to be correctable by the minimum distance. Suppose hard errors are logged as they are detected. Then, if a double error is detected, and if one of the two errors had been previously detected and logged in a register, the effects of that error can be removed from the syndrome corresponding to the double error to create a syndrome for the error which had not been previously detected. Then the syndrome for the remaining error can be used to correct for its effect.

Another technique that can be used when a double error is detected is to complement the word read out of memory and store that complement back into

8.7 SUMMARY

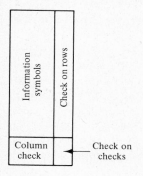

Figure 8.7 Magnetic tape with check bits.

memory. Then read the complemented word. The bit positions corresponding to hard cell failures will be the same, but bits from properly functioning cells will be complemented. Therefore, exclusive-OR the data word and its complement to locate the failed cells, correct the word, and then store the corrected word back in memory. This will not work if two soft errors occurred; at least one of the two errors must be a hard error[11]. This technique can also be used in conjunction with a parity bit to correct hard errors[12]. In either case, whether a single-bit parity error or a double error detected by ECC, the correction procedure can be implemented by having the memory system generate an interrupt whenever an uncorrectable error occurs. A recovery routine residing either in the operating system or in microcode can then be activated to correct bit positions corresponding to hard errors.

8.6.5 Iterated Codes

The use of parity bits on rows and columns of magnetic tapes, Figure 8.7, constitutes an SEC-DEC code[13]. The minimum Hamming weight of the information plus check bits will always be at least four. In addition, a single-bit error in any position complements a row parity bit, a column parity bit, and the check-on-checks parity bit. Therefore, it is possible to correct single-bit errors and detect double-bit errors.

8.7 SUMMARY

Memories must be tested for functional faults, including cells stuck-at-1 or stuck-at-0, addressing failures, and read or write activities which disturb other cells. Memories must also be tested for dynamic faults which may result in excessive delay in performing a read or write. A shipment of memory ICs received from a vendor frequently come from the same lot and were manufactured at the same time. If a sample of the ICs are tested and exhibit certain marginal characteristics, it may be that all of the chips are marginal in that particular characteristic since they were probably all exposed to the same manufacturing processes. It will be desirable to run a thorough test on the shipment. A problem in this regard, with the growing size of memory chips, is to select the algorithm that tests for the characteristics of interest, without consuming excessive test time.

With increasing numbers of memory cells per IC and smaller dimensions, the possibility of failure, both hard and soft, increases. When failure is detected during wafer processing, it is possible to substitute another row and/or column for the row or column in which the failure occurred, if spare rows and/or columns are provided. Recovery from errors when the memory chips are in operation can be achieved through the use of ECCs. Analysis of the problem indicates that significant improvements in reliability can be achieved with the use of ECCs. With memory chips continuing to achieve increased densities, it can be expected that ECCs will gain in popularity.

PROBLEMS

8.1 Given a 64K RAM in which test patterns can be applied at the rate of one every 250 ns. Compute the test time required to apply:

Galpat

Galloping Diagonal

Moving Inversions (assume $B = 32$)

8.2 Create the (8, 4) SEC-DED matrix for the following generator matrix G:

$$G = \begin{bmatrix} 1 & 0 & 0 & 0 & 0 & 1 & 1 \\ 0 & 1 & 0 & 0 & 1 & 1 & 0 \\ 0 & 0 & 1 & 0 & 1 & 0 & 1 \\ 0 & 0 & 0 & 1 & 1 & 1 & 1 \end{bmatrix}$$

8.3 Create the parity check matrix H corresponding to the generator matrix G of the previous problem.

8.4 Using the (8, 4) parity check matrix H of the previous problem, determine which of the following vectors are code vectors and which have errors that are (a) correctable, (b) detectable.

$$\begin{array}{cccccccc} 1 & 0 & 0 & 1 & 0 & 0 & 0 & 1 \\ 0 & 1 & 1 & 1 & 1 & 1 & 1 & 0 \\ 1 & 0 & 1 & 0 & 1 & 0 & 1 & 0 \\ 1 & 0 & 1 & 0 & 1 & 1 & 0 & 0 \\ 1 & 1 & 1 & 0 & 0 & 0 & 0 & 1 \\ 0 & 1 & 0 & 1 & 1 & 1 & 0 & 1 \end{array}$$

8.5 If it is known that bit 3 of the code words has been identified as a solid S-A-0, use that information and the matrix H previously given to correct the following vectors:

$$\begin{array}{cccccccc} 0 & 1 & 0 & 0 & 0 & 1 & 1 & 1 \\ 0 & 1 & 0 & 1 & 1 & 0 & 1 & 0 \\ 1 & 0 & 0 & 1 & 0 & 0 & 1 & 0 \end{array}$$

8.6 For an SEC-DED code, the decoding rules were given for three conditions of the syndrome. However, nothing was said about the condition where the last bit of the syndrome is one, but all other bits are 0. What would you do in that case?

8.7 Prove that the inequality $2^{n-k} - 1 \geq n$ must hold for Hamming codes.

8.8 Prove Theorems 8.2 through 8.5.

8.9 Write memory test programs for the following test algorithms:

> Galloping Diagonal
> Checkerboard
> Moving Inversions

8.10 A register set with 16 registers has an S-A-0 fault on address line A_2 (There are four lines, A_0–A_3). Pick any memory test algorithm that can detect addressing errors and explain, in detail, how it will detect the fault on A_2.

REFERENCES

1. Application Note, "Standard Patterns for Testing Memories," *Electron. Test,* vol. 4, no. 4, April 1981, pp. 22–24.
2. Nowicki, Edwina, "VLSI Board Designs Demand Enhanced Testing Capabilities," *Electron. Test,* April 1984, pp. 72–77.
3. de Jonge, J. H., and A. J. Smulders, "Moving Inversions Test Pattern Is Thorough, Yet Speedy," *Comput. Des.,* vol. 15, no. 5, May 1976, pp. 169–173.
4. Altnether, J. P., and R. W. Stensland, "Testing Redundant Memories," *Electron. Test,* vol. 6, no. 5, May 1983, pp. 66–76.
5. Shannon, C. E., "The Mathematical Theory of Communication," *Bell Syst. Tech. J.,* vol. 27, July and October, 1948.
6. May, T. C., and M. H. Woods, "Alpha-Particle-Induced Soft Errors in Dynamic Memories," *IEEE Trans. Electron. Dev.,* vol. ED-26, no. 1, Jan. 1979, pp. 2–9.
7. Bossen, D. C., and M. Y. Hsiao, "A System Solution to the Memory Soft Error Problem," *IBM J. Res. Dev.,* vol. 24, no. 3, May 1980, pp. 390–398.
8. Khan, A., "Fast RAM Corrects Errors on Chip," *Electronics,* Sept. 8, 1983, pp. 126–130.
9. Levine, L., and W. Meyers, "Semiconductor Memory Reliability with Error Detecting and Correcting Codes," *Computer,* vol. 9, no. 10, Oct. 1976, pp. 43–50.
10. Palfi, T. L., MOS Memory System Reliability," *Proc. IEEE Semiconductor Test Symp.,* 1975.
11. Travis, B., "IC's and Semiconductors," *EDN,* Dec. 17, 1982, pp. 40–46.
12. Wolfe, C. F., "Bit Slice Processors Come to Mainframe Design," *Electronics,* Feb. 28, 1980, pp. 118–123.
13. Peterson, W. W., *Error Correcting Codes,* chap. 5, MIT Press, Cambridge, Mass., 1961.

Chapter 9
Self-Test and Fault Tolerance

9.1 INTRODUCTION

In previous chapters we looked at a number of methods used to create tests for digital logic circuits. The algorithms and heuristic processes can be traced back to the very beginning of the digital logic era. These methods, for the most part, have proved inadequate to the task. With increasing levels of integration, entire processors now exist in a single chip; entire systems exist on a single board.

New methods for testing and verifying physical integrity are being researched and developed. We saw in Chap. 7 that concessions to testability are gaining acceptance. However, even with design-for-testability guidelines, testing difficulties remain. Circuits continue to grow in both size and complexity. When operating at higher clock rates they are susceptible to performance errors that are not well modeled with stuck-at fault models. As a result, there is increasing concern about both the cost and the effectiveness of tests. The problem is compounded by the fact that it is increasingly necessary to develop test strategies for devices when description of internal structure is unavailable.

There is a growing requirement to develop improved test methods for use at a customer's site where test equipment is not readily accessible or where the environment cannot be easily duplicated, as in military avionics subject to high gravity stresses. This has led to the concept of built-in-test (BIT), wherein test circuits are placed directly within the product being designed. They are closer to the functions they must test, hence they have greater control and observability. They can test the device in its normal operating environment, at its intended operating speed, and can therefore detect parametric failures that occur during operation. The BIT can also permit remote access to a device via telephone lines so that a field engineer may already have a good idea as to what problems exist before he responds to a

9.2 SYSTEM TESTING

service request. BIT can reduce the costly investment in test equipment. However, any circuit added during the design of a product adds cost to each unit manufactured. Therefore, if reduction in test development costs is the main motive in considering BIT, cost tradeoff studies must be conducted. Furthermore, it requires either a comprehensive understanding by design engineers of test strategies and goals, or a close alliance between design engineers and test engineers.

In addition to self-test, we can expect to see increasing use of fault tolerance in large, expensive machines or in those where reliable operation is critical. Like BIT, factors that must be considered include cost tradeoffs and availability of design engineers knowledgeable in strategies and goals relating to fault tolerance.

9.2 SYSTEM TESTING

We will first consider functional level testing of systems ranging from microprocessors to large mainframes. In each instance the test strategy is directed toward developing a test for a system composed of functional entities without necessarily having detailed knowledge of the physical structure of the system. Within that context, it is necessary to develop a test program that is thorough while at the same time effective at diagnosing fault locations.

9.2.1 The Ordering Relation

A typical central processor unit (CPU) is illustrated in Figure 9.1. The figure is also typical of the amount of information provided by manufacturers of micro-

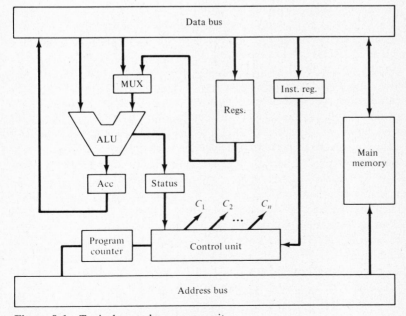

Figure 9.1 Typical central processor unit.

processor devices. It displays a register stack, a control section, an arithmetic logic unit (ALU), a status register, a data bus, an instruction register, and a program counter. Additional information provided by the manufacturer includes a description of the instruction set and the number of machine cycles required to execute the instructions, and signal behavior at I/O pins.

We saw previously (Chap. 5) that one approach to testing a device, when a structural level description is unavailable, is to create a gate equivalent circuit. That, however, is time-consuming and labor-intensive. There is no assurance that automatic test pattern generation methods will work and, furthermore, no assurance that fault simulation of test patterns, automatic or manually generated, will provide an accurate estimate of fault coverage. The difficulty in creating gate-level models has led to a growing interest in the use of functional models. One such approach, described by Robach et al.[1], first partitions a system into macroblocks, which are high-level functional entities such as CPUs, memory systems, interrupt processors, I/O devices, and control sections. The macroblocks are then partitioned into smaller microblocks. Testing is organized at the microblock level; hence it is very detailed and takes into account the characteristics of the individual microcircuits. The objective is to obtain a comprehensive test for the microblock while using the macroblocks to route test information to observable outputs. When testing the microblocks in a given macroblock, all other macroblocks are assumed to be fault-free. Furthermore, the microblocks within a given macroblock are ordered in such a way that a microblock is tested only through modules already tested.

Before discussing the partitioning techniques for microblocks and macroblocks, we discuss the concept of hardcore. Hardcore refers to those circuits used to test a processor, and both first-degree and second-degree hardcore are defined. The *first-degree hardcore* is that hardcore used exclusively for testing. It is verified independently of a processor's normal operational features and is then used to test the operational logic. Examples of first-degree hardcore include such things as a read-only memory (ROM) dedicated to test which is loaded via a special access path not used by operational logic, a dedicated comparator for evaluating results, and watchdog timers which are used to verify that peripherals attached to I/O ports respond within some specified time. A given device may or may not have first-degree hardcore. If it does, then the test strategy dictates that it be tested first. The *second-degree hardcore* is that part of the operational hardware used in conjunction with first-degree hardcore to perform test functions. Examples of this include the writable control store (WCS) used by test microprograms to exercise other operational units as well as the control circuitry and access paths of the WCS.

After first-degree hardcore has been verified, the second-degree hardcore is verified. Then macroblocks are selected for testing. These are chosen such that a macroblock to be tested does not depend for its test on another macroblock which has not yet been tested. Individual microblocks within a chosen macroblock are selected for testing, again with the requirement that microblocks be tested only through other, previously tested, microblocks. To achieve this, two ordering relations are defined. The controllability relation ρ_1 is defined by

$$A \cdot \rho_1 \cdot B \Leftrightarrow A \text{ can be controlled through } B$$

9.2 SYSTEM TESTING

The observability relation ρ_2 is defined by

$$A \cdot \rho_2 \cdot B \Leftrightarrow A \text{ can be observed through } B$$

With these two relations, a priority partial order $\geq$ is defined such that

$$\text{If } B \cdot \rho_1 \cdot a \text{ and } B \cdot \rho_2 \cdot b, \text{ then } B \geq a \cdot b$$

In words, a test of B must follow the test of a AND b, or stated another way, a AND b precedes B. In effect, if B is controlled through a and observed through b, then a and b must both be tested before B is tested. However, it may be that two devices C and D have the property that $C \geq D$ and $D \geq C$. In that case $A \equiv B$ and A and B are said to be *indistinguishable*. This would be the case, for example, if two devices were connected in series and could not possibly be tested individually.

After a complete ordering has been established, the microblocks are partitioned into layers such that each microblock is tested only through microblocks contained in previous layers. A microblock B is contained in a layer L_k if and only if

1. B follows at least one element of L_{k-1}.
2. All elements that precede B are contained in the union $\bigcup_{i=0}^{k-1} L_i$

The layer L_0 is the hardcore; it is directly controllable and observable.

To assist in ordering microblocks, a tree is formed as illustrated in Figure 9.2. In that figure, the dot ($\cdot$) represents the AND operator and the plus ($+$) represents the OR operator. Therefore $B \geq C \cdot D + E \cdot F$ states that the test of B must follow either the test of C AND D, OR it must follow the test of E AND F. In this graph, if an element occurs twice on the graph, with elements in between, then an indistinguishability block is defined that contains all elements joining the two occurrences of the element.

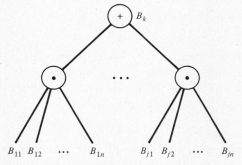

Figure 9.2 Ordering tree.

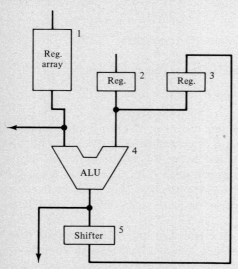

Figure 9.3 ALU circuit.

EXAMPLE ■

We illustrate the ordering algorithm by means of the circuit in Figure 9.3. In the circuit, the various elements are assigned numbers to identify them during the discussion that follows. We first identify the ρ_1 and ρ_2 relations:

Controlled by	Observed through
$1\,\rho_1\,0$	$1\,\rho_2\,0$
$2\,\rho_1\,0$	$2\,\rho_2\,4$
$3\,\rho_1\,5$	$3\,\rho_2\,4$
$4\,\rho_1\,1\cdot(2+3)$	$4\,\rho_2\,0$
$5\,\rho_1\,4$	$5\,\rho_2\,3$

From these relations we obtain the following ordering relations:

$$1 \geq 0$$
$$2 \geq 4$$
$$3 \geq 4 \cdot 5$$
$$4 \geq 1 \cdot 2 + 1 \cdot 3$$
$$5 \geq 3 \cdot 4$$

These relations in turn lead to the tree shown in Figure 9.4.

From the graph we see that 1 does not follow any other microblock, therefore we place it in layer L_1. It is also evident from the ordering relations that $2 \geq 4$ and $4 \geq 2$. That can also be seen from the ordering tree. This

9.2 SYSTEM TESTING

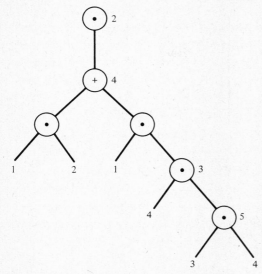

Figure 9.4 Ordering tree for ALU circuit.

implies an indistinguishability between 2 and 4. Therefore, we form a new block $b_1 = \{2, 4\}$ and replace both 2 and 4 by this new block b_1. We get:

$$b_1 \geq b_1$$
$$3 \geq b_1 \cdot 5$$
$$b_1 \geq b_1$$
$$b_1 \geq 1 \cdot b_1 + 1 \cdot 3$$
$$5 \geq b_1 \cdot 3$$

We can apply two reduction properties to the fourth relation; these are:

$$c \geq c \cdot d + e \quad \text{implies} \quad c \geq d + e \tag{r1}$$

$$f \geq g + g \cdot h \quad \text{implies} \quad f \geq g \tag{r2}$$

After applying these properties, the fourth relation becomes:

$$b_1 \geq 1$$

The first and third relations ($b_1 \geq b_1$) are tautologies, so we can eliminate them and the ordering relations become:

$$3 \geq b_1 \cdot 5$$
$$b_1 \geq 1$$
$$5 \geq b_1 \cdot 3$$

A microblock can be put in the present layer only if it appears solely on the right-hand side of the relation operator or if, whenever it appears on the left, the microblocks on the right are in lower-level layers. In this case, microblock 1 is in a lower layer and in all other relations b_1 occurs only on the right. Therefore, it can be put in L_2. If we now let $b_2 = \{3, 5\}$, we get:

$$b_2 \geq b_1 \cdot b_2$$
$$b_1 \geq 1$$
$$b_2 \geq b_1 \cdot b_2$$

The first and third relations can be reduced using reduction relation (r1) to give:

$$b_2 \geq b_1$$
$$b_1 \geq 1$$
$$b_2 \geq b_1$$

Then, b_2 can be placed in L_3.

We have now placed all microblocks into layers. Since the register array is in layer L_1, it should be tested first. This corresponds with the fact, seen in the diagram, that there is a separate path into and out of the register array. After it has been tested, it can be used when testing the ALU and the register denoted as component 2, which were grouped together as indistinguishability block b_1. Finally, the shifter and the register denoted as component 5 can be tested. ■ ■

Ordering a large number of microblocks within a macroblock can be tedious and time-consuming, and indistinguishability classes may become too complex. These complex classes indicate areas in which additional hardware can be used to good advantage to improve controllability and observability.

9.2.2 The Microprocessor Matrix

The microprocessor generally absorbs a great deal of logic into a single IC. There may not be enough information to permit the ordering of microblocks within a macroblock. An alternate strategy[2] employs a matrix which relates the individual instructions within the microprocessor to the physical entities such as registers, ALU, condition code register, and I/O pins. A row of the matrix is assigned for each instruction. Several categories of columns are assigned; these include:

1. Data movement: register-register, register-memory, memory-immediate, and so on
2. Operation type: AND, OR, COMPLEMENT, SHIFT, MOVE, ADD, SUBTRACT, MULTIPLY
3. I/O pins involved: data, address, control
4. Clock cycles involved: a column for each clock cycle
5. Condition codes affected: carry, overflow, sign, zero, parity, and so on

If the ith instruction uses, affects, or is characterized by the property corresponding to column j, then there is a 1 in the matrix at the intersection of the ith row and jth column. A weight is assigned to each instruction by summing the

9.2 SYSTEM TESTING

number of 1s in the row corresponding to that instruction. Another matrix is created in which the rows represent functional units. Columns record such information as the number of instructions using the unit and any other physical information that is available, possibly including: number of gate levels, number of feedback paths, and number of clocks in the unit. Note that the number of instructions that use a given unit may in many cases be a reasonable approximation to an ordering, in the sense in which an ordering was described in the previous subsection.

The test strategy requires that the functional units which have the lowest weight be tested first. Furthermore, the units should be tested using, as often as possible, the instructions of minimum weight. The objective is to obtain a set of programs $\{P_i\}$ which test all of the functional units. Each program P_i has a weight which is equal to the sum of the weights of the individual instructions that make up the program. Because a given program may test two or more functional units, in the sense that two or more units are indistinguishable as defined in the previous subsection, a covering problem exists. The objective, therefore, given a set $\{P_i\}$ of programs which test all of the functional units, is to select the set of programs of minimum weight which cover (test) all functional units. A minimal weight test has the advantage that it can usually be applied more quickly, requires less memory, and reduces the likelihood that a fault will mask symptoms of another fault.

9.2.3 Graph Methods

The LAMP (logic analyzer for maintenance planning)[3] system developed at Bell Laboratories is a comprehensive system containing a number of simulators, test pattern generation programs, and a testability analysis subsystem called COMET (controllability, observability, and maintenance engineering technique)[4] which orders functional units based on controllability/observability dependences. The ordering allows a circuit to be represented graphically. In the graph, nodes represent functional units and edges represent the controllability and/or observability relations. A directed arc emanating at node P and terminating on node Q indicates that functional unit P controls functional unit Q and/or that P is observed through Q. A loop in the resultant directed graph indicates that there is a complex dependence (indistinguishability) among the functional units. Such loops indicate places where testability can be enhanced with the use of additional circuitry to improve controllability or observability, in effect, breaking up the loop.

Graphs can also be used to relate functional units of microprocessors to the instructions which use them[5]. In the method to be described here, we focus our attention on register transfers and their relationship to the instructions that effect these transfers.

The set $R = \{R1, R2, \ldots, Rn\}$ denotes the registers used in the microprocessor, including the program counter and any scratchpad registers not accessible to the programmer. The registers will be represented by nodes in the graph. The set $I = \{I1, I2, \ldots, In\}$ denotes the instruction repertoire of the microprocessor. Data flow during the execution of instruction Ij is represented by means of a directed edge from node Rp to node Rq if data transfer occurs, with or without transformation, from register Rp to register Rq during execution of instruction Ij.

If we denote an I/O port used for input as IN, and an I/O port used for output as OUT, and if we assign graph nodes to IN and OUT, then a directed arc exists from IN to Rp (from Rp to OUT) if data transfer occurs, with or without transformation, from main memory or an I/O port to register Rp (from register Rp to main memory or an I/O port). Instruction Ij may appear several times in the graph since it could affect several registers. For example, an instruction such as

```
        MOV      M1(I1),M2(I2),COUNT
```

might be used to move a number of bytes specified by the argument COUNT from memory location M2, indexed by I2, to memory location M1, indexed by I1. For each byte moved, its source location and destination location in memory must be computed and then the byte must be moved. Some operations can be performed in parallel, others must be performed in correct sequence. In the MOV instruction, the source and destination addresses can be incremented in parallel.

In the discussion that follows, instructions will occasionally be referred to as T, M, or B. The T indicates an instruction which causes data transfer between registers and storage. Instructions which manipulate arguments, such as the arithmetic and logic instructions, are denoted by M, and the branch instructions are denoted by B.

EXAMPLE ■

We will use the following hypothetical microprocessor to illustrate the graph method.

```
REGISTERS
   R1       – Program Counter         (PC)   16 bits
   R2       – Accumulator             (AC)    8 bits
   R3       – Scratchpad register     (AUX)  16 bits
   R4       – Index register          (IX)   16 bits
   R5       – Stack pointer           (SP)   16 bits
   R6       – Condition Code register (CC)    8 bits
                Carry
                Overflow
                Sign
                Zero
                Parity

INSTRUCTIONS
   I1      CALL     TOS  ← PC
                    PC   ← Immed. Addr.
                    SP   ← SP - 2
   I2      RETURN   SP   ← SP + 2
                    PC   ← TOS
   I3      JUMP     PC   ← Immed. Addr.
   I4      JR       PC   ← PC + Disp.
   I5      LJC      PC   ← CC0 · <Immed. Addr.> ∨ $\overline{CC0}$ · <PC + 3>
   I6      SJC      PC   ← CC0 · <PC + Disp.> ∨ $\overline{CC0}$ · <PC + 2>
   I7      PUSH     TOS  ← AC
                    SP   ← SP - 1
```

9.2 SYSTEM TESTING

I8	POP	SP	← SP + 1
		AC	← TOS
I9	LDIM	AC	← Immed. Addr.
I10	LDA	AC	← (Immed. Addr.)
I11	LIX	IX	← Immed. Addr.
I12	LDX	AC	← (IX)
I13	LDXI	AC	← (IX)
		IX	← IX + 1
I14	LDXD	IX	← IX − 1
		AC	← (IX)
I15	LSP	SP	← Immed. Addr.
I16	STA	(Immed. Addr.)	← AC
I17	SIX	(Immed. Addr.)	← IX
I18	STX	(IX)	← AC
I19	STXI	(IX)	← AC
		IX	← IX + 1
I20	STXD	IX	← IX − 1
		(IX)	← AC
I21	LSH	AC	← LEFT(AC)
		CC	← CCO · MASK1 ∨ CCN · $\overline{MASK1}$
I22	ADD	AC	← AC + (Immed. Addr.)
		CC	← CCO · MASK2 ∨ CCN · $\overline{MASK2}$
I23	SUB	AC	← AC − (Immed. Addr.)
		CC	← CCO · MASK3 ∨ CCN · $\overline{MASK3}$
I24	ADC	AC	← AC + (Immed. Addr.) + Carry Flag
		CC	← CCO · MASK4 ∨ CCN · $\overline{MASK4}$
I25	ADI	AC	← AC + Immed. Field
		CC	← CCO · MASK5 ∨ CCN · $\overline{MASK5}$
I26	AND	AC	← AC ∧ (Immed. Addr.)
		CC	← CCO · MASK6 ∨ CCN · $\overline{MASK6}$
I27	OR	AC	← AC ∨ (Immed. Addr.)
		CC	← CCO · MASK7 ∨ CCN · $\overline{MASK7}$
I28	EXOR	AC	← AC ⊕ (Immed. Addr.)
		CC	← CCO · MASK8 ∨ CCN · $\overline{MASK8}$

LEGEND
TOS   Top of Stack
Immed. Addr.   2 byte address field in instruction
Disp.   1 byte argument contained in instruction
( )   specifies that argument is taken from or
                 stored into contents of the indicated
                 address
CCO   old condition code -- the present or
                 existing bit settings in the C.C.
                 register
CCN   new condition code -- the C.C. register
                 bits that are updated by an instruction
LEFT   perform left shift of contents of AC

When an arithmetic or logic operation is performed, some subset of the old condition codes is selected for update with new values and others are inhibited from update by means of a mask vector MASKi, $1 \leq i \leq 8$. The condition code(s) that affect conditional jump instructions I5 and I6 are selected by a field in the instruction.

We now describe three of these instructions in more detail, namely I1, I5, and I22. In the interest of brevity, we have omitted the instruction fetch which would transfer the contents of PC to Memory Addr. and transfer contents of IN to an instruction register IR (not shown).

```
CALL                                    Data Path    Logical Order
     OUT(Addr.) ← PC                        14            1
     AUX[15:8] ← IN                          5            2
     PC ← PC + 1                            19            2
     OUT(Addr.) ← PC                        14            3
     AUX[7:0] ← IN                           5            4
     PC ← PC + 1                            19            4
     OUT(Addr.) ← SP                        12            5
     OUT(Data) ← PC[15:8]                   15            6
     SP ← SP - 1                            17            6
     OUT(Addr.) ← SP                        12            7
     OUT(Data) ← PC[7:0]                    15            8
     SP ← SP - 1                            17            8
     PC ← AUX                                7            9

LJC  (Long Jump Conditional)
     OUT(Addr.) ← PC                        14            1
     AUX[15:8] ← IN                          5            2
     PC ← PC + 1                            19            2
     OUT(Addr.) ← PC                        14            3
     AUX[7:0] ← IN                           5            4
     PC ← PC + 1                            19            4
     PC ← AUX · CC0 ∨ PC · CC0             7, 19          5

ADD
     OUT(Addr.) ← PC                        14            1
     MBR[15:8] ← IN                          1            2
     PC ← PC + 1                            19            2
     OUT (Addr.) ← PC                       14            3
     MBR[7:0] ← IN                           1            4
     PC ← PC + 1                            19            4
     OUT(Addr.) ← MBR                       11            5
     AC ← AC + IN                         8, 21           6
     CC ← CC0 · MASK1 ∨ CCN · MASK1       22, 9           7
```

The data path column refers to arcs on the graph model of Figure 9.5 and the logical order refers to the order in which actions must occur in order to satisfy data dependences. For example, in the CALL instruction, contents of a memory location are transferred over arc 5 to the scratchpad register AUX and the program counter is incremented using arc 19. In an actual implementation these could occur either serially or in parallel. However, for the purpose of analysis, we are only concerned with logical dependence. Other activities must clearly occur in some given sequence; for instance, an ADD cannot be performed until the arguments are available.

Given the instruction Ij, we define its *source registers S(Ij)* as the set of registers that provide its operands and its *destination registers D(Ij)* as the set of

9.2 SYSTEM TESTING

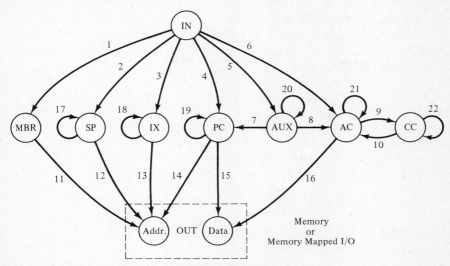

Figure 9.5 Graph model of hypothetical microprocessor.

registers that are changed by Ij during its execution. Given a graph representation of a microprocessor, then the *edge set E(Ij)* of instruction Ij is the set of directed edges defining data flow during execution of instruction Ij. The notation *Read(Ri)* indicates the shortest sequence of class T or B instructions necessary to propagate the contents of register Ri to an output, and *Write (Ri)* denotes the shortest sequence of class B or T instructions required to write register Ri. The *transfer path T(Ij)* is a logical path over which data flows during the execution of instruction Ij. Because we assume that structural implementation details are not available, transfer mechanisms such as buses are mapped into logical transfer paths.

In the graph representation, a label denoted L(Ri) is assigned to each node representing a register and a label denoted L(E(Ij)) is assigned to the edge set E(Ij) of each instruction Ij. The labels are assigned according to the following algorithm:

1. L(OUT) = 0
2. K ← 0
   DO WHILE |{unlabeled nodes}| > 0
       Assign label K + 1 to all unlabeled nodes representing registers whose contents can be transferred (implicitly or explicitly) to any register, I/O device, or main memory represented by a node labeled K by executing a single instruction of type T or B;
       K ← K + 1
   END
3. Assign the label 1 to each edge in the set E(Ij) where Ij is an instruction that reads out a register (implicitly or explicitly) during its execution.
4. If Ij is an instruction whose edge set has not been labeled in step 3, then assign a label K + 1 to each edge in set E(Ij), where L(D(Ij)) = K.

In our hypothetical microprocessor,

$$L(MBR) = L(SP) = L(IX) = L(PC) = L(AC) = 1$$
$$L(AUX) = L(CC) = 2$$

To permit creation of an uncluttered graph, only a single arc is drawn between any two nodes. It represents all occurrences of data transmission from node Ri to node Rj. For example, in Figure 9.5, arcs 5, 14, and 19 are traversed during logical steps 1 through 4 of the CALL instruction. The same three arcs are also each traversed twice during the Long Jump Conditional (LJC). Arc 19 is also traversed during the LJC and is traversed twice more during the ADD instruction. Therefore, the set denoted by arc 19 is defined (partially) by:

$$19 = \{(I1, 2, 1), (I1, 4, 1), (I5, 2, 2), (I5, 4, 2), (I5, 5, 2),$$
$$(I22, 2, 1), (I22, 4, 1), \ldots \ldots\}$$

The third entry in each ordered triple is the label assigned by the labeling algorithm. The CALL instruction causes the stack pointer to be read out and ADD causes MBR to be read out. If we exclude the program counter, which is read out to fetch the jump address, then the LJC does not cause any register to be read out. The complete set for arc 19 contains all ordered triples identifying instructions which cause data flow on arc 19, their logical order, and their label. The set represented by arc 10 is easier to construct. The carry condition code is used only with the ADC (Add with Carry), hence $10 = \{(I24, 3, 1)\}$.

Tests can now be created, using the graph model. Since a basic assumption is that no detailed structural knowledge of the device is available, functional elements must be processed, including:

Register decoding

Data storage

Instruction decoding and control

Data transfer

Data manipulation

Microprocessors

The register decoding and data storage test strategies are similar in approach to memory testing, which we covered in Chap. 8, so we will not comment further on them. Instruction decoding and control tests are concerned with determining, when a specific instruction is selected, that the correct, and only the correct, operations are performed. If a fault has occurred in the instruction decoding and control, then when instruction Ij is selected, one of three possibilities could occur; these are:

$f(Ij/Iq)$—instruction Iq was executed.

$f(Ij/0)$—no instruction was executed.

$f(Ij/Ij + Iq)$—instruction Ij *and* instruction Iq were executed.

9.2 SYSTEM TESTING

Two assumptions are made in order to simplify the test creation process. These are: If $f(Ij/Iq)$ or $f(Ij/Ij + Iq)$ is present, then instruction Iq will be executed correctly, and if $f(Ij/0)$, $f(Ij/Iq)$, or $f(Ij/Ij + Iq)$ is present, then neither $f(Ir/Ij)$ nor $f(Ir/Ir + Ij)$ can be present.

Ordering procedure:

1. Kmax = Max label value
2. K ← 1
3. DO WHILE K ≤ Kmax

Apply tests to detect faults in the following order:

i. $f(Ij/0)$, $f(Ij/Iq)$, $f(Ij/Ij + Iq)$
    where $L(Ij) = L(Iq) = K$
ii. $f(Ij/Ij + Iq)$ where $1 \le L(Ij) \le K$,
    $L(Iq) = K + 1$, and $K < Kmax$
iii. $f(Ij/Ij + Iq)$ where $K + 1 \le L(Ij) \le Kmax$, $L(Iq) = K$

END

EXAMPLE ■

For $f(Ij/0)$, consider the case when $L(Ij) = 2$. We use the following procedure:

1. Store OPERAND1 in $D(Ij)$ and operand(s) in $S(Ij)$ such that Ij produces RESULT1 ($\ne$ OPERAND1) in $D(Ij)$
2. Read out $D(Ij)$
3. Execute Ij
4. Read out $D(Ij)$ ■ ■

In this procedure, a value is initially stored in register $D(Ij)$ using an instruction Iu such that $L(Iu) = 1$. In step 2 the register is read out, using instruction $Ir = READ(D(Ij))$, to verify its contents. In step 3 the contents of $D(Ij)$ should be changed by Ij, and in step 4 register $D(Ij)$ is again read out to determine whether its contents actually changed. A key point in this procedure is the fact that $L(Ir) = |READ(D(Ij))| = 1$. Therefore, instruction Ir is assumed to have been previously verified by virtue of the ordering procedure and can confidently be used in steps 2 and 4 to read out $D(Ij)$. Furthermore, again by virtue of the ordering procedure, it has been verified that Ir does not cause execution of any instruction with label 2.

We give the following procedure, without elaborating on or justifying the steps involved, mainly to illustrate the depth of detail necessary to develop a comprehensive test.

Test generation for f(Ij/Iq), given that $L(Ij) = L(Iq) = K \leq 3$, $D(Ij) = D(Iq)$, and $L(S(Iq)) = K$.

1. $S(Ij) \leftarrow$ OPERAND1,
   $S(Iq) \leftarrow$ OPERAND2, \qquad (OPERAND1 $\neq$ OPERAND2).
2. DO i = 1 to K
   Execute Ij
   Read D(Ij)
   END
3. Execute Iq
4. Read D(Iq)

The interested reader can find details of this procedure and procedures for other cases in the previously cited reference or in the Ph.D. dissertation by Thatte[6]. For a microcomputer with many instructions, it could seemingly become difficult to apply tests to detect all possible combinations such that Iq was executed when Ij was intended. However, in practical cases shortcuts can be taken. Note that in the procedure, it is assumed that S(Ij) and S(Iq) are different registers. Arithmetic and logic operations generally use the same registers and often require the same number of clock cycles. With proper selection of arguments A and B, the correct instruction Ij can be distinguished from many incorrect instructions Iq.

EXAMPLE ∎

Suppose we are testing the ADD instruction. We choose arguments A and B such that every possible binary combination appears in some column of the two operands.

$$A = 00110011$$
$$B = 01010101$$

The various arithmetic and logic operations then produce the following results:

$$A + B = 10001000$$
$$A \wedge B = 00010001$$
$$A \vee B = 01110111$$
$$A \oplus B = 01100110$$
$$B - A = 00100010$$
$$A - B = 11011110$$

∎ ∎

Because the arguments produce different answers for every arithmetic and logic operation, and because two distinct ALU operations cannot operate simultaneously, the arguments can detect any instruction decoding error which converts an ALU operation into another ALU operation. Similar reasoning may apply to other pairs of instructions that use the same resource. Illegal combinations that are more probable include ALU operations occurring in conjunction with I/O operations, interrupt processing, or program counter-related operations.

We have not discussed test procedures for detecting faults in the data transfer and storage functions nor in the data manipulation section. We will address them in a later section.

9.3 SELF-TEST

The ordering relation previously described helps define a systematic test strategy which can first isolate a minimal logic configuration that is testable independent of other logic, and then use it to apply an orderly and exhaustive test to other functional units. The strategy aims for thorough testing of faults and efficient diagnosis. By virtue of first testing the hardcore, and then progressively testing additional functional units, an erroneous response is more easily pinpointed to the failing element. The strategy also gives a test more credibility since, with a high degree of confidence in the hardcore, there is greater confidence in decisions that are made when using the hardcore to test other logic.

Because the ordering only specifies the order in which elements must be tested, it still remains to actually create tests for the functional units. Furthermore, in a microprocessor-based system the microprocessor is normally only one functional element on a circuit board which also contains ROM, RAM, and various LSI peripheral chips. Microprocessors also tend to be bus-oriented. Given that an effective test can be developed for a microprocessor, how can it be integrated into a complete test plan for a bus-oriented microprocessor board?

9.3.1 Signature Analysis

It is often quite easy to generate tests algorithmically. We previously saw that memory tests comprised of simple programs on the order of about 20 lines of software could generate hundreds of thousands of patterns. Similar pattern generators could be employed for random logic. However, evaluating responses to those patterns could require saving a formidable amount of data. A test strategy called signature analysis, developed by Hewlett-Packard[7], addresses the problem of data explosion that occurs during the testing of microprocessor-based boards. The test stimuli consist of both exhaustive functional patterns and specific, fault-oriented test patterns. Regardless of the type of patterns, the output responses are compressed into four-digit hexadecimal signatures.

Consider the circuit of Figure 9.6. It represents a typical microprocessor configuration, a number of devices joined together by address and data buses and controlled by a microprocessor. Included on the diagram are two items not usually

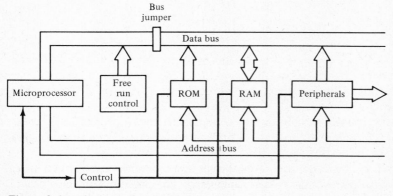

Figure 9.6 Microprocessor-based system.

seen on such diagrams: a free-run control and a bus jumper. When in the test mode, the bus jumper isolates the microprocessor from all other devices on the bus. In response to a test signal or system reset, the free-run control forces an instruction such as a NOP (No Operation) onto the microprocessor data input. This instruction performs no operation, it simply causes the program counter to increment. Since no other instruction can reach the microprocessor inputs while the bus jumper is removed, it will continue to increment the program counter at each clock cycle and put the incremented address onto the address bus. The microprocessor may generate 64K addresses or more, depending on the number of address bits. To evaluate each bit in a stream of 64K bits, for each of 16 address lines, would require storing a million bits of data and then comparing these individually with the response at the microprocessor address output. To avoid this data storage problem, the bit stream from each address line is compressed into a 16-bit signature. For 16 address lines, a total of 256 data bits must be stored.

A number of methods can be used to generate a signature for a bit stream. It is possible, when sampling the bit stream, to count 1s. It is also possible to count transitions. Another approach is to break the incoming stream into a succession of n-bit quantities that can be added to a running sum to create a checksum. The checksum has uneven error detection capability. If a double error occurs, and both bits occur in the low-order column, then the low-order bit is unchanged but, because of the carry, the next higher order bit will be complemented and the error will be detected. If the same double-bit error occurs in the high-order bit position, and if the carry is ignored, the double error will go undetected. In fact, if there is a stuck-at fault on the high-order bit, either at the sending or receiving end, there is only a 50% chance that it will be detected by the checksum that ignores carries. A triple error can also go undetected. A double error in the next-to-high-order position, occurring together with a single-bit error in the high-order position, will again cause a carry-out but have no effect on the checksum. In general, any multiple error which sums to zero, with a carry-out of the checksum adder, will go undetected.

EXAMPLE ■

Given a set of n 8-bit words for which a checksum is to be computed, assume that four of the words are corrupted by errors e_1 through e_4, as shown.

e_1		0	0	1	0	0	0	0	0
e_2		0	1	0	0	0	0	0	0
e_3		1	0	0	0	0	0	0	0
e_4		0	0	1	0	0	0	0	0
	1	0	0	0	0	0	0	0	0

The errors sum to zero, hence they will go undetected if the carry is ignored.
■ ■

Hewlett-Packard chose to use the linear feedback shift register (LFSR). It offers simplicity in implementation together with a high degree of probability that an error in an incoming stream of bits will be detected. The LFSR is made up of

9.3 SELF-TEST

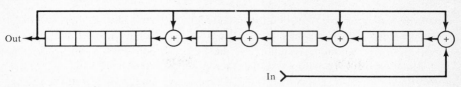

Figure 9.7 Linear feedback shift register.

delays (delay flip-flops), exclusive-ORs, and feedback lines. The exclusive-ORs are actually modulo 2 adders in GF(2), the Galois field containing the two elements 0 and 1 (see Appendix).

Analysis of LFSRs is performed most conveniently by relating them to polynominals in which the coefficients are contained in GF(2). For example, the LFSR in Figure 9.7 is associated with the polynomial

$$p(x) = x^{16} + x^9 + x^7 + x^4 + 1$$

If the incoming binary message stream is represented as a polynominal $m(x)$ of degree n, then the circuit in Figure 9.7 performs a division

$$m(x) = q(x) \cdot p(x) + r(x)$$

The output is 0 until the 16th shift. After n shifts the output of the LFSR is a quotient $q(x)$, of degree $n - 16$. The contents of the delay elements, called the signature, are the remainder. If an error appears in the message stream, such that the incoming stream is now $m(x) + e(x)$, then

$$m(x) + e(x) = q'(x) \cdot p(x) + r'(x)$$
and
$$r'(x) = r(x)$$

if and only if $e(x)$ is divisible by $p(x)$. Therefore, if the error polynominal is not divisible by $p(x)$, the signature contained in the delay elements will reveal the presence of the error.

The LFSR used in the Hewlett-Packard signature analyzer, shown in Figure 9.8, is actually a variation of the LFSR in Figure 9.7. The implementation in Figure 9.8 generates the same quotient as the previous circuit, but does not generally create the same remainder. Regardless of which implementation is used, the following theorem holds[8]:

Theorem 9.1. Let $s(x)$ be the signature generated for input $m(x)$ using the polynomial $p(x)$ as a divisor. For an error polynomial $e(x)$, $m(x)$ and $m(x) + e(x)$ have the same signature if and only if $e(x)$ is a multiple of $p(x)$.

One of the interesting properties of LFSRs is the following[9]:

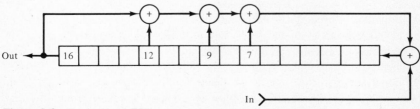

Figure 9.8 Equivalent LFSR.

Theorem 9.2. An LFSR based on any polynomial with two or more nonzero coefficients detects all single-bit errors.

Binary bit streams with 2 bits in error can escape detection. One such example occurs if

and
$$p(x) = x^4 + x^3 + x + 1$$
$$e(x) = (x^6 + 1) \cdot x^n$$

It can also be shown that, if the polynomial has an even number of terms, then it will detect all odd numbers of errors. In addition, all single bursts of length less than the degree of the polynomial will be detected.

After a signature has been generated on the output of the program counter, the signature generator can be applied to the ROM by running through its entire address space and generating a signature for each of its output pins. The ROM, like the program counter, is run through its address space by putting the board in the free-run mode and generating the NOP instruction. After the ROM has been checked, the bus jumper is replaced and a diagnostic program in ROM is run to exercise the microprocessor and other remaining circuits on the board. Note that either the diagnostic tests can reside in the ROM that contains the operating system and other functional code, or that ROM can be removed and replaced by another ROM which contains only test sequences.

After the ROM has been tested, the microprocessor assumes control. When in control, it can exercise the RAM using any of the methods of Chap. 8. In fact, register sets within the microprocessor itself can also be tested with the same types of tests used to test the RAM. These tests tend to be compact, most of them being programmed with just a few lines of software. Tests for the I/O peripherals and other LSI devices can, like the microprocessor tests, be developed by using methods described in the previous section.

Some common functional structures can be tested using algorithmic tests. One such example is the fixed-point ALU. A test which is quite effective is the test which checks propagate and generate signals between adder stages. The equations for these signals are:

$$P = P_{n-1} \cdot P_{n-2} \cdot \cdots \cdot P_0$$
$$G = G_{n-1} + P_{n-1}G_{n-2} + P_{n-1}P_{n-2}G_{n-3} + \cdots + P_{n-1}\ldots P_0 G_{-1}$$

where $P_i = A_i + B_i$
$G_i = A_i \cdot B_i$

and A_i and B_i, $0 \leq i \leq n-1$, are the binary inputs to the ith stage of an n-stage adder and G_{-1} is the carry-in to the adder. The test described here is a variation of the equivalent normal form described in Sec. 2.9.2. We dispense with the subscripts and rely solely on the Boolean equations to generate the tests.

The propagate circuits can be tested by putting all 0s on one port and all 1s on the other port. Then walk a 0 through the field of 1s to test each P_i. Next, test the individual terms in the generate equation. To test the rightmost term, again assign all 0s to one port and all 1s to the other port and set the carry-in to 1. All

9.3 SELF-TEST

of the other product terms in the equation will be zero since the G_i, $i \geq 0$, are all zero. Walk a 0 through the set of P_i, $i \geq 0$. Then set G_{-1} to 0, set G_0 to 1, set the P_i to 1 for $i \geq 1$, and again walk a 0 through the P_i, $i \geq 1$. Continue until all terms have been completely tested.

EXAMPLE ■

For an 8-bit ALU, the following tests will check the generate circuits:

A_7	A_6	A_5	A_4	A_3	A_2	A_1	A_0	B_7	B_6	B_5	B_4	B_3	B_2	B_1	B_0	C_{IN}
0	0	0	0	0	0	0	0	1	1	1	1	1	1	1	1	1
0	0	0	0	0	0	0	0	1	1	1	1	1	1	1	0	1
0	0	0	0	0	0	0	0	1	1	1	1	1	1	0	1	1
0	0	0	0	0	0	0	0	1	1	1	1	1	0	1	1	1
0	0	0	0	0	0	0	0	1	1	1	1	0	1	1	1	1
0	0	0	0	0	0	0	0	1	1	1	0	1	1	1	1	1
0	0	0	0	0	0	0	0	1	1	0	1	1	1	1	1	1
0	0	0	0	0	0	0	0	1	0	1	1	1	1	1	1	1
0	0	0	0	0	0	0	0	0	1	1	1	1	1	1	1	1
1	1	1	1	1	1	1	1	0	0	0	0	0	0	0	1	0
1	1	1	1	1	1	0	1	0	0	0	0	0	0	0	1	0
1	1	1	1	1	0	1	1	0	0	0	0	0	0	0	1	0
1	1	1	1	0	1	1	1	0	0	0	0	0	0	0	1	0
1	1	1	0	1	1	1	1	0	0	0	0	0	0	0	1	0
1	1	0	1	1	1	1	1	0	0	0	0	0	0	0	1	0
1	0	1	1	1	1	1	1	0	0	0	0	0	0	0	1	0
0	1	1	1	1	1	1	1	0	0	0	0	0	0	0	1	0

....
....
....

A_7	A_6	A_5	A_4	A_3	A_2	A_1	A_0	B_7	B_6	B_5	B_4	B_3	B_2	B_1	B_0	C_{IN}
0	0	0	0	1	0	0	0	1	1	1	1	1	0	0	0	0
0	0	0	0	1	0	0	0	1	1	1	0	1	0	0	0	0
0	0	0	0	1	0	0	0	1	1	0	1	1	0	0	0	0
0	0	0	0	1	0	0	0	1	0	1	1	1	0	0	0	0
0	0	0	0	1	0	0	0	0	1	1	1	1	0	0	0	0
1	1	1	1	0	0	0	0	0	0	0	1	0	0	0	0	0
1	1	0	1	0	0	0	0	0	0	0	1	0	0	0	0	0
1	0	1	1	0	0	0	0	0	0	0	1	0	0	0	0	0
0	1	1	1	0	0	0	0	0	0	0	1	0	0	0	0	0
0	0	1	0	0	0	0	0	1	1	1	0	0	0	0	0	0
0	0	1	0	0	0	0	0	1	0	1	0	0	0	0	0	0
0	0	1	0	0	0	0	0	0	1	1	0	0	0	0	0	0
1	1	0	0	0	0	0	0	0	1	0	0	0	0	0	0	0
0	1	0	0	0	0	0	0	0	1	0	0	0	0	0	0	0
1	0	0	0	0	0	0	0	1	0	0	0	0	0	0	0	0

■■

The test is quite effective at testing propagate and generate circuits, as well as the circuits which add bits A_i and B_i, and can also detect shorts between adjacent bit positions. The A and B port arguments can be reversed, as shown in the example, when the test is advanced to the next term in the generate equation to check for sensitivity to large numbers of simultaneously switching signals. It is easily programmed, requires very few lines of software, and therefore occupies very few

ROM or RAM cells. In addition, it is rather brief, requiring $(n + 2) \cdot (n + 1)/2$ addition operations. After tests for arithmetic operations are completed, tests for logic operations can be performed simultaneously at all bit positions. For a negative logic ALU, such as the 74181, it is necessary to complement the arguments.

The signature analyzer that is used to compute signatures has several inputs, including START, STOP, CLOCK, and DATA. The DATA input is connected to a signal point that is to be monitored in the logic board being tested. The START and STOP define a window in time during which the DATA input is to be sampled, while the CLOCK determines when the sampling process occurs. All three of these signals are derived from the board under test and can be set to trigger on either the rising or falling edge of the signal. The START signal may come from a system reset signal, or it may be obtained by decoding some combination on the address lines, or a special bit in the instruction ROM can be dedicated to providing the signal. The STOP signal, which terminates the sampling process, is likewise derived from some signal in the logic circuit being tested. The CLOCK is usually obtained from the system clock of the board being tested.

For a signature to be useful, it is necessary to know what signature is expected. Therefore, documentation must be provided listing the signatures expected at the IC pins being probed. The documentation may be a diagram of the circuit with the signatures imprinted adjacent to the circuit nodes, much like the oscilloscope waveforms found on television schematics, or it can be presented in tabular form, where the table contains a list of ICs and pin numbers with the signature to be found at each signal pin for which a meaningful signature exists. This is illustrated for a hypothetical circuit in Table 9.1.

During test the DATA probe of the signature analyzer is moved from node to node. At each node the test is rerun in its entirety and the signature registered by the signature analyzer is checked against the value listed in the table. This operation, analogous to the guided probe used on automatic test equipment (see Chap. 6) traces through a circuit until a device is found which generates an incorrect output signature but which is driven by devices that all produce correct signatures on their outputs. Note that the letters comprising the signature are not the expected 0–9 and A–F. The numerical digits are retained but the letters A–F have been replaced by ACFHPU, in that order, for purposes of readability and compatibility with seven-segment displays[10].

TABLE 9.1 SIGNATURE TABLE

IC	Pin	Signature	IC	Pin	Signature
U21	2	8UP3	U41	3	37A3
	3	713A		5	84U4
	4	01F6		6	F0P1
	7	69CH		8	1147
				9	8P7U
U33	9	77H1		11	684C
	11	10UP		15	H1C3
	14	1359			
	15	U11A			

9.3 SELF-TEST

A motive for placing stimulus generation within the circuits to be tested, and compaction of the output response, is to make field repair of logic boards possible. This, in turn, can help to reduce investment in inventory of logic boards. It has been estimated that a manufacturer of logic boards may have up to 5% of its assets tied up in replacement board kits and "floaters," boards in transit between customer site and a repair depot. Worse still, repair centers report no problems found on up to 50% of some types of returned boards[11]. A good test, one which can be applied successfully to help diagnose and repair logic boards in the field, even if only part of the time, can significantly reduce inventory and minimize the drain on a company's resources.

The use of signature analysis does not obviate the need for sound design practices. Signature analysis is useful only if the bit streams at various nodes are consistent. If even a single bit is susceptible to races, hazards, uninitialized flip-flops, or disturbances from asynchronous signals such as interrupt inputs, then false signatures will occur with the result that either correctly operating devices may be replaced or confidence in the signature diminishes. Needlessly replacing nonfaulted devices in a microprocessor environment, where devices frequently have 40 or more pins, can negate the advantages provided by signature analysis.

9.3.2 Random Patterns

It is important to bear in mind that the purpose of signature analysis is to compress long bit strings into hexadecimal signatures. The quality of the test is still dependent on the stimuli used to detect faults. The stimuli must induce a fault to create an error signal in the output stream.

In Chap. 2 we mentioned the method of exhaustive testing in which all possible patterns are applied to a circuit. If a combinational circuit responds correctly to all input combinations, then it has been satisfactorily tested not only for the traditional stuck-at faults, but for most multiple faults as well. Some multiple faults may still escape detection, including those that change a combinational circuit into a sequential circuit. In addition, if tests are applied at a rate slower than normal circuit operation, parametric faults which affect circuit response time may go undetected. Within integrated circuits the possibility of faults affecting seemingly unrelated logic gates, due to such things as mask defects and shorts caused by metal migration or breakdown of insulation between layers, makes it desirable to exhaustively test for a more comprehensive set of fault classes.

In circuits where the number of inputs is large, as in an ALU with two 32-bit input ports, a carry-in, and several control lines, exhaustive testing is not feasible. The values on the high order bits and the carry-out are affected by logic values on as many as 70 inputs. Two alternatives exist. It may be possible to partition the circuit into smaller subcircuits using one or more of the techniques discussed in Chap. 7. The smaller subcircuits can then be individually tested. Additional tests can be added to test the signal paths that were blocked from being tested by the partitioning circuits. The second alternative is to accomplish self-test using random patterns. A small, randomly selected subset of the entire set of patterns is applied to the circuit.

Self-test, using random patterns, is subject to constraints imposed by the design environment. For example, when designing with microprocessors and related peripheral chips, there is little control over internal operations, so the test must be tailored to the architecture of the microprocessor and peripherals. These devices are characterized by a great deal of internal logic that, once set in operation by an instruction, frequently runs for several clock cycles independent of external stimuli, other than arguments for the instructions being executed. In such cases it may be possible to repeatedly apply control signals that force a device through a predetermined sequence of states, and within that sequence, apply random signals to the data ports at the appropriate times.

For the purpose of generating random patterns, the LFSR is quite effective. Particularly attractive are LFSRs corresponding to *primitive* polynomials over GF(2) (see Appendix). Given a primitive polynomial of degree m, the corresponding LFSR can, when initialized to a nonzero value, run autonomously and generate a sequence of outputs of length $2^m - 1$ before repeating any subsequence of m consecutive bits. This sequence is quite random compared to a counter, which, starting at value 0, will not generate a 1 in the high-order bit until half the patterns have been exhausted. The shift register generator of Figure 9.9 is one such pseudo-random pattern generator (PRG). It can be initialized or "seeded" with a nonzero value by performing a set operation on one or more of the flip-flops.

The $2^{16} - 1$ consecutive distinct 16-bit values can be obtained by taking outputs simultaneously from all 16 flip-flops. The circuit is easily implemented using standard commercially packaged serial/parallel shift register parts and a four-input parity generator with output equal to 1 whenever an even number of inputs are at 1. The serial/parallel shift register can be one that performs functional tasks in parallel mode and creates test stimuli in serial mode to exercise the functional unit connected to the register outputs. Signatures are computed downstream from the PRG. Such a configuration is illustrated in Figure 9.10. The register, when in self-test mode, generates test stimuli. At some other register connected to the output of the combinational logic, signatures are generated. The START signal puts the register into test mode and "seeds" it with an initial value. The STOP signal can be generated from the last pattern for which a signature is to be computed. For example, it could be configured so that the last signal has 0s in $n - 1$ of the flip-flops. The OR of all the flip-flops having 0 on the last pattern is then used as the stop signal.

From use of a serial/parallel register, it is a small step to the BILBO (built-in logic block observer)[12]. The BILBO, shown in Figure 9.11, operates as follows: when $B_1, B_2 = 0, 0$ it is reset. When $B_1, B_2 = 1, 0$ it can be loaded in parallel and

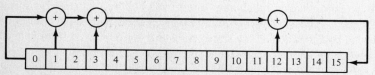

Figure 9.9 Maximal length LFSR.

9.3 SELF-TEST

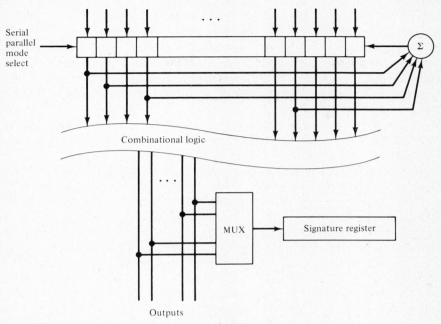

Figure 9.10 Test configuration using maximal LFSR.

used as a conventional register. When $B_1, B_2 = 0, 1$ it can be loaded serially and incorporated as part of a serial scan path. When $B_1, B_2 = 1, 1$ it can be used as a MISR (multiple input signature register) to sum the incoming data I_1–I_n, unless the data are held fixed, in which case it creates pseudo-random sequences of outputs.

The MISR is a feedback shift register which forms a signature on n inputs in parallel. After an n-bit word is added, modulo 2, to the contents of the register, the result is shifted one position before the next word is added. The MISR can be implemented with combinational logic[13] in such a way that the generated signature is identical to that obtained with serial compression. The equations are computed for a given LFSR by assuming an initial value c_i in each register bit position r_i, serially shifting in a vector $(b_0, b_1, \ldots, b_{n-1})$, and computing the new contents $(r_1, r_2, \ldots, r_n)$ of the register following each clock. After n clocks the contents of each r_i are specified in terms of the original register contents $(c_1, c_2, \ldots, c_n)$ and the new data that was shifted in. These new contents of the r_i define the combinational logic required for the MISR to duplicate the signature in the corresponding LFSR.

EXAMPLE ■

We use a register corresponding to the polynomial $p(x) = x^4 + x^2 + x + 1$. The register is shown, in equivalent form, in Figure 9.12. We assume that initially flip-flop r_i contains c_i. The data bits enter serially starting with bit b_0. The contents of the flip-flops are shown for the first two shifts. After two

9/SELF-TEST AND FAULT TOLERANCE

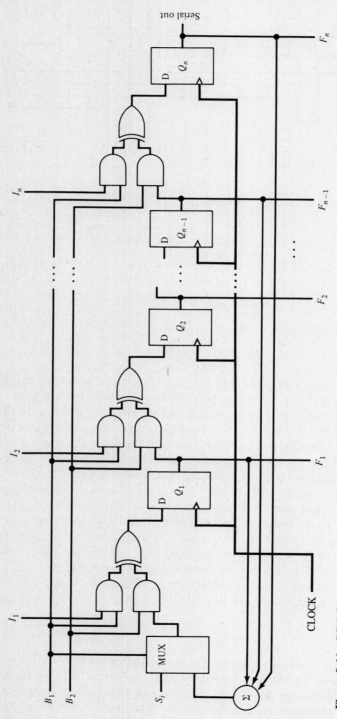

Figure 9.11 BILBO.

9.3 SELF-TEST

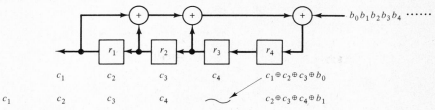

Figure 9.12 Fourth-degree LFSR.

more shifts, and making extensive use of the fact that $a \oplus a = 0$ and $a \oplus 0 = a$, the contents of the flip-flops are:

$$r_1 = c_1 \oplus c_2 \oplus c_3 \oplus b_0$$
$$r_2 = c_2 \oplus c_3 \oplus c_4 \oplus b_1$$
$$r_3 = c_1 \oplus c_2 \oplus b_0 \oplus b_2$$
$$r_4 = c_1 \oplus b_0 \oplus b_1 \oplus b_3$$

■ ■

For the purpose of creating effective signatures, it is not necessary that parallel data compression generate a signature which matches the signature generated using serial data compression. What is of interest is the probability of detecting an error. As it turns out, the MISR has the same error detection capability as the serial LFSR when they have an identical number of stages. In the discussion that follows, we informally demonstrate the equivalence of the error detection capability.

Using serial data compression, and an LFSR of degree r, and given an input stream of k bits, $k \geq r$, there are $2^{k-r} - 1$ undetectable errors since there are $2^{k-r} - 1$ nonzero multiples of $p(x)$ of degree less than k which have a remainder $r(x) = 0$.

When analyzing parallel data compression, it is convenient to use the linearity property. It permits us to ignore message bits in the incoming data stream and focus on the error bits. When clocking the first word into the register, any error bit(s) can immediately be detected. Hence, as in the serial case, when $k = r$ there are no undetectable errors. However, if there is an error pattern in the first word, then the second word clocked in is added (modulo 2) to a shifted version of the first word. Therefore, if the second word has an error pattern that matches an error pattern in the shifted version of the first word, it will cancel out the error pattern contained in the register and the composite error contained in the first and second words will go undetected. For a register of length r, there are $2^r - 1$ error patterns possible in the first word, each of which could, after shifting, be cancelled by an error pattern in the second word. When compressing n words, there are $2^{(n-1)r} - 1$ error patterns in the first $n - 1$ words. Each of these error patterns could go undetected if there is an error pattern in the nth word which matches the shifted version of the error pattern contained in the register after the first $n - 1$ words. Consequently, after n words, there are $2^{(n-1)r} - 1$ undetectable error patterns. Note that an error pattern in the first $n - 1$ words which sums to zero is vacuously canceled by the all-zero "error" in the nth word. The number of errors matches the number of undetectable errors in a serial stream of length $n \cdot r$ being processed by a register of length r.

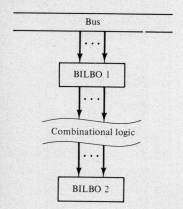

Figure 9.13 BILBO used to test circuit.

EXAMPLE ■

Using Figure 9.8, if the first word contains the error pattern $e_1 = 0000000001000000$ superimposed on the message bits, then after one shift of the register the error pattern becomes $e_2 = 0000000010000001$. Therefore, if the second word contains an error pattern matching e_2, it will cancel the error in the first word, causing the error to go undetected. ■ ■

There are a number of ways in which the BILBO can be used. One such example is to convert any register connected to a bus into a BILBO. Then, as depicted in Figure 9.13, either BILBO1 can generate stimuli for combinational logic while BILBO2 generates signatures, or BILBO1 can be configured to generate signatures on the contents of the bus. In that case, the stimulus generator can be another BILBO or a ROM whose contents are being read out onto the bus. After the signature has been generated, it can be scanned out by putting the BILBOs into the serial scan mode. Then, assuming the results are satisfactory, the BILBOs are restored to operational mode. In a complex system employing several functional units, there may be several BILBOs, and it is necessary to control and exercise them in correct order. Hence, a controller must be provided to assure orderly self-test in which the correct units are generating stimuli and forming signatures, scanning out contents and comparing signatures to verify their correctness.

9.4 MAINTENANCE PROCESSORS

In a number of applications, including large mainframes and complex electronics systems for controlling avionics and factory operations, where a system may be comprised of several hundred thousand gates, the maintenance processor is becoming an indispensable part of the overall test strategy. Maintenance processors may have different names and have somewhat different assignments in different systems, but one thing they all have in common is the responsibility to respond to error symptoms and to help diagnose faults more quickly. The maintenance processor may use some or all of the methods discussed in this and previous chapters, and

9.4 MAINTENANCE PROCESSORS

some methods that we will discuss in following sections. The general range of functions performed by the maintenance processor includes the following[14]:

- System start-up
- Communication with operator
- System reconfiguration
- Performance monitoring
- System testing

A typical system configuration is depicted in Figure 9.14. During system start-up the maintenance processor must initialize the main processor, set or reset specific flip-flops and indicators, clear I/O channels of spurious interrupt requests, load the operating system, and set it into operation. Communication with the operator may result in operator requests to either conduct testing of the system or make some alterations in the standard configuration. A system reconfiguration may also be performed in response to detection of errors during operation. Detection of a faulty I/O channel, for example, may result in that channel being removed from operation and I/O activities for that channel being reassigned to another channel. The maintenance processor may perform some or all of the reconfiguration in conjunction with the main processor.

Performance monitoring requires observing error indicators within the system during operation and responding to error signals. If an error signal is observed, an instruction retry may be in order. If the retry results in another error indication of the same nature, then a solid failure exists and a detailed test of some part of the system is necessary. The maintenance processor must determine what tests to select and it must record the internal state of the system so that it can be restarted, whenever possible, from the point where the error was detected. After applying tests, decisions must be made concerning the results of the tests. This may involve communicating with a field engineer either locally or, via remote link, at some distant repair depot. If tests do not result in location of a fault, but the error persists, then the field engineer may want to load registers and flip-flops in the system with

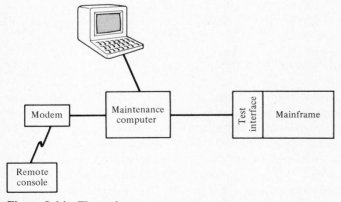

Figure 9.14 The maintenance processor.

specific test data via the maintenance processor, run through one or more cycles of the system clock, and read out results for evaluation.

In conjunction with remote diagnosis, it is possible to maintain a data base at a depot to assist the field engineer in situations where the error persists but a fault cannot be located. The RETAIN (remote terminal access information network)[15] system is one such example. It is a data base which maintains symptoms of faults which have proved difficult to diagnose. It includes the capability for structuring a search argument for a particular product symptom to provide efficient and rapid data location. The data base is organized both on a product basis and on a symptom basis.

Because the maintenance processor is part of the hardcore for a mainframe, it must be thoroughly tested before the mainframe is tested. It is worth noting that the maintenance processor, frequently a microprocessor or small minicomputer which has been previously designed and debugged, is known to be correctly designed and can be quite useful as an aid in debugging the design of a new processor. In addition, it is usually of much higher reliability than the main processor.

In microprogrammable systems implemented with writable control store, the maintenance processor can be given control over loading of control store. This can be preceded at system start-up time by first loading diagnostic software which operates out of control store. Diagnostics written at this level generally provide greater control over internal hardware. Tests can run more quickly since they can be designed to exercise functional units without making repeated instruction fetches to main memory. In addition, control at this level makes it possible to incorporate hardware test features such as BILBOs and similar built-in-test devices, and directly control them from fields in the microcode.

Maintenance processors can be given control over a number of resources, including power supplies and system clocks[16]. This permits power margining to stress logic components, useful as an aid in uncovering intermittents. Intermittents can also occasionally be isolated by shortening the clock period. With an increased system clock period, the system can operate with printed circuit boards on extender cards. Other reconfiguration capability includes the ability to disconnect cache and address translation units to permit operation in a degraded mode if errors are detected in those units.

The maintenance processor must control the overall operation of the mainframe and must be flexible enough to respond to a number of different situations, which implies that it must be programmable. However, speed of operation of the maintenance processor is usually not critical, hence microprocessor-based maintenance processors can be used. One such system uses a Z80 microprocessor[17]. The maintenance processor can trace the flow of activity through the CPU, which proves helpful in writing and debugging both diagnostic and functional routines in writable control store. Furthermore, the maintenance processor can reconfigure the system to operate in a degraded mode wherein an internal processor unit (IPU) that normally shares processing with the CPU can take over CPU duties if the CPU fails.

Another interesting feature of the maintenance processor is its ability to intentionally inject fault symptoms into the main processor memory or data paths to

9.4 MAINTENANCE PROCESSORS

verify the operation of parity checkers and error detection and correction circuitry[18]. The logging of relevant data is an important aspect of the maintenance processor's tasks. Whenever indicators suggest the presence of an error during execution of an instruction, an instruction retry is a normal first response because the error might have been caused by an intermittent condition which may not occur again during instruction retry. However, before the instruction retry, all relevant data is captured and stored. This includes any registers and/or flip-flops in the unit that produced the error signal. Other important information which may be relevant includes such factors as temperature, line voltage, and time and date[19]. If intermittents become too frequent, it may be possible to correlate environmental conditions with frequency of occurrence of certain types of intermittent errors. If a given unit is prone to errors under some stressful conditions, and if this is true in a large number of the units in use at customer sites, the recorded history of a product may indicate an area where the product might benefit from redesign.

The inclusion of internal buses in the mainframe to make internal operations visible is supported in the VAX-11/780[20]. An interesting addition to this architecture is the cyclic redundancy check (CRC) instruction, which enables both the operational programmer and the diagnostic programmer to generate signatures on data buffers or instruction streams.

The scan path can also be integrated with the maintenance processor, as in the DPS88[21]. In this configuration the maintenance processor has access to test vectors stored on disk. The tests may be applied comprehensively at system startup or may be applied selectively in response to an error indication within some unit. The tests are applied to specific scan paths selectable from the maintenance processor. The scan path is first addressed and then the test vectors are scanned into the addressed serial path. Addressability is down to specific functional unit, board, and micropack (assembly on which 50 to 100 unpackaged dice are mounted and soldered).

The random pattern and signature features can be used in conjunction with the maintenance processor[22]. This is illustrated in Figure 9.15. The PRG of length n which corresponds to a primitive polynomial creates a pseudo-random pattern of length $2^n - 1$. For each test, enough bits from the maximal length bit stream are clocked in to completely update all shift register latches (SRLs). If L is the length

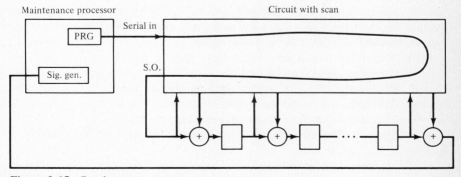

Figure 9.15 Random pattern scan.

TABLE 9.2 FAULT COVERAGE WITH RANDOM PATTERNS

	No. of gates:	No. of random patterns			Fault coverage with ATPG
		100	1000	10000	
CHIP1	926	86.1	94.1	96.3	96.6
CHIP2	1103	75.2	92.3	95.9	97.1

of the scan path plus the number of primary inputs, and if L and $2^n - 1$ are relatively prime, then it is possible to apply $2^n - 1$ unique patterns of length L.

A pseudo-random pattern generator of arbitrary n can create a nonrepeating sequence of great length, and the length L may be on the order of 100 to 200 latches. To apply every possible pattern to a scan of that length, in a machine where there may be a hundred or more scan paths, can take an inordinate amount of time. Consequently, it may be desirable to terminate the set of patterns applied to a given scan path well before the upper limit of patterns is reached. The objective is to apply sufficient patterns to obtain acceptable fault coverage, recognizing that there are faults that will escape detection. The data in Table 9.2 shows the improvement in fault coverage, determined via fault simulation, as the number of random test vectors applied to two circuits increases from 100 to 10 thousand[22]. For the purpose of comparison, fault coverage obtained with an automatic test pattern generator is also listed. The numbers of test patterns generated by the ATPG are not given, but another ATPG under similar conditions, that is, combinational logic tested via scan path, shows that fault coverage ranging between 99.1 and 100% for circuits with gate counts ranging from 2900 to 9400 can be obtained with 61 to 198 test vectors[23].

Some of the difficulties in successfully testing combinational logic with random patterns result from large fan-in and fan-out. To overcome these problems, controllability and observability points can be added by inserting shift register latches into the logic, just as test points can be added to nonscan logic. These latches are used strictly for test purposes. Being in the scan path, they do not add to board or chip pin count. In Figure 9.16(a), the AND gate with large fan-in will likely have a low probability of creating a 1 on its output, adversely affecting observability of the OR gate; therefore an SRL is added for the sole purpose of improving observability of the OR gate. An SRL together with an OR gate, as shown in Figure 9.16(b), can force a 1 to improve controllability. During normal operation the SRL is at its noncontrolling value.

The benefits of random pattern generation and signature analysis are evident. Test pattern generation and response evaluation are virtually automatic. Simply add the LFSRs and, after building the first working prototype system, run the system in test mode and record the signatures generated. However, the user must have some confidence that fault coverage is adequate or, the related problem, there must be a means for determining how many random patterns to apply. There must be a highly efficient and accurate means, such as SCOAP, for detecting troublesome nodes from a controllability/observability standpoint. There must also be some means for determining a fault location after detection of an erroneous signature.

9.4 MAINTENANCE PROCESSORS

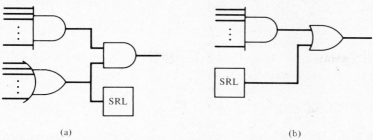

Figure 9.16 Enhancing random test.

If an erroneous signature has been recorded, fault isolation can be achieved by running the test and forming a signature on individual bit positions in the scan path. When scanning out the data, all data is ignored except the data in a selected bit position. That single bit is shifted into the signature generator from each response pattern. After the entire test has been run, the test is rerun to form a signature on the next bit position. This is continued until a signature has been formed for each bit in the scan path. Then, after identifying the SRL(s) with erroneous signatures, the logic in the cone fanning into that SRL(s) is suspect. Note that with addressable flip-flops (see Chap. 7), it would be possible to initialize the machine to some specified start state by loading the flip-flops, then simply run the machine in operational mode for some number of clock cycles while the addressing mechanism is held fixed at a particular flip-flop, in order to record the signature for that flip-flop. This sequence is repeated for each flip-flop until all have been recorded. Because of the addressing mechanism and the general randomness of combinational logic, it may not be necessary to reload the flip-flops after each operational clock.

We consider briefly the data compaction obtained using signature analysis and the pseudo-random generator in conjunction with scan path. We consider a very simple scenario, making assumptions that may not always be true in practice; that is, we assume that all data for a given circuit can be scanned in via a single scan-in path and read out via one scan-out path. Suppose that we have a circuit with 100 flip-flops in the scan path and that 200 test vectors are to be applied to the circuit. The test vectors come from the pseudo-random generator so there is no storage requirement for test vectors. The 200 test vectors generated by the ATPG will require 20,000 bits of storage (assuming 1 bit for each input, that is, only 0 and 1 values allowed). Assuming also a 100-bit response vector that is scanned out, a total of 40,000 bits must be saved for this one scan test. If a 16-bit signature generator is used, then 16 bits are required for each of the 200 responses. For diagnostic purposes, when generating a signature for each of the 100 flip-flops, a signature is required for each flip-flop, hence there are $16 \times 200 + 16 \times 100 = 4800$ bits of response data that must be saved. There is almost a 10-to-1 data compression using signatures for both fault detection and diagnosis for a single scan path. Considering that a typical mainframe has many scan paths, and that the numbers used here are conservative, a tremendous savings in disk storage space for test data can be obtained by use of PRG and signature analysis.

9.5 T-FAULT DIAGNOSIS

Up to this point we have considered diagnosis of systems wherein we start with a small kernel or hardcore and work outward, embracing more of the system with each phase of the test. That strategy can be quite effective when a system yields to the required partitioning. Sometimes, however, it is not so obvious how to partition for such testing. Some systems can only be partitioned so that each unit is a peer unit; that is, each is equal in decision-making capability to all others. Furthermore, it may be that each is capable of autonomous behavior, and that each is capable of communicating with some subset of the others. When there are several autonomous units capable of testing and diagnosing a subset of the remaining units, one problem immediately becomes evident: How do we know which units are telling the truth?

Consider the following: two units, u_1 and u_2, are connected so as to be able to test one another. When unit u_1 tests u_2, the outcome, denoted $a_{1,2}$, is 0 if u_1 and u_2 are both fault-free; it is 1 if u_1 is fault-free and u_2 is faulty; and it is inconclusive, denoted x, if u_1 is faulty. Similar considerations hold for the outcome $a_{2,1}$ when u_2 tests u_1. What can we conclude about the results when the two units test one another? We analyze the possibilities with the following table:

Condition	$a_{1,2}$	$a_{2,1}$
Both good	0	0
u_1 faulty	x	1
u_2 faulty	1	x
Both faulty	x	x

When u_1 is faulty, and it tests u_2, the results are unreliable. It could, in fact, report u_2 as faulty. A diagnosis can be made only if some a priori knowledge exists of fault probabilities. A significant amount of research has been conducted which addresses the problem of fault diagnosis for systems capable of correctly diagnosing one another without replacement. In what follows, we are not concerned with how u_i tests u_j. We assume the tests are accurate and thorough, and attempt to determine what conclusions can be drawn from the test results.

9.5.1 One-Step Diagnosis

Given a system S which is decomposed into n units $u_1, u_2, \ldots, u_n$, a graph model is defined in which the units u_i represent nodes. A directed arc $b_{i,j}$ exists if unit u_i tests unit u_j. Each $b_{i,j}$ is assigned a weight $a_{i,j}$, where $a_{i,j} = x$ if u_i is faulty and $a_{i,j} = 0 (= 1)$ if u_j is fault-free (faulty). Given a set of n units, with test links $b_{i,i+1}$ and $b_{n,1}$, the vector $(a_{1,2}, \ldots, a_{j,j+1}, \ldots a_{n,1})$ is called the syndrome. If only a single unit u_j is faulty, then the syndrome becomes $(0, \ldots 0, 1, x, 0, \ldots, 0)$, where $a_{j-1,j} = 1$, $a_{j,j+1} = x$, and all others are 0. With three or more units, a single defective unit can always be diagnosed simply by connecting the test links in a ring.

We now want to determine some general conditions necessary for diagnosing multiple faults in a system. However, first we point out that faults can be diagnosed

9.5 T-FAULT DIAGNOSIS

completely in a single step, or they may be diagnosed sequentially, such that a fault is diagnosed and repaired, and then an additional fault is identified and repaired, until all faults have been repaired. A system of n units is *one-step (sequentially) t-fault diagnosable* if all units (at least one unit) within the system can be identified without replacement provided the number of faulty units does not exceed t.

Given a system with n units, the following relationship holds between n and t:

Theorem 9.3. Let system S be one-step, t-fault diagnosable. Then $n \geq 2t + 1$. Conversely, if $n \geq 2t + 1$, it is always possible to provide a connection to form a system S such that S is one-step, t-fault diagnosable.

Proof. The converse follows from a maximal connection in which every unit tests every other unit. In such a configuration it is possible to find a loop of $t + 1$ or more units in which all test signals are 0. Then either all units are faulty or fault-free. However, they cannot all be faulty since that would violate the premise.

For necessity, assume $n < 2t + 1$. Partition the units into sets P_1 and P_2, each with t_0 units, $t_0 \leq t$, where all units in P_1 are faulty and all units in P_2 are fault-free. All tests between units in P_2 are 0 and all tests from units in P_2 to units in P_1 will have value 1. Because all units in P_1 are faulty, tests between units in P_1 may all be 0 and tests from units in P_1 to units in P_2 may be 1. Hence, with that partition, it is impossible to distinguish which set contains the faulty units and which contains the fault-free units.

Theorem 9.4. In a one-step, t-fault diagnosable system S, a unit is tested by at least t other units.

Proof. Let $F_1 = \{u_1, u_2, \ldots, u_k\}$, $k < t$, be a set of faulty units. Let $u \notin F_1$ be tested by all of the $u_j \in F_1$ but not by any other units. Furthermore, suppose that u_i is faulty. The units in F_1 may all diagnose u_i as fault-free, in which case it is impossible to distinguish between fault sets F_1 and $F_2 = F_1 \cup \{u_i\}$.

A set of test links is *optimal* if $n = 2t + 1$ and each unit is tested by exactly t units. An optimal test has $n \cdot t$ test links. An optimal design for $n = 5$ is shown in Figure 9.17.

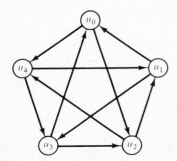

Figure 9.17 Design for two-fault diagnosable system.

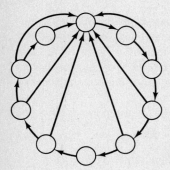

Figure 9.18 Sequential diagnosis connections for $n = 10$, $t = 4$.

9.5.2 Sequential Diagnosis

It is easily shown that Theorem 9.3 holds for sequential t-fault diagnosable systems. However, it is not necessary that each unit be diagnosed by k other units. The object of sequential diagnosability is to reduce the number of test links. However, when there are fewer test links, the problem of identifying faulty and fault-free links becomes more of a challenge. A class of designs which is sequentially t-fault diagnosable, and for which a fault-free unit can be identified, is given in the following (see Figure 9.18):

Theorem 9.5. Given a set $S = \{u_i\}, 0 \leq i \leq n - 1$, assign test links $b_{i,i+1}$ and $b_{n-1,0}$. Select a subset S_1 of $2t - 2$ units from $\{u_i\}, 1 \leq i \leq n - 2$, and add test links from each unit in S_1 to u_0. This design, with $n + 2t - 2$ links, is sequentially t-fault diagnosable.

Proof. Denote the number of test signals from S_1 to u_0 which have the value 0 (1) to be n_0 (n_1). Then:

Case 1: $n_1 > t$. There can be at most t faulty units. Therefore at least one test signal must reliably be reporting u_0 to be faulty.

Case 2: $n_1 < t$. Then u_0 must be fault-free else there would be $n_0 > t - 1$ other units which incorrectly diagnose it as fault-free.

Case 3: $n_1 = t$. The set $S^* = S_1 \cup \{u_{n-1}, u_0\}, 0 \leq i \leq n - 1$, contains $2t$ units. If u_0 is not faulty then S^* contains $n_1 = t$ faulty units. If u_0 is faulty, then S^* contains an additional $n_0 = t - 1$ faulty units which incorrectly diagnosed u_0. In either case, S^* contains t faulty units. Since $n \geq 2t + 1$, the set $S - S^*$ therefore contains at least one fault-free element.

We locate a faulty element as follows: in case 1 u_0 is faulty. In cases 2 and 3 u_0 is fault-free. Therefore, starting at u_0, follow the directed arcs along the loop until arriving at a test link with a 1. It identifies a faulty unit.

Designs for sequential diagnosability have been found having fewer than $n + 2t - 2$ test links, but they become rather involved. Some results also exist in which a faulty unit can be identified in single loop systems, that is, those in which the only test links are $b_{n-1,0}$ and $b_{i,i+1}, 0 \leq i \leq n - 2$. Perhaps of more interest is the fact that the diagnosability theory described here need not be confined to autonomous systems. As seen in previous sections, functional units in the computer

9.6 FAULT TOLERANCE

If we distinguish between the logic machine, which is an abstract specification defining tasks to be performed and algorithms to perform them, and the host, which is the physical implementation of that abstract machine, then fault tolerance can be defined as the architectural attribute of a digital system which permits the logic machine to continue performing its specified tasks when its physical host suffers various kinds of component failures[25]. We have already seen, in Chap. 8, the use of error detecting and correcting codes, which are perhaps the most widely used form of fault tolerance. We will look at other forms of fault tolerance, but first we distinguish between *active* and *passive* fault tolerance. Active fault tolerance is the ability to recover from error signals by repeating an operation, such as instruction retry, or rereading a data buffer or file, or requesting that a device retransmit a message. Passive fault tolerance is the ability to detect and correct errors without intervention by the host. These are somewhat arbitrary distinctions since even in the error detection and correction (EDAC) circuits that we examined in Chap. 8, an error signal resulted in logic activity in the hardware circuits of the host physical machine to correct the data, activity that would not have occurred if the error signal had not been detected. Perhaps a useful distinction is that active fault tolerance requires attention at the architectural level while passive fault tolerance contains errors before the symptoms are detected at the architectural level. In this text we will refer to active fault tolerance as *performance monitoring* since it more closely suggests the nature of the activities that take place.

The object of fault tolerance is either to prevent data contamination or to provide the ability to recover from the effects of data contamination. Applications range from data bases to industrial processes and transportation control. Consequences of faulty operation range from negligible to catastrophic. Hence the cost impact of fault-tolerant options employed may range from minor to significant. In some applications, such as space probes, it is simply not possible to repair faulty machines, hence cost for fault tolerance must be balanced against cost for failure of a critical part, which in turn may be equated with cost for failure of the entire mission.

9.6.1 Performance Monitoring

Performance monitoring involves the observation and evaluation of data during the course of normal operation. The monitoring may take advantage of information redundancy in the data or it may take advantage of structural characteristics of particular functional units.

Parity Bit A parity bit is an example of monitoring information redundancy. It is claimed that in most digital systems, parity checking accounts for 70 to 80% of error detection coverage[26]. It can be applied to memory, control store, data and address buses, and magnetic tape storage. Parity bits can be appended to data transmitted between I/O peripherals and memory as well as to data transmitted via radio waves.

Signatured Instruction Stream A concept which requires an integrated software/hardware approach is the *signatured instruction stream*[27]. This approach, which can be applied to both microcode and program instructions, requires that a signature be generated on the stream of instructions coming out of memory or control store. Any branch or merge point in a set of instructions is accompanied by a signature generated by an assembler or compiler. The merge and branch points are illustrated in Figure 9.19. Each node represents an instruction. A merge node is any instruction that can be the successor to two or more instructions. In an assembler language program, most labeled instructions represent merge nodes. A branch node is an instruction, such as a conditional jump, that can have more than one successor.

The hardware computes a signature and then compares the computed signature against the signature provided by the assembler or compiler. If the signatures do not agree, there is an error in the instruction flow, either hardware or programming error, since self-modifying code is not permissible in this environment. When generating the signatures, it is necessary to reset the signature prior to a merge node since the value in the signature generator will normally be different for two paths converging at a common node. This is illustrated by instruction j, which could be executed in-line from instruction h or could be reached from instruction e via a branch. Therefore, if j has a label, permitting it to be reached via a branch instruction, it is preceded by a special instruction which signals a check on the

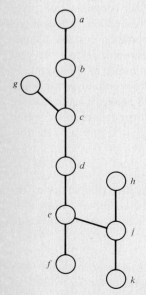

Figure 9.19 Graph representation of instruction stream.

9.6 FAULT TOLERANCE

signature. Likewise, a branch instruction at e must cause the signature to be checked and reset.

The signature, being part of the instruction stream, must be designed in at the architectural level. Hardware and software must be designed and developed jointly. The signature is incorporated into the instruction stream by the assembler or compiler, which inserts an unconditional branch to location PC + 2 that causes the machine to skip the following 2 bytes during operation. A 16-bit embedded signature is inserted following the branch instruction. The special hardware recognizes the unconditional jump as being a signal that the next 16-bit word contains the signature. It can actually contain the inverse, so that the sum of the hardware-calculated signature and the software-calculated signature is zero. Then a nonzero value signals the presence of an error. Conditional jumps must also be considered. Since the instruction at node e may pass control to instruction f or instruction j, the signature generator must be resynchronized when going to instruction f.

A related scheme, called *branch address hashing*[28], incorporates a signature into the branch address by performing a bit-by-bit exclusive-OR of the computed signature and the branch address. This permits a significant savings in program space and execution time. The branch address must, of course, be recovered before being used.

Diagnostic Programs In computers where priority scheduling and time sharing is employed, a maintenance program can reside in a part of memory and obtain a time slice of the CPU and other resources like any other user program. When it receives control of the CPU, it executes special diagnostic procedures designed to test out as much of the machine as possible at the program level. If an error is detected during its performance, it can generate an interrupt to signal the operating system to load special diagnostic programs to further isolate the cause of the error. To avoid tying up resources during periods of peak computational demand, it can be a low-priority task which runs only during off-peak time periods when resources are relatively inactive, or during times when the program mix in memory is I/O-intensive, permitting access to the CPU.

Test Data Injection It was previously pointed out that some maintenance processors are designed to inject test data into a circuit to specifically test parity checkers and other error detection devices. Some architectures are particularly well suited to that operation. A single-instruction, multiple-data (SIMD) array processor, which performs an identical calculation process on multiple streams of incoming data, is one such example. During design of the hardware, time slots can be allocated for insertion of predetermined data samples into the data streams. The processing hardware then checks the received test data for correctness, knowing in advance what results to expect. This can verify most, if not all, hardware between the data capturing end and the processor.

9.6.2 Self-Checking Circuits

In some functions the output response can be analyzed for correctness based on the observation that some responses are simply not possible in a correctly func-

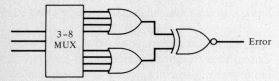

Figure 9.20 Self-test decoder.

tioning circuit and, if they occur, are indicative of a malfunction. One such example is the 3-to-8 decoder. As designed, only a single output can be active for any 3-bit input. If two or more outputs are simultaneously active, it is indicative of an error in operation. If two OR gates are added to the outputs as shown in Figure 9.20, then the circuit becomes self-testing relative to many of the faults which either inhibit selection of an output line or cause two or more outputs to be selected simultaneously[25]. In general, a circuit is *self-testing* if any modeled fault eventually results in a detectable error.

If a circuit is designed so that during normal operation any modeled fault either does not affect the system's output or is indicated no later than when the first erroneous output appears, then the circuit is said to be *fault-secure*. A majority logic decoder implemented with three AND gates and one OR gate, for which the output $M(a, b, c) = ab + bc + ac$, is fault-secure against opens on inputs since, during normal operation, all three input variables a, b, and c are identical. Therefore, a single open on a gate input will not affect the majority function output. The 3-to-8 decoder becomes fault-secure if the outputs are monitored so that an error signal occurs whenever more than one output is active. In fact, since it is both self-testing and fault-secure, it is said to be *totally self-checking*[29].

The multiplexer can be designed with self-testing features that take advantage of the function. The multiplexer must produce a logic 1 on its output if all inputs are at 1 and it must produce a 0 if all inputs are at 0, regardless of which input port was selected. In the 2-to-1 MUX shown in Figure 9.21, five gates are used to check for correct output from a three-gate circuit. However, only half of the

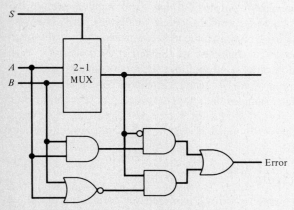

Figure 9.21 Multiplexer with self-test.

9.6 FAULT TOLERANCE

input combinations can enable the error circuitry. For values of $n > 2$, the checking circuitry is more efficient in usage of components, since it still requires only five gates, although requiring more inputs, but it is less efficient in percentage of input combinations which can enable the error detect circuitry.

The use of self-testing and fault-secure circuits is growing because of growing system complexity and because of the decreasing cost of hardware relative to other system development costs, including testing. In complex systems, it becomes virtually impossible to recreate an error that has corrupted data. Because of shorter clock cycles and closer proximity of circuits to one another in VLSI, intermittents are a growing problem. If these can be detected when they occur, by means of self-testing circuits, then instruction retry can be employed to help contain the effects of transient errors. In addition, if error signals are logged whenever they occur, then statistics can be compiled to indicate areas where inordinately large numbers of errors are occurring, causing large numbers of instruction retries. This information may point to areas where redesign can improve reliability and availability of a machine.

9.6.3 Burst Error Correction

The EDAC codes used with semiconductor memories are an example of passive fault tolerance. If an error is detected it is repaired "on-the-fly" by the EDAC circuits so that the processor is not aware that an error was detected and corrected. Error-correcting codes can also be used in an active fault-tolerant role to correct burst errors in data transmitted from disk drives to main memory[30]. Disk packs have an extremely thin coating of magnetic material. Errors occurring as a result of imperfections on a disk frequently take the form of bursts. A type of code called Fire codes, based on irreducible polynomials over GF(2), can correct single bursts in extremely long input streams.

Consider the equation

$$x^{n-k}M(x) = G(x) \cdot Q(x) + R(x)$$

where $M(x)$ = a message polynomial, of degree $k - 1$
$G(x)$ = the code generator polynomial
$Q(x)$ = quotient
$R(x)$ = remainder

By virtue of modulo 2 arithmetic

$$G(x) \cdot Q(x) = x^{n-k}M(x) + R(x)$$

therefore $x^{n-k}M(x) + R(x)$ is a code vector. But the coefficients of $x^{n-k}M(x)$ of degree less than $n - k$ are zero, and $R(x)$ has degree less than $n - k$. Therefore, in the codeword $x^{n-k}M(x) + R(x)$, $x^{n-k}M(x)$ is the original set of bits and $R(x)$ is the set of check bits.

In a field of polynomials over GF(2), y is a *root* of $P(x)$ if $P(y) = 0$. The *order* of a polynomial y is defined to be the smallest integer e such that $y^e = 1$, and a polynomial $P(x)$ is irreducible in GF(2) if there are no non-trivial polynomials $P_1(x)$ and $P_2(x)$ with coefficients in GF(2) such that $P(x) = P_1(x) \cdot P_2(x)$.

EXAMPLE ■

Consider the residue class of polynomials modulo $G(x)$ over GF(2). If $a(x) = b(x) \cdot G(x) + c(x)$, then $a(x) \equiv c(x)$. Since $G(x) = a \cdot G(x) + 0$ for $a = 1$, x is a root of $G(x)$.

Let $G(x) = x^3 + x + 1$ over GF(2). The order of x is 7 since

$$x^7 = G(x) \cdot [x^4 + x^2 + x + 1] + 1 = 1 \bmod[G(x)]$$

and no other power of x of degree less than 7 has remainder equal to 1 when divided by $G(x)$. ■ ■

A Fire code is defined by its generator polynomial

$$G(x) = P(x) \cdot (x^c - 1)$$

where $P(x)$ is an irreducible polynomial over GF(2), of degree m, whose roots have order $e = 2^m - 1$. It is also required that c not be divisible by e. The length n of the code is the least common multiple LCM(c, e) of c and e. We then have:

Theorem 9.6. A vector that is the sum of a burst of length b or less and a burst of length d or less cannot be a code vector in a Fire code if

$$b + d - 1 \leq c$$

and m is at least as large as the smaller of b and d.

Proof. We represent a burst of length b by a polynomial $x^i \cdot B(x)$, where degree$[B(x)] = b - 1$. We do likewise for $D(x)$. Then $F(x) = x^i \cdot B(x) - x^j \cdot D(x)$, where we assume, without loss of generality, that $i \leq j$. We use the Euclidean division algorithm, $j - i = cs + r, 0 \leq r < c$, to get

$$F(x) = x^i[B(x) - x^r D(x)] - x^{i+r}[D(x)(x^{cs} - 1)]$$

We assume $F(x)$ is a codeword, so $j - 1 < n$, and $F(x)$ is divisible by $x^c - 1$. Therefore, the first term on the right is divisible by $x^c - 1$ so

$$B(x) - x^r D(x) = (x^c - 1) \cdot H(x)$$

where we assume $H(x)$ is nonzero. Then we get $r + d - 1 = c + h$, where h is the degree of $H(x)$. Using the inequality in the theorem, we get the result that $r \geq b + h$. We also have that $b \geq 1$ and $h \geq 0$, so $r \geq b$ and $r > h$. $D(x)$ has a zero-degree term, hence a term on the left has degree r and there is no term on the right with degree r because $h < r < c$. Hence we conclude that $r = 0$ and $H(x) = 0$, so $B(x) = D(x)$ and

$$F(x) = x^i D(x) \cdot (x^{cs} - 1)$$

Now, for $F(x)$ to be divisible by $P(x)$, it is necessary that $P(x)$ divide $B(x)$, $x^{cs} - 1$, or x^i. $P(x)$ cannot divide $B(x)$ since degree $[P(x)] = m \geq b >$ degree $[B(x)]$ and $P(x)$ is relatively prime to x, therefore $P(x)$ divides $F(x)$ if and only if $P(x)$ divides $x^{cs} - 1$. We have that e is the smallest number such that a root y of $P(x)$ satisfies $y^e = 1$. Therefore cs must be a multiple of e. But c is not divisible by e,

9.6 FAULT TOLERANCE

and n is the least common multiple of c and e; therefore cs is a multiple of n, which is impossible.

The number of check bits in this code is $c + m$ and the number of information bits is $n - c - m$. The code can correct any burst of length $b \le m$ and simultaneously detect any burst of length $d \ge b$, where $c \ge b + d - 1$.

The burst error processor is able to detect bursts because the factor $x^c - 1$ causes an evenly spaced interlacing of parity checks, so that the message symbols involved in any single parity check bit are spaced c symbols apart. None of the c parity bits will be affected by more than a single error in any burst of length c or less. Hence, a single burst of length c or less will be reproduced in the check bits.

EXAMPLE ■

Consider the AmZ8065 Burst Error Processor[31]. It has several different user-selectable polynomials, including:

$$g(x) = p(x) \cdot (x^c - 1) = (x^{11} + x^2 + 1) \cdot (x^{21} - 1)$$
$$e = 2^{11} - 1$$
$$\text{LCM}(21, 2^{11} - 1) = 42{,}987$$
No. of check bits $= 11 + 21 = 32$
No. of information bits $= 42{,}987 - 32 = 42{,}955$
Correctable burst length $= 11$

The same LSI chip has other polynomials that can correct single bursts of 11 bits in streams of up to 585,442 bits. The register length for correction is equal to the number of check bits; in this example 32 flip-flops are required. The check symbols are generated by shifting a message polynomial $M(x)$ into a divider circuit such as the one shown in Figure 9.8, high-order bit first. After n shifts, k for the information symbols and $n - k$ for the low-order zeros, the remainder is in the shift register. It is the modulo 2 inverse of the check symbols. The check symbols replace the low-order zeros to form the code vector. ■■

When a data stream is received, the nature of the received data

$$x^{n-k}M(x) + R(x) = G(x) \cdot Q(x)$$

implies that, after the complete data stream has been shifted through the decoding register, the parity bits $R(x)$ should be zero. If not zero, then an error has occurred. A correctable burst error $B(x)$ is of the form:

$$E(x) = x^j B(x) = G(x) \cdot S(x) + R(x)$$

from which we get

$$x^i R(x) = x^i x^j B(x) - x^i G(x) \cdot S(x)$$
$$= (x^n - 1) \cdot B(x) - x^i G(x) \cdot S(x) + B(x)$$

where we use $n = i + j$. Since $G(x)$ divides $x^n - 1$, $B(x)$ is the remainder after division of $x^i R(x)$ by $G(x)$.

This suggests the following decoding algorithm. Shift the received bits, including the parity bits, through a register identical to the encoder. If the register contains all zeros after the shift, there is no error. Otherwise, load the remainder $R(x)$ and shift until a burst $B(x)$ of length b or less occurs in the register. If such a burst does not occur, the error is uncorrectable. If a burst of length b or less occurs in the low-order b bits, and all zeros occur in the remaining bits, then the number of shifts that were applied to the remainder $R(x)$ to form the burst indicate where the burst occurred. At that point, the burst is added to the received message to correct it.

When data is required from a disk, the CPU normally initiates an I/O request and continues with other operations while an I/O processor supervises the read operation and, if required, the error correction. This illustrates the difficulty in classifying a method of fault tolerance as active or passive. The burst error correction may appear as passive fault tolerance to the CPU and as active fault tolerance to the I/O processor.

9.6.4 Triple Modular Redundancy

When designing a system, the cost of reliability must be balanced against the cost of system failure or occasional transients. On a video display, an occasional glitch may be totally acceptable. On a deep space probe, where maintenance is not possible, errors are intolerable. Enhanced reliability may be a matter of using more reliable parts or it may be a matter of incorporating redundancy into a system. The cost for reliable parts tends to be a nonlinear function. To extend the mean time to failure (MTTF) of some components by 5% may cause the price of those components to increase an order of magnitude. In such cases, adding redundant parts may produce a significant increase in availability of the system at less cost than would be had by purchasing more reliable parts.

The most obvious approach to fault tolerance is the use of triple modular redundancy (TMR). Using three identical computers, if R_m is the reliability of one machine, and all have the same reliability value, then

$$1 = [R_m + (1 - R_m)]^3 = R_m^3 + 3R_m^2(1 - R_m) + 3R_m(1 - R_m)^2 + (1 - R_m)^3$$

The reliability of the system, then, assuming perfect voter circuits, is

$$R = \Pr(\text{no failures}) + \Pr(\text{one failure})$$

where $\Pr(x)$ denotes the probability of occurrence of x. From the previous equation we then get

$$R = R_m^3 + 3R_m^2(1 - R_m) = 3R_m^2 - 2R_m^3$$

Now, let the reliability of the machine be a decaying exponential of the operating time:

$$R_m(t) = e^{-ft} = e^{-t/\text{MTTF}}$$

9.6 FAULT TOLERANCE

where f is the failure rate and MTTF, the reciprocal of f, is the mean time to failure. Then

$$R(t) = 3e^{-2t/\text{MTTF}} - 2e^{-3t/\text{MTTF}}$$

When $t > \text{MTTF}$, we get $R < Rm$, hence triple modular redundancy may actually degrade performance of the computer, that is, unreliable parts only make the situation worse. Since the computer is made up of functional units, each of which is more reliable than the entire computer, it suggests employing TMR at the level of functional units to obtain enhanced reliability.

The equations are generated on the assumption of perfect voter circuits. This is not unreasonable since the voter circuits are relatively simple circuits compared to the circuits whose outputs they are evaluating. It can be shown[32] that, since voter circuits are imperfect, maximum reliability can be had by using TMR on circuits at that functional level where reliability of the parts equals the reliability of the voters. However, this implies voting on circuits of approximately the same reliability as the voters themselves, which implies that the operational circuits being voted upon are of about the same size as the voters, which would lead not to a tripling but to a sixfold increase in the amount of logic needed to implement a machine.

The benefits of TMR can be enhanced by periodic maintenance, just as we saw earlier in the discussion of error-correcting codes (Chap. 8). However, as with error-correcting codes, conventional testing and fault tolerance are at odds with one another. The test is oriented toward making a fault visible, while fault tolerance is oriented toward masking the effects of a fault. Therefore, when testing a unit it is necessary to disengage a module from its TMR environment or sample the outputs of the functional unit at some point prior to the voter circuits.

An example of fault tolerance with self-diagnosis capability is the Stratus/32, in which four processors run concurrently[33]. There are two processors on a board and two of each kind of board, including CPU, disk controller, and bus controller. If the pair of processors on a board disagree during operation, they remove themselves from operation and signal the system that they have failed. Maintenance software then runs tests on the board. If it passes, the maintenance routines assume that the error resulted from an intermittent and the board is restored to service. If the board fails again, it is removed from service until further, more extensive, service can be provided. The other pair of processors takes over its tasks. The maintenance routines store all information concerning failures in a log for inspection by a field engineer. A key requirement of the system is that failed boards be capable of being removed and replaced while the system is on line.

9.6.5 Software-Implemented Fault Tolerance

Fault tolerance can be implemented at the software level. The software-implemented fault tolerance (SIFT) system achieves fault tolerance by replication of tasks among processing units[34]. The primary method for detecting errors is the corruption of data. Interfaces between units are rigorously defined in order to deduce the effects on a unit when it receives erroneous signals from a faulted unit.

The unit of fault detection and reconfiguration is a processor, memory module, or bus.

In the SIFT system, operation proceeds by execution of a set of tasks, each of which consists of a sequence of iterations. After executing an iteration, a processor places the results in its memory. A processor using the results will examine and compare the results from all processors performing that iteration. Discrepancies are recorded or analyzed by the executive system. An interesting concept is that of "loose coupling." Different processors executing the same iteration are not in lockstep synchronization, and may in fact be out of step by as much as 50 microseconds. Therefore, a transient in the system is not likely to affect all processors in the same way, thus increasing the likelihood that an error in the data caused by a transient will be discovered.

The number of processors performing an iteration can vary, depending on the importance of the task. A global executive determines relative importance of tasks. Spread of contaminated data is prevented by allowing a processor to write only into its own local memory. A processor reading data from the faulted unit will, when comparing that data, discover the error. Further protection from error is achieved by enabling a processor to acquire data, not only from different processors, but via different buses as well.

In order to prevent incorrect control signals from producing wrong behavior in a nonfaulted unit, each unit is autonomous. In addition, the system has been designed to be immune to the failure of any single clock. The clock algorithm can be generalized such that a system can be made immune to failure of M out of N clocks when $N > 3M$.

9.7 SUMMARY

The pervasive nature of digital electronics and its growing complexity have spurred interest in the development of new testing techniques. One approach is to investigate functional units and general architectures, including microprocessor-based architectures. The object is to find invariants that permit development of test methods which have applicability to many similar architectures.

Another approach incorporates the tester into the logic, thus testing smaller functional units, testing exhaustively, and testing at normal operational speed. This also permits testing to be performed on site so that faults can be detected and repairs accomplished with a lower investment in expensive test equipment and spares inventory.

For large processors, comprehensive approaches must also address the additional problem of delivering maximum system availability. This requires reliability in order to prevent breakdowns and maintainability to get a system back into operation quickly after it has failed. Maintainability, in turn, requires the ability to detect faults and diagnose their location quickly.

Because availability requires confidence in the correctness of a system's operation, it is necessary that intermittent errors be minimized and traced to their origin. This has produced growing interest in fault-tolerant operation, including self-checking circuits, EDAC, and other performance-monitoring techniques. The use of remote monitoring in conjunction with centralized data bases such as the RETAIN data base discussed previously is becoming an important adjunct to other methods. These data bases seem a potentially rich area for artificial intelligence methods, relating symptoms of an unreliable processor with knowledge incorporated into the data base describing the machine architecture and other relevant information.

Reliable delivery of system performance has been largely a hardware effort. The SIFT system, however, is one example of software fault tolerance. The signatured instruction stream is an example of hardware and software working together to provide reliable computing. While it is not within the scope of this book to address the subject of software correctness, nevertheless it was noted that the signatured instruction stream can occasionally detect software errors caused by unauthorized writing into program areas during program execution. Architectures can be developed specifically to enhance the detectability of software errors[35]. Programming errors exist which are not detectable at compile time but which can be detected at execution time. It seems safe to predict that, with the increased availability of IC fabrication facilities and declining cost of hardware, more reliable architectures will ultimately evolve.

Much of what has been presented in this chapter has been a cursory overview. In practice, there are many additional factors which must be considered when implementing a particular technique. The primary objective of this chapter was to describe the available options and give a brief explanation of their salient features so that the reader knows what choices are available and can thus make an informed judgment as to which of them will fit his need. Successful application of any of the methods first requires an understanding of one's objectives, whether it be fast repair, reduction of spares inventory, less downtime, fewer field returns, reducing test expense, or some combination of the above as budget allows. Then, knowing one's objectives, and knowing what techniques are available, it is possible to make an informed judgment as to what methods are best suited to one's particular problem and what the cost will be for that solution.

PROBLEMS

9.1 Prove the reduction properties (r1) and (r2).

9.2 Modify Figure 9.3 as follows: place a multiplexer at the node where registers 2 and 3 are tied together in a tri-state configuration. Label the multiplexer as unit 6. The shift circuit which goes to register 3 now also goes to another register, labeled register 7, which is directly observable. Rework all calculations as a result of these changes.

9.3 For the ALU circuit of Figure 9.3 draw a graph which shows the controllability/observability dependences of the functional units.

9.4 Create a matrix of instructions versus functional units and status bits for the hypothetical microprocessor.

9.5 Write as short a computer program as you can, in the programming language of your choice, to create the test patterns given in the text for the fixed-point ALU propagate and generate circuits. If your simulator developed in Chap. 4 can perform fault simulation, check the tests against a common ALU such as the 74181. Try positive and negative logic arguments.

9.6 Show that $x^4 + x^3 + x + 1$ divides $(x^6 + 1) \cdot x^n$, where the coefficients are over the field GF(2).

9.7 When using checksum to detect errors, how many double errors go undetected when n words are being checksummed? How many triple errors?

9.8 Prove: if the generator polynomial has an even number of terms, then all odd numbers of errors will be detected.

9.9 Prove: an LFSR corresponding to a polynomial of order n will detect all bursts of degree less than the polynomial.

9.10 When using random pattern scan, as in Figure 9.15, if the scan length plus the number of PIs is not relatively prime to the sequence length $2^n - 1$ generated by the PRG, how many unique patterns can be applied to the circuit?

9.11 Delete the exclusive-OR between r_2 and r_3 in Figure 9.12. Then recompute the equations for the circuit which transforms it into an MISR.

9.12 Using the MUX in Figure 5.5a, set the enable input E to 0, set the data inputs to 1, and cycle the MUX select line through the values 0 and 1. Identify all stuck-at faults that can be detected on the MUX with the self-test circuit of Figure 9.21. Set the data inputs to 0, and repeat the exercise. Extend your results to n-to-1 MUX.

9.13 Define a rule for detecting pairs of faulty units in the design of Figure 9.17. Hint: consider separate cases for failures in adjacent units and nonadjacent units (u_i and u_j are adjacent if $j = i + 1 \bmod 5$).

9.14 Determine the order of the root x of the polynomial $F(x) = x^5 + x^2 + 1$ over GF(2). Determine the length of the Fire code generated by

$$g(x) = F(x) \cdot (x^{17} - 1)$$

REFERENCES

1. Robach, C., G. Saucier, and J. Lebrun, "Processor Testability and Design Consequences," *IEEE Trans. Comput.*, vol. C-25, no. 6, June 1976, pp. 645–652.
2. Srini, V. P., "Fault Diagnosis of Microprocessor Systems," *Computer*, vol. 10, no. 1, Jan. 1977, pp. 60–65.
3. Chang, H. Y., G. W. Smith, Jr., and R. B. Walford, "System Description," *Bell Syst. Tech. J.*, vol. 53, no. 8, Oct. 1974, pp. 1431–1450.
4. Chang, H. Y., and G. W. Helmbigner, "Controllability, Observability and Maintenance Engineering Technique (COMET)," *Bell Syst. Tech. J.*, vol. 53, no. 8, Oct. 1974, pp. 1505–1534.
5. Thatte, S. M., and J. A. Abraham, "Test Generation for Microprocessors," *IEEE Trans. Comput.*, vol. C-29, no. 6, June 1980, pp. 429–441.
6. Thatte, S. M., "Test Generation for Microprocessors," Ph.D. dissertation, Univ. Illinois Rept. UILU-ENG-78-2235, May 1979.
7. Hewlett-Packard Corp., "A Designer's Guide to Signature Analysis," Application Note 222, April 1977.
8. Meggett, J. E., "Error Correcting Codes and Their Implementation for Data Transmission Systems," *IRE Trans. Inf. Theory*, vol. IT-7, Oct. 1961, pp. 234–244.
9. Smith, J. E., "Measures of the Effectiveness of Fault Signature Analysis," *IEEE Trans. Comput.*, vol. C-29, no. 6, June 1980, pp. 510–514.
10. Nadig, H. J., "Testing a Microprocessor Product Using a Signature Analysis," *Proc. Cherry Hill Test Conf.*, 1978, pp. 159–169.

REFERENCES

11. White, E., "Signature Analysis-Enhancing the Serviceability of Microprocessor-based Industrial Products," *Proc. IECI*, March 1978, pp. 68–76.
12. Konemann, B., et al., "Built-in Logic Block Observation Techniques," *Proc. 1979 IEEE Int. Test Conf.*, pp. 37–41.
13. Nebus, J. F., "Parallel Data Compression for Fault Tolerance," *Comput. Des.*, April 5, 1983, pp. 127–134.
14. Liu, Tze-Shiu, "Maintenance Processors for Mainframe Computers," *IEEE Spectrum*, vol. 21, no. 2, Feb. 1984, pp. 36–42.
15. Hsiao, M. Y., et al., "Reliability, Availability, and Serviceability of IBM Computer Systems: A Quarter Century of Progress", *IBM J. Res. Dev.*, vol. 25, no. 5, Sept. 1981, pp. 453–465.
16. Wallach, S., and C. Holland, "32-Bit Minicomputer Achieves Full 16-Bit Compatibility," *Comput. Des.* Jan. 1981, pp. 111–120.
17. Hawk, R. L., "A Supermini for Supermaxi Tasks", *Comput. Des.*, Sept. 1983, pp. 121–126.
18. Boone, L., et al., "Availability, Reliability and Maintainability Aspects of the Sperry Univac 1100/60," *Proc. 10th Fault Tolerant Comput. Symp.*, Oct. 1980, pp. 3–8.
19. Frechette, T. J., and F. Tanner, "Support Processor Analyzer Errors Caught by Latches," *Electronics*, Nov. 8, 1979, pp. 116–118.
20. Swarz, R. S., "Reliability and Maintainability Enhancements for the VAX-11/780," *Proc. 8th Fault Tolerant Comput. Symp.*, June 1978, pp. 24–28.
21. Miller, H. W., "Design for Test via Standardized Design and Display Techniques," *Electron. Test*, vol. 6, no. 10, Oct. 1983, pp. 108–116.
22. Eichelberger, E. B., and E. Lindbloom, "Random-Pattern Coverage Enhancement and Diagnosis for LSSD Logic Self-Test," *IBM J. Res. Dev*, vol. 27, no. 3, May 1983, pp. 265–272.
23. Laroche, G., D. Bohlman, and L. Bashaw, "Test Results of Honeywell Test Generator," *Proc. Phoenix Conf. Computers and Communication,* May 1982.
24. Russell, J. D., and C. R. Kime, "System Fault Diagnosis: Closure and Diagnosability with Repair," *IEEE Trans. Comput.*, vol. C-24, no. 11, Nov. 1975, pp. 1078–1089.
25. Avizienis, A., "Fault-Tolerance: The Survival Attribute of Digital Systems," *Proc. IEEE*, vol. 66, no. 10, Oct. 1978, pp. 1109–1125.
26. Bossen, D. C., and M. Y. Hsiao, "Model for Transient and Permanent Error-Detection and Fault-Isolation Coverage," *IBM J. Res. Dev.*, vol. 26, no. 1, Jan. 1982, pp. 67–77.
27. Sridhar, T., and S. M. Thatte, "Concurrent Checking of Program Flow in VLSI Processors," *Proc. 1982 Int. Test Conf.*, pp. 191–199.
28. Shen, J. P., and M. A. Schuette, "On-Line Self-Monitoring Using Signatured Instruction Streams," *Proc. 1983 Int. Test Conf.*, pp. 275–282.
29. Smith, J. E., "A Theory of Totally Self-Checking System Design", *IEEE Trans. Comput.*, vol. C-32, no. 9, Sept. 1983, pp. 831–844.
30. Lignos, D., "Error Detection and Correction in Mass Storage Equipment," *Comput. Des.,* Oct. 1972, pp. 71–75.
31. "AmZ8065 Product Specification," Advanced Micro Devices, Sunnyvale, CA 94086.
32. Lyons, R. E., and W. Vanderkulk, "The Use of Triple-Modular Redundancy to Improve Computer Reliability," *IBM J.*, April 1962, pp. 200–209.
33. Hendrie, G., "A Hardware Solution to Part Failures Totally Insulates Programs, "*Electronics*, Jan. 29, 1983, pp. 103–105.
34. Wensly, J. H., et al., "SIFT: Design and Analysis of a Fault Tolerant Computer for Aircraft Control," *Proc. IEEE*, vol. 66, no. 10, Oct. 1978, pp. 1240–1255.
35. Myers, G. J., *Advances in Computer Architecture*, chap. 13, Wiley, New York, 1978.

Chapter 10

Functional Test and Other Topics

10.1 INTRODUCTION

In the previous chapter we looked at methods for self-test. There are applications where such methods are clearly necessary. Self-test is attractive for military systems in remote areas which are not accessible to test equipment. Military aircraft may fail in operation under high-stress conditions which cannot be duplicated on the ground. Built-in-test is essential to detect these faults. Large, fast commercial systems, when implemented in technologies such as emitter-coupled logic (ECL) which generate large amounts of heat, tend to be more susceptible to failure, hence they may require more frequent testing. Frequent testing, in turn, makes it more desirable that a test be performed as quickly and as conveniently as possible.

However, many digital products in the home entertainment category, although quite complex, represent modest investments by their owners. They use standard off-the-shelf components and tend to be quite reliable. In fact, as complexity increases, failure rates actually decrease. One experiment, based on 5 billion device hours, showed LSI circuits to have an effective failure rate which was one-seventh that of equivalent functions implemented in SSI circuits[1]. Thus, ironically, the increasing level of integration makes testing more difficult but reduces the number of failures. We may then want to test these products only once, before they leave the factory, to insure that they are operating correctly. These "throwaway" computers will likely be discarded before they fail. However, the test must be thorough to insure that the product works correctly before it is shipped.

In this chapter we look at methods which employ alternative representations of circuits, including functional and behavioral descriptions, for generating tests. Some of the ideas presented here are in formative or research stages, and may ultimately prove ineffective or impractical. Others may provide clues to effective

10.2 HARDWARE DESIGN LANGUAGES

test pattern generation, and perhaps design verification as well, since the two bear similarities in their approaches and in their goals.

10.2 HARDWARE DESIGN LANGUAGES

We have been concerned up to this point with structural descriptions of circuits (this assumes that the logic symbols for AND gates, OR gates, and inverters are an accurate reflection of structure). Two major problems in testing are the fact that structural descriptions simply are not always available and, even when available, it is not always obvious how to use the information to control the circuit. Consequently, researchers have been looking at ways to use functional and behavioral models in test and simulation. These models are described in hardware design languages (HDLs), also called register transfer languages (RTLs). The HDLs are quite similar in appearance to high-level languages (HLLs) such as FORTRAN, PL/I, and Pascal. The flexibility of the languages permits modeling a circuit at a level of detail ranging from structural to functional or behavioral. A design may even be expressed as a composite of elements, some of which are expressed in terms of structure and others expressed in terms of function or behavior. The HDLs are used for a number of purposes, including:

Specify an architecture.

Partition an architecture into smaller modules.

Verify that a structural implementation corresponds to the architectural design.

Check out microcode.

Serve as documentation.

Other emerging applications include the use of HDLs as input to automatic hardware compilers and formal verification systems[2].

A design expressed in an HDL consists of a declarative and a behavioral section. The declarative section defines inputs, outputs, and internal variables. The internal variables may be scalars, vectors, or arrays, representing flip-flops, registers, and $M \times N$ memories. The behavioral section contains statements describing the flow of data and control and the transformations that take place on data.

A block diagram for a typical computer is shown in Figure 10.1. In the figure, state transitions and the conditions under which they occur are determined by the contents of an instruction register (IR), by various status signals from the data flow section and other miscellaneous controls, and by the present state of the machine. The data flow section is made up of three basic kinds of devices:

Data routing switches

Storage elements

Transformation devices

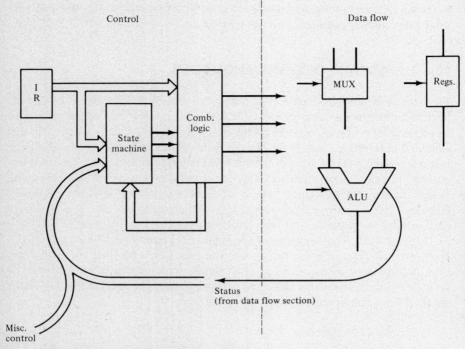

Figure 10.1 Functional view of CPU.

The storage elements are the latches, flip-flops, registers, and memories. These are sometimes referred to as *carriers*. Another entity included in the category of carrier is the *terminal*, which is a variable without memory. They correspond to nets in combinational logic. Data routing switches include multiplexers and tri-state controls gating data onto buses. A number of transformation devices exist, including arithmetic logic unit (ALUs), encoders, decoders, and virtually any other circuit that transforms a digital signal into another signal. The switches and transformation devices are sometimes called *operators*.

Since the purpose of an HDL is to describe hardware, the HDL language constructs must be translatable into corresponding hardware constructs. For basic logic functions it is easy to see the hardware equivalents; for example, a three-input AND gate can be represented by the expression

```
F := A.AND.B.AND.C
```

A functional device such as the 2-to-1 multiplexer can be represented by the expression

```
IF S THEN A ELSE B
```

10.2 HARDWARE DESIGN LANGUAGES

where S is a control variable corresponding to the select line and A and B are the input ports and may be scalars or vectors. A behavioral expression for an n-bit adder would be

X := U + V

The behavioral representation frequently provides little, if any, clue as to the underlying structure of the device. The adder may be implemented with ripple carry or carry look-ahead. If it is necessary to verify the structural model, then the behavioral expression is replaced by a more detailed structural description and the previous expression is replaced by an expression such as

X := SUM(U,V)

This results in a call to a subroutine in which the sum is computed using a structural description.

Although the description of a device in an HDL may appear identical to a program written in an HLL, the two need not necessarily give the same results. For example, the pair of statements

A := B
B := A

may indicate that the contents of registers A and B are to be exchanged in one clock cycle in order to implement the register swap instruction EXCHG A, B. However, in a high-level language, the contents of variable B are transferred to register A and the next line is superfluous. If it is desired to swap the contents of A and B, it is first necessary to save the contents of variable A in a temporary variable, then transfer the contents of B to A, and then load B from the temporary variable. The HDL can correctly exchange the contents of the two variables by using READ and WRITE arrays (see Sec. 4.4). The new value for variable A is retrieved from the contents of a READ array entry for variable B and stored in the WRITE array location corresponding to variable A. The next line is implemented similarly. After all lines of code have been executed, contents of the WRITE array are transferred to the READ array. One complete pass through the HDL code therefore corresponds to a single state change or clock pulse in a synchronous circuit, and the order in which the statements are executed is immaterial.

Note that the two statements above are order-sensitive in an HLL. If the two statements occur in the opposite order, the results will differ. The HDL is usually insensitive to the order. A language which attaches no significance to the order of occurrence is called *nonprocedural*; statement execution is either absolute, occurring in every pass through the code, or conditional upon the value of some control variable. A *procedural* language is one which imposes an order on activities. Activities are initiated conditional upon completion of preceding activities. Concurrent activities are described in time blocks which group together those actions that are to be performed concurrently. A succession of time blocks is used to define sequential operations[3].

To implement the register exchange instruction EXCHG A, B in a procedural language, terminals S and T can be defined. Then the code is written

```
S := A
T := B
A := T
B := S
```

Since terminals S and T physically correspond to wires, they have no memory, their contents can be changed immediately, rather than conditioned upon arrival of a clock signal.

The control section is responsible for the orderly progression of a machine through a series of state transitions. It must control the flow of data between flip-flops, registers, and other storage devices, and the transformations that take place en route. In an HDL it is not always obvious how the control section translates into a physical implementation.

Consider the circuit in Figure 7.18. Its behavior may be expressed as:

$$Q_2 := IN \cdot Q_2 \cdot \overline{Q_1} + \overline{IN} \cdot Q_2 \cdot Q_1 \cdot Q_0 + IN \cdot Q_2 \cdot \overline{Q_0} + IN \cdot \overline{Q_2} \cdot Q_1$$
$$Q_1 := \overline{IN} \cdot Q_2 + \overline{Q_1} \cdot Q_0 + IN \cdot Q_1 \cdot \overline{Q_0} + Q_2 \cdot Q_1 \cdot \overline{Q_0}$$
$$Q_0 := \overline{IN} \cdot \overline{Q_2} + Q_1 \cdot \overline{Q_0} + \overline{IN} \cdot Q_0 + IN \cdot \overline{Q_0}$$
$$OUT := IN \cdot Q_1 + \overline{Q_2} \cdot Q_1 \cdot \overline{Q_0} + \overline{IN} \cdot \overline{Q_2} \cdot \overline{Q_1} \cdot Q_0 + IN \cdot Q_2 \cdot \overline{Q_1} \cdot \overline{Q_0}$$

The same state machine can be written as:

$$0: OUT \leftarrow 0$$
$$1: OUT \leftarrow \overline{IN}$$
$$\rightarrow (2 \cdot IN) + (3 \cdot \overline{IN})$$
$$2: OUT \leftarrow IN$$
$$\rightarrow (3 \cdot IN) + (5 \cdot \overline{IN})$$
$$3: OUT \leftarrow IN$$
$$\rightarrow (1 \cdot \overline{IN}) + (4 \cdot IN)$$
$$4: OUT \leftarrow \overline{IN}$$
$$\rightarrow (2 \cdot \overline{IN}) + (5 \cdot IN)$$
$$5: OUT \leftarrow 0$$
$$\rightarrow (3 \cdot \overline{IN}) + (6 \cdot IN)$$
$$6: OUT \leftarrow IN$$
$$\rightarrow (3 \cdot \overline{IN}) + (7 \cdot IN)$$
$$7: OUT \leftarrow IN$$
$$\rightarrow (7 \cdot \overline{IN}) + (0 \cdot IN)$$

where we arbitrarily associate state S_i with its equivalent binary value on flip-flops Q_2, Q_1, Q_0.

This second version, written in AHPL (A Hardware Programming Language)[4], is procedural. Correct operation is dependent on statements being executed in the correct order. For example, state 0 implicitly transfers to state 1 when complete. Whereas a pass through all statements corresponds to a single state change in the first example, each statement in the AHPL representation is associated with a single state. Each statement either implicitly passes control to the next statement or transfers, conditionally or unconditionally, to some statement other than the next one. The second description is more abstract, more indicative of behavior, whereas the first description more closely reflects the physical structure.

10.3 FUNCTIONAL SIMULATION

Functional descriptions have been used for many years; some of the more widely known and used languages include CDL[5] (Computer Design Language), DDL[6] (Digital Design Language), and AHPL[7]. Along with these languages, there are simulators and simulator control languages provided so that the hardware description, once compiled, can be used in conjunction with input stimuli to evaluate the design. A typical control language specifies the input stimuli to be used, the output pins and internal nets to be observed, and the format in which output responses are to be printed. The inputs, both the input pins and the values to be applied to them, are normally expressed symbolically to relieve the designer of the tedium of keeping track of such details. The duration of the simulation may be specified to be some fixed time period or it may be contingent upon simulation results. For example, the designer may simulate the arrival of an interrupt at an input and, via the control language, look for some output condition to indicate that the interrupt was correctly processed. If after n clock pulses, the output is not as expected, simulation is halted to permit analysis of simulation results. Output control includes request for printout of values at specific I/O pins or internal nets, including registers and flip-flops. Conditional control statements may also be used to specify time intervals during which output is to be examined. A SAVE facility which records all internal memory states at the end of simulation is useful so that simulation can be restarted at the point where it was halted. A COMPARE facility permits comparison of output response for two different versions of a model, such as a functional and structural, to aid in the detection of any discrepancies in response between the two models.

10.3.1 Functional Fault Simulation

Although functional and behavioral descriptions of logic circuits are usually aimed at specifying and evaluating correct behavior, they can also be used for faulted machine simulation. Applications range from use of a few functional entities in an otherwise gate-level description to a complete functional or behavioral description.

Functional fault simulation has been incorporated into a parallel fault simulator[8]. Normally, the parallel fault simulator packs computer words so that many machines can be simulated in parallel. This works well when the elements being simulated, AND, OR, exclusive-OR, and inverters, have corresponding computer instructions. When simulating more complex elements, it is necessary to unpack the individual bits when the simulator comes to a functional level element. When unpacked, the bits are encoded into an integer or character format. Each of the faulted machines, as well as the nonfaulted machine, is then individually simulated. After processing, the result is packed back into the bit-encoded form again to be processed at the gate level. While cumbersome, the functional models themselves are simulated rapidly so there is an overall gain in processing speed.

At least one simulator has been developed which employs functional models in conjunction with deductive simulation[9]. However, concurrent fault simulation seems most amenable to functional fault modeling[10–12]. Because of the manner in which the concurrent fault simulator creates and processes separate copies of an element for the unfaulted model and faulted models which disagree with the good model at a given node, processing for gate and functional level models is essentially the same. Functional faults which cannot be described by the stuck-at model, as well as delay faults, can be handled in concurrent simulators in the same way as logic faults.

Major motives for the use of functional level simulation include the ability to deal with devices for which there is no detailed structural knowledge and the ability to model faults other than the stuck-at faults. However, because it can be instrumental in providing memory and CPU time savings, another variation is to use functional models for all parts of a design except for a particular function. The function under consideration is inserted into the design as a gate-level model and gate-level faults are inserted into the model. The faulted behavior is simulated at the gate level to the function terminals, and from there the simulation is carried out using functional models. Functional and gate-level representations can be swapped so that different functions can be chosen to be simulated at the gate level[13].

A frequent criticism of function level simulation is the fact that there is loss of detail in the model, hence loss of accuracy in the simulation. As was shown in Chap. 4, spikes can result when signals arrive close together in time at a latch. Simulating a latch as a functional primitive rather than as an interconnection of gates normally will not reveal the presence of that spike. On the other hand, with the functional element it is usually easier to determine whether a manufacturer's specified setup and hold times for a flip-flop have been satisfied[10].

10.3.2 Functional Fault Modeling

The subject of fault modeling in the context of functional models is one on which there is some disagreement. After all, the purpose of testing is to detect the presence of structural defects and perhaps to diagnose their location as well. How does one detect a structural defect when not even using structural information? The subject is, however, not entirely unexplored. Universal test sets have been shown to exist which detect all stuck-at faults in combinational logic functions, without having access to structural information[14]. In sequential logic, the sequential path sensitizer (see Chap. 3) employs functional models of flip-flops and defines functional tests for these devices. The checkpoint arcs described in Chap. 5 demonstrate the equivalence of signal paths with a subset of the stuck-at faults. More recently, S. M. Thatte, in his Ph.D. dissertation[15], examined several functions and defined conditions under which fault behavior could be determined independent of implementation details. A clearer understanding is emerging as more investigations are conducted, largely in response to the continuing emergence of LSI and VLSI parts for which detailed structural knowledge is unavailable.

10.3 FUNCTIONAL SIMULATION

One approach to modeling has been to design gate-level models whose behavior matches that of the device in question. A major motive for this approach is the fact that many automatic test pattern generators (ATPGs) and simulators simply cannot work on anything but gate-level models. It is generally accepted that an adequate test for stuck-at fault coverage in the gate equivalent model will provide a similar fault coverage for the real device, although this is not completely true, as was demonstrated in Sec. 5.6.

A problem with gate-level modeling for functional primitives is that different technologies employ different basic building blocks. The NAND gate is natural for TTL and the NOR gate is natural for ECL. The NAND conveniently implements a sum of products whereas the NOR more conveniently implements a product of sums. The circuit in Sec. 5.6 is implemented as

$$F = (\overline{x_1} + x_2 + \overline{x_3} + x_4)(\overline{x_1} + \overline{x_2} + \overline{x_3} + \overline{x_4}) \cdot$$
$$(x_1 + \overline{x_2} + x_3 + \overline{x_4})(x_1 + x_2 + x_3 + x_4)$$

or
$$F = x_1 \cdot \overline{x_3} + \overline{x_1} \cdot x_3 + x_2 \cdot \overline{x_4} + \overline{x_2} \cdot x_4$$

depending on which technology is chosen by the designer to implement his function.

Although small circuits can be found for which the fault coverage between different implementations varies greatly, experience suggests that for large circuits the fault coverage for an actual circuit is reasonably close to the fault coverage of its gate equivalent circuit. This is not surprising since the stuck-at model has good granularity, it effectively models most, if not all, signal paths, and for a large circuit it is an effective random sample. There are more practical objections to gate-level modeling. It is extremely time-consuming, labor-intensive, prone to error, and not easy to correct if an error is discovered. The alternative is to create functional models. However, that leads to the next question: "How do we fault these models?"

One suggestion is to simply fault the I/O pins of a device. However, even for logic boards comprised mainly of SSI and MSI, an acceptable fault coverage, say 95%, for pin faults usually provides only in the neighborhood of 70–75% fault coverage for internal faults[13,16]. Pin faults are apparently not a good statistical sample. For LSI and VLSI they would be expected to be even less effective since the ratio of gates to pins is much greater.

If we are to create tests for faults internal to a function, then it might be useful to look at a typical function to determine whether any approaches are suggested by that function. Consider the sum-of-products and product-of-sums implementations of a frequently occurring device, the 2-to-1 multiplexer. The test vectors in Table 10.1 will detect all single stuck-at faults in the NAND gate multiplexer shown in Figure 10.2. However, the table also shows that the set of vectors will only detect four of six checkpoint faults in the NOR gate product-of-sums implementation. Input 2 of NOR gate 1 and input 1 of NOR gate 2 may or may not be detected, depending on which value is assigned to the don't cares. An alternative way to view the multiplexer as a functional entity is shown in Figure 10.3.

TABLE 10.1 FAULT DETECTION TABLE FOR THE 2-TO-1 MULTIPLEXER

				Faults detected	
A	B	S	F	(NAND)	(NOR)
0	1	0	0	1.1, 2.1 S-A-1	3.1 S-A-0
1	0	1	0	1.2, 2.2 S-A-1	3.2 S-A-0
x	1	1	1	3.2 S-A-1	2.2 S-A-0
1	x	0	1	3.1 S-A-1	1.1 S-A-0

Faults in the functional unit can be classified as control faults or data faults. The data faults are:

1. Cannot propagate 0 through A
2. Cannot propagate 1 through A
3. Cannot propagate 0 through B
4. Cannot propagate 1 through B

The control faults are:

5. Select A, got B
6. Select A, got both ports, that is, $A + B$
7. Select B, got A
8. Select B, got $A + B$.

The eight functional faults can be detected with the following set of four test vectors:

A	B	S	F	Faults detected
0	1	0	0	1, 5, 6
1	0	0	1	2
1	0	1	0	3, 7, 8
0	1	1	1	4

Comparing this table with Table 10.1 suggests that the don't cares in Table 10.1 should be set to 0. If we set them to 0 and again check the faults in the NOR gate

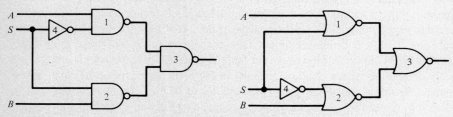

Figure 10.2 Two implementations of the 2-to-1 multiplexer.

10.3 FUNCTIONAL SIMULATION

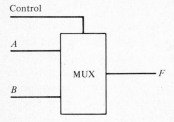

Figure 10.3 Functional view of multiplexer.

model of the multiplexer, we find that the previously undetected faults have now been detected. Using a NAND model as the logic model for a circuit implemented in ECL, whose basic building block is the NOR gate, results in a less accurate fault simulation detection estimate. Using the functional model, we generate a set of tests, akin to the primitive D-cubes of failure (PDCFs) of the D-algorithm, which can detect structural defects in a TTL or ECL implementation, and perhaps in a metal oxide semiconductor (MOS) circuit as well. While not a proof that functional fault modeling is better than, or even as good as, gate-level stuck-at fault modeling, it does suggest that functional fault modeling merits further study.

The above results can be generalized to any multiplexer. For an n-to-1 MUX, $2n$ tests verify that 0 and 1 can be propagated through the n ports. Selection of the wrong port is detected by using the same $2n$ vectors and putting values on the other ports that are opposite to the value on the selected port. With the single-fault assumption it is not necessary to put opposing values on all ports. For a 4-to-1 MUX with two select lines, S_1 and S_0, port 1 is selected by setting $S_1, S_0 = (0, 0)$. A single select line fault is likely to select either port 2 ($S_1, S_0 = 0, 1$) or port 3 ($S_1, S_0 = 1, 0$) but not port 4 ($S_1, S_0 = 1, 1$).

Other functional entities can be similarly processed. The objective is to try to identify invariant properties common to all or most physical realizations. Then, effective tests can be created without detailed structural descriptions. There is the added advantage that test pattern generation can be begun before a complete description of the design has been put in place. The basic types of functional entities include:

Elementary gates: AND, OR, Invert, simple combinations

Latches, flip-flops: JK, D, T

Multiplexers

Encoders and decoders

Comparators

Parity checkers

Registers

ALUs: logic, arithmetic—fixed point, binary coded decimal (BCD), floating point

Memory arrays

In the final analysis, the purpose of fault models is to provide a means for evaluating the effectiveness of test vectors relative to the detection of physical defects in logic circuits. To that end, the modeling of faults for the functional primitives should reflect the types of physical defects that can occur and their effect on functional behavior. As an example, a binary counter with parallel load capability must be capable of performing a parallel load, it must be capable of advancing the count to the next higher binary stage, and it must be resettable. A physical defect which alters any of these functional capabilities must be modeled in terms of its effect on the function; the fault model must reflect the behavior of the device when the fault is present. If the output of the ith flip-flop is S-A-1, then whenever performing a reset, the counter begins counting with an initial value of 2^i rather than 0. In normal operation, when counting up, bit position i resets to zero when bit position $i + 1$ switches to 1. To simulate faulted operation it must be forced to remain at 1.

Fault insertion in functional models can be done in many ways, depending on the capabilities of the language in which the functional models are written. A straightforward way, for a single fault, is to introduce a fault variable v into an expression such that the expression evaluates correctly if $v = 0$, indicating that the fault is not present, and incorrectly when $v = 1$, indicating that the fault is present. Notationally, this can be expressed as

$$F = \bar{v} \cdot f_g + v \cdot f_f$$

In terms of the amount of computation, when a function has many possible faults, it will usually require less CPU time if, whenever possible, a single multivalued fault variable can be used to specify either the unfaulted function or one of n faulted models. Then the fault variable is set before the function is evaluated. Upon entering the function, the fault variable is evaluated once. For the 2-to-1 MUX, the following CASE statement determines whether the fault-free code or code corresponding to a faulted condition is executed.

```
CASE FAULT_#(N) OF
   0:
   1:    A:=1
   2:    A:=0
   3:    B:=1
   4:    B:=0
   5:    A:=B
   6:    A:=A+B
   7:    B:=A
   8:    B:=A+B

CASE S OF
   0:    F:=A
   1:    F:=B
   X:    IF A=B THEN F:=A ELSE F:=X
```

The fault number determines which case statement is to be executed. Case 0 corresponds to the fault-free machine. After the fault is inserted, the simulation code is executed. If the control signal is indeterminate, but the inputs match, then the output is set equal to the inputs, otherwise it is set equal to X. If A and B are m-bit-wide ports, then a more detailed bit-by-bit comparison is necessary.

For a more complex function, such as the fixed-point ALU, additional difficulties arise. Normally, when one is only interested in good machine simulation at a behavioral level, the simulation will be performed with input arguments encoded like those used in high-level languages. Then the native ADD instruction of the host machine can be used. For simulation in an environment where indeterminates may occur, a more detailed simulation is required and it becomes necessary to convert the input arguments to vector form. Then iterative computations are performed to compute the sum $A_i + B_i$ at any bit position, as well as the propagate P_i and generate G_i (see Sec. 9.3.1). The sum begins at the carry-in and proceeds from low-order to high-order bit position. If an input bit position is indeterminate in one vector, but the other input and the carry-in are both 0s, then the indeterminate will not propagate to the next higher position. The iterative method lends itself to any argument size since the number of iterations can be an argument in a loop control statement.

For fault simulation this iterative approach permits simulation of faults internal to the ALU. However, all the P_i and G_i must first be computed, based on the inputs A_i and B_i. Then, as with the MUX previously described, an individual A_i, B_i, P_i, or G_i is faulted, based on the fault number. The ALU result is then computed for either the fault-free or some faulty circuit. The sum at position i is computed using A_i, B_i, and a carry C_i into position i where

$$C_i = G_{i-1} + P_{i-1}C_{i-1}$$

The G_i and P_i can be computed once. Then the faulting of individual parameters and computation of the effect can be performed in a loop until all fault effects have been analyzed.

There are many unanswered questions when considering functional fault modeling. Fault effects in control logic are difficult to analyze and model because of the diverse nature of the design possibilities for such logic. One approach to this problem is to distinguish between external and internal faults[17]. An *external* fault is one which is associated with inputs to or outputs from a control section. For example, a HOLD signal is an external signal which causes some predictable action of a processor and, if faulted, the effects of that fault are usually predictable. If that signal is faulted, the processor may go into a HOLD state when it should not, and then wait for a subsequent signal enabling it to leave the HOLD state, or it may not go to the HOLD state when it should. An *internal* fault is any fault, other than an external fault, which causes an erroneous state transition. The effects of internal faults are extremely difficult to predict because these effects are implementation-dependent and a particular control section usually can be implemented in many different ways. A successful traversal of all arcs in a state transition graph will probably detect most faults, but even that may fail to detect all physical defects.

10.4 THE FUNCTIONAL ATPG

If a device or system is defined as an interconnection of the basic functions previously listed, then well thought out fault models for the basic functions provide an orderly way to define faults for an entire device. These fault models are then useful for evaluating the effectiveness of test vectors applied to the device; that is, the functional faults provide a figure of merit for sets of test vectors. Furthermore, with a representative set of functional faults, it may be possible to create test vectors which are invariant with respect to structural implementation.

As useful as functional models are for functional simulation, there are better arguments for the use of functional models in conjunction with an ATPG. Perhaps the most compelling reason is the fact that gate-level ATPGs simply have not worked. Something as elementary as a four-stage binary counter, capable of counting from 0 to 15, can frustrate most ATPGs, since it could require as many as 16 time frames to propagate or justify an argument. An additional argument for functional level test generation stems from the fact that, with shrinking dimensions of ICs, problems such as pattern sensitivity become more severe. Primitive function test pattern sets can be made more flexible to address the specific pattern sensitivity characteristics of a particular device.

Functional modeling techniques, while analogous to gate-level methods, must of necessity be more flexible. It is simply not practical to store D-cubes and PDCFs for a large function such as a 32-bit ALU. Therefore, the test must be expressed in terms of algorithms, augmented with tables[18]. To accomplish this, an algorithm for a functional primitive can be partitioned so as to be expressed in terms of its data inputs and its control inputs. The control inputs define the operation to be performed, select the input port(s), and possibly select a destination.

To illustrate this functional modeling approach, let us suppose that we are to develop processing routines for an ALU. It may have several capabilities, including fixed-point addition and subtraction, AND, OR, Invert, Complement, All 0s, and All 1s. In addition, it may be able to pass an argument straight through from any input port to the output without it being altered. Each of these operations requires a specific setting on the control lines. During justification, if an all-0s or all-1s vector is required on the output, then the appropriate control line values are selected and the input ports are ignored. If a specific value is required on an output port, and if the control lines can be set to pass a value from an input port directly to the output, then that setting is used for justification. If a straight pass-through does not exist, then a logic operation is used. The desired value is placed on one port, the all-0s (all-1s) is placed on the other port, and the OR (AND) operation is selected.

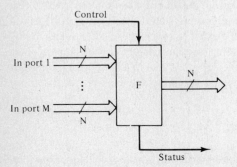

Figure 10.4 General functional model.

10.4 THE FUNCTIONAL ATPG

The entire process just described can be structured as a sequence of IF...THEN statements. The possibility exists that one or more operations may be blocked by virtue of the fact that the machine, as designed, does not use the operation. The all-0s operation may exist in the ALU but the control section, whether microcoded or hardwired, may not be able to support that operation. In that case, the operation should be marked as BLOCKED; then the ATPG does not attempt to again use that operation for justification or propagation operations. This is not the same as a conflict. A BLOCKED operation occurs in a microcoded machine if the particular operation does not appear in any microinstruction. A conflict occurs in a microcoded machine if two or more device operations exist but never in the same microinstruction.

Another significant difference between gate-level and functional primitives lies in the fact that the propagation and justification rules for sequential devices can, and usually will, be sequences of operations rather than single operations. As a rather simple example, the test for an edge-triggered flip-flop may be expressed as a sequence in which a 1 on the Data line is clocked in, and then the Clear line is exercised to confirm that it will, indeed, reset the flip-flop output to 0.

A more complex example of sequential devices is the serial/parallel shift register or the counter. Processing is complicated by the fact that these devices may be required to be inactive, that is, in a Hold state. The range of functional operations can be described symbolically as:

L left shift
R right shift
P parallel load
H hold (do nothing)
U count up
D count down
C clear

A particular operation that must be performed is expressed in terms of these operators.

EXAMPLE ■

A 1 can be justified on the ith output of a counter by means of the following sequence:

$$CH*\{UH*\}^{2^i}$$

The notation indicates that a 1 is obtained by performing a clear, followed optionally by one or more hold operations (the asterisk denotes an arbitrary number of consecutive operations of the type specified by the operator to its immediate left). Then, the entire sequence in braces, which is a count up optionally followed by zero or more hold operations, is repeated the number of times indicated by the superscript, in this case 2^i times. ■ ■

These abstract operations must be related to real counters, either those available from semiconductor manufacturers, or those designed by the user. The operations correspond to I/O pins which perform the operation. It is also necessary to relate the operations to such things as rising or falling clock edge, depending on which edge enables the activity.

Rules can also be defined for propagation and implication. Again, the rules are expressed functionally. The signal values are analogous to D-cubes in that they specify, for a D or $\bar{D}$ on an input, exactly what signal(s) must appear on other inputs to make the output sensitive to the D or $\bar{D}$. For a shift register which has a clear line K and control lines S_1, S_0 which may select hold, parallel load, shift left, and shift right, the following table expresses some of the propagation rules:

Composite	K	S_1	S_0	$y(i-1)$	$y(i)$	$y(i+1)$	A	$Y(i)$
C/H	D	—	—	—	1	—	—	$\bar{D}$
L/H	0	D	0	1	0	—	—	$\bar{D}$
P/H	0	0	D	—	1	—	0	$\bar{D}$
L/L	0	1	0	D	—	—	—	D
H/H	0	0	0	—	D	—	—	D
R/R	0	1	1	—	—	D	—	D
P/P	0	0	1	—	—	—	D	D

In this table $y(i)$ represents the present value in the flip-flop at position i and $Y(i)$ represents the new value in flip-flop i after clocking the register. For entry u/v in the composite column, u denotes the action taken by the fault-free machine and v denotes the action taken by the faulted machine. The first line defines the conditions for propagating a D on the Clear line to output i. It requires first clocking a 1 into register flip-flop $y(i)$. The faulted machine will perform a hold operation. To propagate a D through control line S_1 to output $y(i)$ requires a 1 in register bit position $y(i-1)$ and a 0 in position $y(i)$. The above table is not a complete list. For example, propagation through a control line could also be accomplished with a 0 in $y(i-1)$ and a 1 in $y(i)$.

Implication tables can also be expressed functionally. They can be created via composition; that is, if a D ($\bar{D}$) occurs on one or more lines, then the results can be computed individually by first setting $D = 0$ ($\bar{D} = 1$) and performing the computation, then setting $D = 1$ ($\bar{D} = 0$) and again performing the computation. After computing each case individually, set the output or internal state variable to 0, 1, or x if it has value 0, 1, or x for both good machine and faulted machine. If it assumes value 0 (1) for the good machine and 1 (0) for the faulted machine, set it to D ($\bar{D}$). If it assumes value x for the good machine, set it to x. If it assumes x only for the faulted machine, then its value depends on whether the user wants to consider possible detects or only absolute detects.

A functional test pattern generation algebra has been developed for use in conjunction with HDLs[19]. We first define $U = \{0, 1, D, \bar{D}\}$. Then, if S_i is a subset of U, x^{S_i} denotes the fact that $x \in S_i$, and the following equations hold:

$$x^{S_i} \cdot x^{S_j} = x^{S_i \cap S_j} \tag{10.1}$$

$$x^{S_i} + x^{S_j} = x^{S_i \cup S_j} \tag{10.2}$$

10.4 THE FUNCTIONAL ATPG

$$x^{S_i} + x^{\overline{S_i}} \cdot y = x^{S_i} + y \tag{10.3}$$

$$x^{S_i} + x^{S_i} \cdot y = x^{S_i} \tag{10.4}$$

For the AND operation, the following equations define all of the combinations on the inputs a and b which will produce the indicated value on the output c:

$$c^0 = (ab)^0 = a^0 + b^0 + a^D b^{\overline{D}} + a^{\overline{D}} b^D \tag{10.5}$$

$$c^1 = (ab)^1 = a^1 b^1 \tag{10.6}$$

$$c^D = (ab)^D = a^D b^1 + a^1 b^D + a^D b^D \tag{10.7}$$

$$c^{\overline{D}} = (ab)^{\overline{D}} = a^{\overline{D}} b^1 + a^1 b^{\overline{D}} + a^{\overline{D}} b^{\overline{D}} \tag{10.8}$$

The Invert function is obtained by complementing the superscript; that is, if $b = \overline{a}$, then $b^i = a^{\overline{i}}$ where $\overline{S}_i$ is obtained by complementing each of the individual elements contained in S_i. The equations for the OR gate can be computed from the equations for the AND gate and the inverter. Eq. 10.5 states that if c is an AND gate with inputs a and b, then a 0 is obtained on the output by setting either a or b to 0 or by putting D and $\overline{D}$ on its inputs. Note from Eq. 10.8 that a $\overline{D}$ on both inputs does not put a 0 on the output but, rather, a $\overline{D}$.

These basic equations for the logic gates can be used, together with the four rules, to compute D cubes for more complex functions.

EXAMPLE ■

The JK flip-flop behavior can be represented by

$$Q = J\overline{q} + \overline{K}q$$

where Q is the next state and q is the present state. The D-cubes are computed using

$$\begin{aligned}
Q^D &= (J\overline{q} + \overline{K}q)^D \\
&= (J\overline{q})^0 (\overline{K}q)^D + (J\overline{q})^D (\overline{K}q)^0 + (J\overline{q})^D (\overline{K}q)^D \\
&= J^0 K^0 q^D + J^D K^0 q^D + K^{\overline{D}} q^1 + J^1 K^{\overline{D}} q^D + J^D K^{\overline{D}} q^D \\
&\quad + J^1 K^1 q^{\overline{D}} + J^D K^1 q^{\overline{D}} + J^D q^{\overline{D}} + J^1 K^{\overline{D}} q^{\overline{D}} + J^D K^{\overline{D}} q^{\overline{D}}
\end{aligned}$$

The result can be used to create a table of propagation and justification cubes for both D and $\overline{D}$ values. ■ ■

The basic operators can also be used to create tables for more complex functions. The adder can be created a bit position at a time. The sum and carry tables are created from the exclusive-OR and the AND, respectively. These are then used to build up one complete state of a full adder. That, in turn, is used to build an n-stage adder. The multiplexer can be expressed in equation form as

$$F = S \cdot A + \overline{S} \cdot B$$

where S is the select line and A and B are the input ports and F is the output port. Since it is now expressed in terms of ORs and ANDs, the cubes for the equation can be generated.

10.5 ARTIFICIAL INTELLIGENCE METHODS

In Chap. 2 we described the PODEM test pattern generator. It made use of a "branch-and-bound" technique borrowed from the field of artificial intelligence (AI) to reduce the problem of combinatorial explosion. Specific assignments of logic values to primary inputs were evaluated and either accepted or rejected based on whether or not they blocked a potential solution to the problem. Rejection of a value on an input significantly reduced the problem domain. Interestingly, PODEM seeks to find the right solution by avoiding wrong choices. That is effective because there are usually many input combinations which will test a particular fault.

We now look at two test pattern generation systems which use techniques borrowed from AI: SCIRTSS and Hitest. The first of these, SCIRTSS, uses heuristics, or learning mode, to search through state space. When searching, the number of choices grows exponentially and it is not practical to enumerate them all. Two assumptions in AI applications that influence SCIRTSS are:

1. A limited n-level search may have a greater payback than an exhaustive $n - 1$ level search.
2. Self-modifying methods, based on previous results, can improve the probability of pursuing the correct path in a search.

Hitest is an example of an expert system. It has knowledge of functional devices used in a circuit and it can use specific behavioral knowledge, provided by the test engineer, on how those devices can be controlled or observed in a particular circuit. Expert systems, in general, tend to apply techniques on the premise that the techniques have worked in the past, or that the techniques resemble those used by humans to solve the problem. We will also examine some related research activities which may ultimately help to solve test generation problems. However, first we examine one of the circuit configurations that has always proved to be a stumbling block for ATPGs, the state machine.

10.5.1 The State Machine Problem

Automatic test pattern generators have generally been quite ineffective when applied to sequential circuits. The problems caused when the inherent delay of combinational logic is used to create clock pulses for edge-triggered flip-flops have been previously discussed. ATPGs traditionally have operated strictly in the logic domain and circuits that operate in the time dimension are simply beyond the capability of such ATPGs. However, even when a circuit is completely synchronous, it still is capable of thwarting the ATPG. Consider again the circuit of Figure 7.18. We pointed to a fault on an AND gate which required that the flip-flops Q_2, Q_1, Q_0 contain the values $1, 0, 0$. If the ATPG performs a reset on the circuit, then it would appear that the problem is solved by simply driving Q_2 to a 1. Seemingly an easy solution is to be had.

Suppose now we represent the circuit by the state transition graph of Figure 10.5. In this graph, the first number of the pair at each arc represents the value of

10.5 ARTIFICIAL INTELLIGENCE METHODS

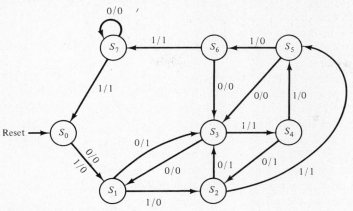

Figure 10.5 State transition graph.

IN. The second number represents the value of *OUT*. Therefore, when the machine is in state S_1 and $IN = 1$, the machine will have an output value of 0 and will go to state S_2 in response to a clock pulse. More importantly for our purposes, we see that it is necessary to pass through state S_3 to get to state S_4. But state S_3 corresponds to $Q_2, Q_1, Q_0 = 0, 1, 1$. In other words, after performing a reset on the circuit, the ATPG must be clever enough to get a $Q_2, Q_1, Q_0 = 1, 0, 0$ by first driving the flip-flops to $Q_2, Q_1, Q_0 = 0, 1, 1$. But does it seem reasonable to implement an ATPG such that the ATPG tries to put a 1 on the output of a flip-flop by driving it to the 0 state?

10.5.2 SCIRTSS

The testing problem in state machines occurs because the typical ATPG does not take a global view of a circuit; it processes components in isolation. There are interrelationships between flip-flops in a circuit that cannot be seen at the gate level. However, from a graph it is a trivial exercise to determine how to get from the reset state S_0 to the objective state S_4. This observation is the basis for the system called SCIRTSS (sequential circuit test search system)[20]. The SCIRTSS system employs two models of a circuit; it uses a detailed gate-level circuit description and a register transfer level description expressed in AHPL.

SCIRTSS assumes that a state machine is in a known starting state. Within that state, a conventional *D*-algorithm attempts to find a partial test for some selected fault. The test is propagated forward and justified back either to I/O pins or to stored state devices, including latches, flip-flops, or register elements. When the fault is propagated to a stored state element, then we say that it is *trapped* in that element.

The *D*-algorithm is strictly combinational; it does not attempt to create multiple time images in order to propagate a fault through sequential elements. Once the test has been extended to boundaries defined by primary inputs, primary outputs, and memory elements, the *D*-algorithm is done. At that point, the state of those memory elements which must be changed from their present value in order to satisfy the justification requirement is recorded. These values define a state to which the machine must make a transition from its present state in order to cause the fault to either appear at a primary output or become trapped in a memory element.

To drive the machine from its present state to the objective state, SCIRTSS enters the *sensitization search* phase in which it employs the AHPL description of the circuit. The AHPL description implicitly specifies a transition directly to the next state or it specifies several possible next states and the conditions that determine which of these next states is selected. The sensitization search is essentially a tree search in which SCIRTSS, starting at the present state, tries to find a sequence of inputs which will drive a machine to the objective state.

If the sensitization search is successful, then SCIRTSS has computed a sequence which, starting at the present state, either makes the fault visible at an output or has trapped the fault in a memory element. If the latter has occurred, then SCIRTSS enters the *propagation search*. In this phase, SCIRTSS attempts to drive the machine through a sequence of states which will make the fault visible at an output. This phase, like the sensitization phase, tries to control the behavior of a state machine by using the AHPL description to compute state transitions.

When a complete test has been achieved, including both a sensitization and propagation search, SCIRTSS again reverts to a gate-level description. This time, the gate-level description is used to perform gate-level simulation. The simulation must confirm that the fault was, indeed, detected. In addition, the simulation determines what other faults were detected and, finally, the simulation will also indicate whether or not any other faults were trapped in bistables by the computed sequence. If faults are trapped, then one of them is selected for processing and SCIRTSS again goes into the propagation search. If there are no trapped faults, then SCIRTSS goes into the sensitization search. The entire process is shown in flowchart form in Figure 10.6.

The tree search conducted by SCIRTSS is subject to data explosion. With m inputs, a search depth of k states could produce a tree with 2^{mk} sequences, resulting in a need for massive amounts of memory and CPU time. To address this problem a machine is viewed, as in Figure 10.1, as a control part and a data flow part. State changes in registers contained in the data flow part are viewed strictly as data transfers, not as state changes. A second simplification, when searching for sequences of state changes, is the use of heuristics. The heuristic assigns a value to each node according to the following formula:

$$H_n = G_n + w \cdot F_n$$

where G_n is the distance from the starting node to node n, F_n is a function of any information available about node n as defined by the user, and w is a constant that determines the extent to which the search is to be directed by F_n. The object is to find the easiest or shortest path to an objective state.

EXAMPLE ■

We consider the eight-state machine which caused testing difficulties with the *D*-algorithm but which, in Chap. 7, we were able to test using a scan

10.5 ARTIFICIAL INTELLIGENCE METHODS

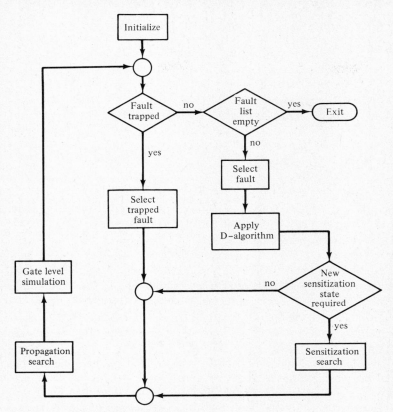

Figure 10.6 SCIRTSS flowchart.

path. We first apply a reset, which puts the machine in state S_0. From the fault list we select the third input to gate 23 S-A-1, and apply the combinational D-algorithm. We find that we can sensitize a path to the output if we set $Q_2, Q_1, Q_0 = 1, 0, 0$. Therefore, we wish to go from state S_0 to state S_4. Using the AHPL description given in Sec. 10.2, we see that when in state S_0 we will always go to state S_1. From there, we may go to state S_2 or S_3, depending on what value we assign to *IN*. Continuing one additional step, as shown in the tree of Figure 10.7, we see that either of the two sequences $0, 0, 1$ or $1, 0, 1$ will put the machine in state S_4. We arbitrarily pick $0, 0, 1$ and drive the machine to S_4. Then, the input IN = 1 is applied. Finally, the entire sequence is simulated at the gate level to confirm its effectiveness and to determine whether other faults are tested. Note that when creating a tree, previously occurring states are crossed off. No new information is gained by pursuing those paths.

At the conclusion of gate-level simulation, one or more faults may be trapped in a flip-flop. If the output of gate 16 is S-A-1, it will not be detected originally when the reset is applied and the first two transitions to S_1 and S_3 will not distinguish between the good machine and the machine with the

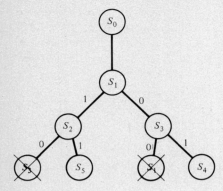

Figure 10.7 Sensitization search tree.

fault. However, in the transition to S_4, the faulted machine goes to S_5, hence a $\overline{D}$ is trapped in Q_0. From the state graph, Figure 10.5, it is seen that $IN = 0$ causes an output of 1 from the good machine and an output of 0 from the faulted machine. Furthermore, it was not even necessary to clock the circuit. If there are no faults trapped when in state S_4 and if the output of gate 12 S-A-0 is selected from the fault list, then the D-algorithm would start by assigning the values 1, 1 to the inputs of gate 12. The fault would be propagated into Q_0 by assigning $IN = 0$, and $Q_2, Q_1, Q_0 = 0, 0, 1$ and then clocking the machine. From the graph, it is seen that there are a number of ways that the sensitization search can go from S_4 to S_1. The signal IN can be set to 0 or 1, but it is also possible to reset the machine and go immediately to state S_0, hence there are three possible successor states to S_4. Furthermore, the transition from S_0 to S_1 is trivial to compute. However, it may be desirable to force the machine through states S_5–S_6–S_7 from state S_4 in order to exercise additional logic and attempt to pick up faults that might otherwise require individual processing. This can be done with the heuristic. The w term and the F_n term in the heuristic can be chosen to force SCIRTSS to go through those additional states rather than directly to S_1. It may also be desirable to modify the heuristics after the process has run for some time in order to force state transitions through other logic. This modification on-the-fly requires that intermediate results be available for inspection. ■ ■

The trapped fault in Q_0 can be processed by the D-algorithm or it can be processed directly from the graph. The D-algorithm can propagate the D in Q_0 to OUT (through gate 22) simply by setting $IN = 0$. The value $IN = 0$ could also have been determined from the graph. The fault-free machine is in state S_1 and the faulted machine is in state S_0. Therefore, it is easily determined from the graph that $IN = 0$ causes different outputs from the two states.

We pointed out that SCIRTSS employs two models, a gate-level model and an AHPL model. This observation is rather interesting because many systems implicitly assume that the ATPG and the simulation models must be identical. This assumption has the effect of locking in a gate-level model for both ATPG and

10.5 ARTIFICIAL INTELLIGENCE METHODS

simulation in order to satisfy a user requirement for precise simulation. However, for an ATPG to be successful, it must make use of more information than simply the gate-level interactions. As is shown with SCIRTSS, this does not rule out the use of gate-level simulation to confirm the success of the ATPG, or to uncover serious timing problems with the pattern developed by the ATPG.

Some observations concerning SCIRTSS:

> It must be possible to correlate the abstract states S_i with values on the flip-flops; for instance, it is necessary to know that state S_4 corresponds to $Q_2, Q_1, Q_0 = 1, 0, 0$.
>
> The heuristics can be updated to reflect successful state transition sequences. In the example given, a transition from S_4 to S_6 or S_7 is always performed more quickly when the first transition is directly to state S_5 rather than to S_2.
>
> It is possible to give up on a fault and succeed later when the sensitization search is started from some other state.
>
> Unlike sequential D-algorithms, the sensitization search is forward in time; in fact all sequential searching is forward in time. This is possible because the state machine is a circular construct.
>
> It is not necessary to have a completely specified objective state. If the D-algorithm leaves one or more flip-flops in the state machine unassigned, then the objective may be a set of states determined by setting the xs to 1 and 0. The sensitization search is successful if any state in the set is reached.
>
> During gate-level simulation it is necessary to keep track of the fault effects of all faults of interest since a fault may, over time, affect both data registers and one or more control flip-flops. This could cause the fault to mask its own symptoms.
>
> Arguments required in the data flow section to effect a propagation may originate in other registers; therefore it may be necessary to derive sensitization sequences in which an argument is first loaded from a data port into a register or accumulator, and then used to propagate the fault to an output or flip-flop.
>
> As the implementation was reported, SCIRTSS employs trial vectors at the data ports. These may include such typical vectors as the All 1s, the All 0s, 100...0, 010...0, 000...1, etc. However, with functional test generation employing symbolic propagation, it should not be necessary to restrict the data at input ports to user-suggested trial vectors.
>
> SCIRTSS can also be applied to circuits controlled by microcode,

It is particularly interesting that SCIRTSS attempts to avoid backtracking. This is in some sense analogous to PODEM which avoids having to perform conflict resolution by working forward from primary inputs. Backtracking to justify assignments in previous time images can be especially ill-advised when applied to incompletely specified sequential machines. Once the backtrack operation enters an illegal or unused state which cannot be reached from a forward in time traversal of the machine states, all computations from that point became superfluous.

State transition graphs do not model well the activities that take place in asynchronous circuits. In an asynchronous circuit two or more processes may occur simultaneously, and some activities may be inhibited from proceeding until a signal is received from another circuit indicating that it has completed its task. Some hardware design languages provide constructs to accommodate these asynchronous features. Typical among these are such constructs as COBEGIN, which activates a task to run concurrent with the task that spawned it, and WAITFOR, which specifies that a task cannot proceed until a signal arrives to indicate that some other task has been completed.

The Petri net is a directed graph that is capable of expressing asynchronous activities. It consists of two kinds of nodes, called *places* and *transitions*, that are connected by directed arcs[21]. The transitions, represented on a diagram by bars, can only be connected to places, represented by circles, and the places can only be connected to transitions. The places from which arcs emanate are called input places of a transition and the places on which an arc terminates are called output places. A Petri net in which every transition has exactly one input place and one output place is a state machine.

A place may have a token (sometimes called a marker) or it may be empty. If all of the input places to a transition have tokens, then the transition is enabled, and this permits the transition to fire. In the process of firing, the transition moves one token from each input place and puts one token into each output place. A Petri net is illustrated in Figure 10.8. The place designated p_1 has a token. Upon firing, the token is transferred to p_2. When transition t_2 fires, a token is placed in both p_3 and p_4. The transition t_4 cannot fire unless there are tokens in both p_3 and p_5. A token in place p_6 will enable both t_5 and t_6 but only one of them will fire.

The Petri net has been successfully used in conjunction with SCIRTSS[22]. It is used to reduce the search cost required to reach a goal state and also to generate input vectors used to expand the state space nodes. Because even a synchronous design frequently requires that several events be properly set up before an operation can occur, the Petri net representation can sometimes be more successful than a state machine representation when trying to control a synchronous machine. Consider as a simple example a CPU controlled by a state machine which implements multiplication by performing a succession of shift and add operations. The host CPU must reach some specific state S_c and the multiplier must be reset and then cycled through a series of computations. A counter must be initialized with some value indicating the number of shift and adds to be performed. The multiplier and multiplicand must be loaded into the appropriate registers. Then, the controlling state S_c must be held fixed while the product is being formed.

If this operation is represented by a Petri net in which the goal state is the actual multiplication, then the conditions that must be fulfilled before the beginning of multiplication represent places which require tokens. Each of these places must therefore be individually analyzed in order to compute sequences of firings which move tokens into these places. These places, such as loading of a multiplicand,

10.5 ARTIFICIAL INTELLIGENCE METHODS

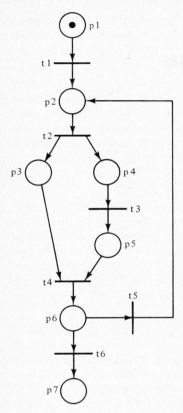

Figure 10.8 Petri net.

become subgoals. Ultimately, from the original Petri net graph of the machine, a kind of inverse tree is formed in which each path leads into the goal state. These paths specify starting states and transitions which contribute to moving tokens forward in order to satisfy the subgoals and enable firing of the multiplication transition.

10.5.3 Hitest

The Hitest ATPG system was created by Cirrus Computers, Ltd., a company which had considerable experience developing board test programs for printed circuit boards. The system architects based its development on observations of successful test engineers generating tests for those printed circuit boards. Hitest attempts to capture the essential characteristics of their endeavors, and embody them in a system software environment. In that sense it can be called a knowledge-based artificial intelligence system. The system permits test engineering knowledge to be stored in usable form by the system and it is also interactive so that a test engineer can take control and apply guidance and corrections if needed[23].

Overview When analyzing the approaches used by successful test engineers, it is immediately apparent that these engineers do not decompose a circuit into interconnections of elementary gates and write tests based on some particular algorithm. The successful engineer begins by identifying high-level features. These include buses, microprocessors, peripherals, state machines, frequency dividers, and counters. The engineer can also identify and attach meaning to some signal names such as RESET, CLOCK, S-100, and RS-232. He uses the knowledge gained from these observations to define a test strategy, which usually includes plans for individually observing and controlling each of the various functional entities. However, unlike the gate-level ATPG, which solves the same control and observability problems repeatedly, the test engineer solves the problem once and exploits the solution as long as it is successful in deriving tests for untested faults[24].

Successful test strategies and specific knowledge items useful to the test engineer are stored in an extensible knowledge base where they can be represented and used by both algorithmic and heuristic techniques. The driving force is to constrain the test generator to look for tests among useful sequences, in contrast to gate-level test generators, which usually consider all possible sequences. Although some gate-level ATPGs have in the past used constraints to prohibit illegal combinations on stored state variables, the number of possibilities is usually too great to permit specifying all such illegal combinations. Specifying *positive constraints*, that is, specifying valid input sequences or state transitions, appears to be more manageable.

To assist in building test plans around high-level features, Hitest brings the test engineer into the loop by providing an interactive system. The engineer can take control when the test generator is stuck in a maze of details. In addition, he can update test strategies for a particular circuit. In this way, on a difficult to test board, the engineer performs the difficult tasks, such as analyzing the logic to compute sequences of state transitions, while the test generator performs tedious bookkeeping chores associated with test generation.

Since it was observed that test engineers work extensively with circuit diagrams, a graphics package was developed. This package provides the engineer with the ability to put selected parts of a design onto a video display terminal with the amount of detail controlled by the user. Different colors on the screen denote different logic values. These visual cues help the test engineer to quickly identify logic values at the various components.

The graphics package operates in tandem with a simulator which permits the engineer to make assignments to primary inputs, simulate them, and then immediately see the results on the screen and determine whether the assignments accomplished the desired effect. This system has no page boundaries; that is, the entire schematic is one large page, and when viewing the results of simulation, the user can follow changing signals across this page through a window. This window can move across the page and zoom in or out to let the user focus in on particular nodes or get a more global view of activity. Since the engineer can visually follow color-coded logic signals across the diagram, he can more quickly spot blocking signals on gates or functions which inhibit a sensitized path from propagating to an output.

A timing analyzer display format (see Chap. 4) is also provided so that simulation results can be visualized temporally. In addition to the usual features

10.5 ARTIFICIAL INTELLIGENCE METHODS

found in these kinds of display formats, Hitest can also put up two waveforms of the same node; one waveform shows the good machine response at that node and the other shows the response at that node when some specified fault is present in the circuit. The schematic display and the analyzer display complement each other. The engineer can use the timing analyzer display to identify the time frame when an event either occurred or failed to occur, and he can use the schematic diagram to help identify why the event did or did not occur.

The Hitest system employs Cirrus Circuit Language (CCL) to describe the circuit[25]. It provides:

Structural and functional modeling

Uni- and bidirectional buses

Grouping of wires into vectors

Multiple technologies (MOS, ECL, TTL, and others)

Expression of complex sequences of events in stored state logic

Control of the test environment is accomplished through the Cirrus Waveform Language (CWL). It permits specification of the logic values for primary inputs and also permits separate specification of timing associated with the pins. Therefore, timing relationships at the pins can be changed independent of the logic values. This timing information makes it possible to reproduce the timing set (TSET) structure found on dynamic testers. CWL is also used to specify the values on outputs.

Cirrus Waveform Language contains statements to control both the simulator and the test generator. In a typical scenario, the user may be simulating input stimuli, and he may briefly interrupt simulation activity to request that the ATPG perform some task. The task may be a request that the ATPG justify a value at some node or a request to propagate an internal state to an output. The waveform fragments created by the ATPG are edited into the waveform and the simulator continues either until another ATPG statement appears or until all waveform statements have been executed.

The CWL waveforms can be specified using high-level constructs and looping facilities. Postprocessors convert the waveform into corresponding constructs for the test equipment on which the test is to run. It is possible that the test may not run on a tester for which it was intended because the tester is not fast enough or flexible enough. In those cases the waveform may have to be manually converted by a test engineer into a sequence that can be implemented on the tester. Therefore, comments and structure of CWL are preserved throughout the process of developing the test, so that the test is decipherable by a test engineer. A preprocessor assists in converting tests from some specific tester types into the CWL format. As an additional bonus, the availability of both a preprocessor and postprocessor allows conversion of tests for one particular tester into the correct format for another tester.

The Test Generator The Hitest test generator is based on the premise that successful test pattern generation must somehow be constrained to state transition sequences that are both useful and realizable in the controlling device(s) of the circuit under test in order to efficiently test that circuit. Test pattern generators must be inhibited from random, unproductive searches. A simple example can illustrate this situation. Consider a microprocessor driving a serial input/output (SIO). The test engineer views the SIO as a separate entity. To test the SIO, he uses a small subset of the microprocessor instruction set to develop a loop or small series of loops. This test program is used repetitively to drive test stimuli through the SIO to its output pin(s) and to read data which the microprocessor may then redirect to observable outputs. This is done until such time as the test engineer feels confident that the SIO has been adequately tested or until fault simulation indicates sufficient fault coverage.

Given a circuit of this type, the gate-level ATPG would attempt to drive tests through the microprocessor, one logic gate at a time. If ever it were successful, all traces of its success would literally be swept away at the conclusion of the test and it would attempt to repeat the entire process for the next fault selected in the SIO.

If the ATPG were designed to emulate the process employed by the human test engineer, it would use a repetitive sequence to control the microprocessor. The sequence would configure the SIO for some specific mode of operation and freeze that mode until it had treated all faults that were accessible from that mode. The microprocessor would, of course, have to loop through some sequence of instructions to send new data to the SIO when the SIO signaled that it was ready for new data, and the SIO would convert this to serial form for transmission. Similar considerations hold when receiving data. When it was determined that no more faults were accessible from a given mode, another mode would be selected and the process continued. Eventually, all testable faults in all modes in the SIO will have been processed.

In the above scenario, suppose that a knowledge base existed which defined how the microprocessor transmits data over a bus to a peripheral and how the microprocessor receives data from the bus. With that knowledge an ATPG could be directed to process a gate-level model of the SIO for the purpose of generating structural tests, and use the knowledge base to complete the exchange of data between the microprocessor and the SIO. The knowledge required for retrieving instructions from a RAM or off-chip and using them to control the SIO is relatively low-level. In some microprocessors, it consists of performing a Load Accumulator or Load Register instruction, followed by Store Accumulator or Store Register, and then a wait loop in which the microprocessor waits for a signal from the SIO indicating that it is ready for the next byte of data.

In this scenario, the microprocessor is constrained not by specifying illegal combinations on flip-flops but rather by a paradigm. The paradigm, a model of permissible behavior, guides the test generator in its attempt to use the microprocessor when testing other devices. Since the microprocessor may be very complex, and only a very small subset of its capabilities are needed to drive other devices, the constraints are positive, stating what the microprocessor is permitted to do, rather than negative, stating what the microprocessor is prohibited from doing. This can be done by providing a stored sequence of instructions which take

10.5 ARTIFICIAL INTELLIGENCE METHODS

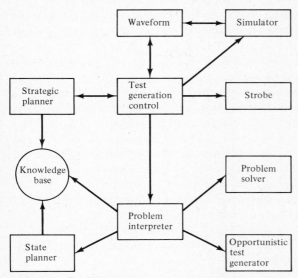

Figure 10.9 Hitest automatic test pattern generator.

the microprocessor through a complete I/O operation. Another approach is to provide a scaled-down model of the microprocessor for use when testing other structural or functional entities. This reduced microprocessor contains only a few very elementary operations, a subset of the original set of operations, so that the ATPG has few, if any, choices when using the microprocessor to test other devices. This does not, of course, obviate the need for a complete model of the microprocessor for testing of its structure.

A block diagram of Hitest is shown in Figure 10.9. The CWL is in overall control of the system. Commands are contained in an *Outline Waveform*, which emits orders to the simulator and to Test Generation Control. In fact, the test generator is activated by waveform statements in what could otherwise be a normal simulation job. The Outline Waveform may request that the test generator find a test for a particular fault or set an internal net to some value. It may define a basic looping structure as a framework within which the test generator has freedom to exercise some primary input pins. It may define some timing rules and impose other restrictions on some primary inputs.

When Test Generation Control gets a command from the waveform, it first determines whether it is a reserved word. The reserved words activate the Strategic Planner or the Strobe. The Strobe uses current simulated output values to produce waveform statements which specify values on the primary outputs in CWL format. These are then edited into the waveform. If the Strategic Planner is called it chooses test generation tasks and test generation problems.

If a command from the waveform does not contain a reserved word, then the Problem Interpreter is invoked. It uses the definition of the problem to drive the Problem Solver, the Opportunistic Test Generator and the State Planner. In solving problems, the contents of stored state devices are considered to be pseudo-primary inputs and pseudo-primary outputs. Objectives can be specified to change the contents of these devices or to make the present state of one or more of them observable at an output.

The Opportunistic Test Generator attempts to find tests from the present state of the ATPG. Its job is to drive fault effects to primary outputs or to stored state devices which are about to be clocked; in other words, it looks for opportunities to test faults from the present state of the circuit. To accomplish this objective, it uses the PODEM algorithm, modified to handle tri-state and bidirectional signals. To find untested faults which may be testable from the present state, Xs are assigned to uncontrollable primary inputs and internal nodes, U to unassigned inputs, and the logic values 0 and 1 to input and stored state devices which have these values presently assigned. These values are then propagated through the circuit. If fault effects are trapped in a register or blocked at a logic gate, then it may be possible to propagate these fault effects to an output or to another stored state device if a propagation path exists with unassigned values on all its nets. The Opportunistic Test Generator converts the Us to 0s and 1s by means of assignments on primary inputs, using the PODEM algorithm.

If the Opportunistic Test Generator cannot find faults which can be driven to outputs merely by converting Us to 0s and 1s, then the Problem Solver must find a new machine state which provides opportunities for the Opportunistic Test Generator. If it can decide on a next state which is useful, that is, one which causes faults to be either visible or trapped in stored state variables, then it resorts to the PODEM algorithm to attempt to achieve that state. If PODEM cannot achieve the objective within some predetermined number of decisions on the primary inputs, then it is necessary to resort to use of the State Planner.

The State Planner is the area where interactive test generation becomes important. The human test engineer can enter the loop. He provides guidance so that the system makes a transition to the desired state efficiently, without going up blind alleys. Additional capabilities in the State Planner include the ability to use stored sequences of inputs which drive a machine through specific sequences of states. These include sets of control and observability waveforms. Of major importance to the State Planner is the concept of the *well-known signal* (WKS), which is a signal or set of signals that can be observed as a single unit, usually a register or scan path. These signals are defined by the test engineer and recorded in the knowledge base. Associated with the WKS is an observability waveform. When fault effects are driven into a WKS, the observability waveform is retrieved by the State Planner and used to drive fault effects to observable outputs. For the scan path, it is simply a sequence which puts the scan path into test mode and then clocks out the scan path contents. For a register in a microprocessor, it may be a Store Register instruction which makes the register contents visible on the data bus.

A typical starting sequence for a simple combinational circuit may appear as follows:

```
begin
    testgen(next_task);
end
```

10.5 ARTIFICIAL INTELLIGENCE METHODS

When the waveform routine encounters the testgen statement, the Test Generation Control is invoked. The next_task problem name is a reserved work. The Strategic Planner comes into play at that point to find a new test generation task (TG_Task). The TG_Task consists of an Outline Waveform together with definitions of any test generation problems (TG_Problem) to be solved during execution of the task.

For a combinational circuit, a rather simple Outline Waveform is used, which may appear as follows:

```
begin
    testgen(spot_them);
    testgen(strobe);
    testgen(next_task);
end
```

Since it is a combinational circuit, the first statement following the begin statement activates the Opportunistic Test Generator. The test generator is instructed to spot faults. Therefore it selects a fault and produces input values to drive the fault to an output. The input values are added to the waveform and then simulated. The next statement, strobe, accesses the simulated results and uses them to add expected output values to the waveform. The next_task statement causes the Strategic Planner to be again invoked. This continues until there are no more faults in the circuit that have not been processed. After the first vector, the Outline Waveform might appear as follows:

```
begin
    S0:=1 S1:=0
    DATABUS:=00101110;
    DATAOUT:=11111110;
    testgen(next_task);
end
```

where S0, S1, and DATABUS are input signal names and DATAOUT is an output signal name.

For sequential circuits the Outline Waveform becomes more complex. It contains information which constrains the State Planner, it may define basic looping structure for the logic, and it may define timing rules for the circuit. It may also define times at which the test generator can make decisions and it may request that some objective be achieved. The objective may be a request that the State Planner drive the circuit to some internal state or a request that some stored state be made observable. Features also exist which make the test engineer an active part of the loop.

Sequential Circuit Test: Analysis There are some interesting similarities between SCIRTSS and Hitest. In both systems a gate-level combinational ATPG is used. Its purpose is to drive fault effects through the combinational logic either to primary outputs or to stored state devices which are treated like pseudo-primary inputs and outputs. The contents of stored state devices are held fixed until it is determined that no more faults can be tested or until no more fault effects can be propagated from stored state devices to primary outputs simply by toggling primary inputs.

Then the gate-level ATPG may be used to drive fault effects into stored state devices.

When faults are trapped in stored state variables, they must be propagated by means of state changes. At that point, both SCIRTSS and Hitest make a drastic shift in mode of operation. In SCIRTSS a high-level register transfer level description is used to drive trapped faults to primary outputs. In Hitest the State Planner is invoked and it makes use of a knowledge base to compute state transitions. In either case, a gate-level simulator then evaluates fault coverage. Note the similarities to SOFTG (Sec. 4.8.4), which, after each test pattern is simulated, looks for fault effects trapped in stored state devices. SOFTG was biased to look for faults in bistables that are "close" to primary outputs, in the sense that they can be driven to primary outputs without going through more bistables. If input values are selected to drive them to outputs, then frequently more faults are propagated forward into bistables, where they can be rather easily driven to primary outputs, so the process can be self-sustaining for several clock cycles.

Although SCIRTSS as implemented uses an RTL description of the complete circuit, written in AHPL, its methods can be adapted to other RTL languages and confined to descriptions of the control logic. Configured in that way it could then be incorporated as part of the State Planner in a system such as Hitest. Another point of interest, when examining SCIRTSS and Hitest, is the fact that SCIRTSS goes forward in time when performing state searches, and Hitest uses PODEM, which goes forward from primary inputs, and its State Planner seems tailored to forward state searches. In other words, both of these systems avoid, as much as possible, backtracing through time and space, one of the more difficult and time-consuming operations.

Because the State Planner is both flexible and extensible, virtually any specific information that is useful in testing a circuit can be stored in the knowledge base and used to test the circuit. The key is to represent the knowledge in some useable form. The knowledge may be classified as controllability, or observability, or complete cycles of state transitions, or constraints (unattainable states or illegal input combinations). The representation can be in graph form, specifying state transitions, or in graphs representing transfer of information between registers, as described in the previous chapter. It can also be in the form of a matrix which relates internal resources to the instructions which use them. If it is necessary to propagate the contents of a bistable or register to the data bus, then the matrix is useful for indicating which instruction(s) may help to accomplish that objective. An RTL description of the instructions can then help refine the knowledge. For example, if the sign bit in the status register has a trapped fault, then the matrix may indicate that it is used by a conditional branch instruction. The RTL instruction appears as:

```
PC ← (Sign) · (PC+1) + (Sign) · (Eff. Addr.)
Address Out ← PC
```

The conditional branch causes either the next program counter address or a branch address to be placed on the address bus, depending on the value of the sign bit.

10.5 ARTIFICIAL INTELLIGENCE METHODS

This immediately provides information as to what machine instruction would be useful in propagating the trapped fault in the status register to an output.

10.5.4 Extracting Behavior from Structure

The conventional view that the ATPG takes of a circuit is that it consists of an apparently amorphous collection of gates, wires, and possibly transistors. There may on occasion be a slightly higher level of abstraction, such as flip-flops, but essentially, functional or behavioral information is ignored. Perhaps we became locked in because printed circuit board designs have traditionally been recorded in data bases at a structural level. That level of detail then becomes available without effort, simply by writing an easily defined program that accesses the data base and converts the board design into internal computer format for the ATPG. However, suppose we could deduce some behavior from that information?

Some behavioral clues are fairly obvious. In Figure 10.10(a), a simple inverter is shown. The relevance of the circuit lies in the fact that there is an implied

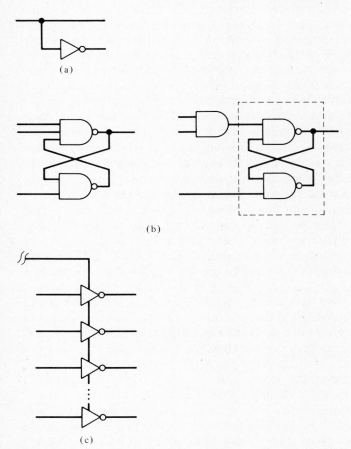

Figure 10.10 Identifying basic structures.

choice by virtue of the fan-out and the existence of both a and $\bar{a}$. That implies the structure

IF a THEN b ELSE c

Figure 10.10(b) is an RS latch. However, it is not immediately obvious that it is such a latch. If we modify the circuit slightly, we can put the latch into a form from which it can be immediately recognized. Bus structures, as in Figure 10.10(c), may also be recognizable. If a gate has fan-out to numerous other gates, and if all of them are identical, this may imply the existence of a bus.

A motive for abstracting behavior is to perform design verification. The purpose of design verification is to determine whether a physical implementation of a logic design corresponds to a behavioral specification. The specification may be a programmer's manual, a principles of operation manual, or a detailed breakdown of each instruction or operation of the device with a description of what must occur during that operation.

The traditional approach to design verification applies stimuli specified by the designer, or by the individual who originally created the specification, to the structural design. The stimuli are simulated at one or more levels of detail and/or run on the actual device. The results are then evaluated to determine whether the device responded correctly. One problem with this approach is that the designer may misunderstand the specification and, if he creates stimuli based on this misconception, his simulation results may only confirm that his design functions according to his understanding, which was initially wrong. A second problem is that there is no criterion for determining when enough simulation results have been obtained, that is, no measure of the goodness or thoroughness of the simulation.

An alternative approach constructs a behavioral description from a structural description. This can be done by several techniques, including the recognition of key physical constructs[26]. The methods used depend on the rather simple expedient of whether or not they work. For example, the constructed behavior for a given primary output is compared with the specified behavior by considering both the logic driving the output and the internal state variables upon which it depends. Output expressions for constructed and specified behavior are equated and an attempt is made to prove that the equation is an identity. If the constructed and specified behaviors are identical at all outputs and state variables, then the modules are correct.

The comparison starts by trying to determine whether the equation is a trivial identity. If it cannot be determined that it is an identity, then the domain space is examined. It may be possible to enumerate the possibilities if the domain space is small. If that fails, the next step is symbolic manipulation, including evaluation, simplification, expansion, canonicalization, and case analysis.

Evaluation may permit replacing at least part of an output expression by a constant. If two or more constants appear in conjunction with some arithmetic or Boolean operators, then the expression is replaced by the evaluated result. Simplification involves replacing a Boolean expression or subexpression by a simple expression. For example, if the expression $a + \bar{a}b$ appears in the output, then it can be replaced by $a + b$.

10.5 ARTIFICIAL INTELLIGENCE METHODS

Expansion occurs if a function occurs on only one side of the equation. It is replaced by its definition. Canonicalization tries to standardize the representation of expressions. Nested expressions are replaced by their "flattened" equivalents. The associative and commutative properties are used and expressions are lexically ordered. Conditional expressions are replaced by their evaluated value whenever the value is known. An interactive mode is employed so that, when all else fails, the user can guide or influence the program. Some interesting results have been obtained from these programs. They have been able to identify functional constructs such as adders, multiplexers and multipliers when starting with only a structural model at the transistor level.

These methods, suitably adapted, may eventually prove useful in test pattern generation as well. As we have seen, ATPGs operating at the gate level have fared rather poorly with LSI and VLSI circuits. A major stumbling block has been the inability to drive fault symptoms through sequential logic, or to control sequential logic in such a way as to make faults in the data flow circuits visible at an output. The SCIRTSS approach requires an AHPL description. It may be possible to employ the SCIRTSS concepts without requiring a formal behavioral description if behavioral properties of key parts of a circuit can be recognized.

In using the concepts derived from artificial intelligence, it is necessary to recognize a very important distinction between design verification and structural test. Design verification attempts to confirm that a design conforms to intent, while structural test attempts to confirm that a physical structure conforms to the design. Design verification assumes structural integrity, while structural test operates on the premise that the design is correct (perhaps, more accurately, the structural test is indifferent to the correctness of the design).

Because of this fundamental difference, a structural test must take into account the pathological behavior of a design which is faulted and may have to provoke such behavior from the design. This can be illustrated by referring back to the edge-triggered delay flip-flop (DFF) in Figure 3.6. Assume we want a test for an S-A-1 fault on the bottom input to gate N_4 and that present state Q is at 1. Then, while holding $\overline{\text{Preset}}$ high, toggle the $\overline{\text{Clear}}$ line, first setting it to 0 and then returning it to 1. If the $\overline{\text{Clear}}$ is toggled while either the Clock or Data is low, a faulted input will not be detected. If the Clock and Data are high, then the faulted DFF will initially clear but will revert to the set state when the $\overline{\text{Clear}}$ is returned to 1; hence it will be detected (assuming a sensitized path to some output). Because detectability requires both Clock and Data lines to be high, the fault may appear intermittent to the test engineer who is not aware of this requirement on the Clock and Data lines and lets them assume random values.

The correct approach may be to analyze several different physical implementations of a function to find effective test sets for physical defects at the functional level. This may lead to technology-dependent modeling of faults wherein identical functional models have different fault models, contingent on the technology in which they are implemented. Then a behavioral description of faulted behavior becomes a fault model equivalent to the PDCF; it becomes the primitive function test pattern (PFTP). Whereas the gate-level simulator determines what percentage of PDCFs were successfully applied and propagated to outputs, the functional simulator determines what percentage of PFTPs were successfully applied. A major distinction lies in the fact that the PFTP may extend over several time images. Note that there is nothing to preclude gate-level primitives; the PDCFs are, in fact, a subset of the PFTPs.

10.5.5 Expert Systems for Diagnosis

In this section we examine the use of expert systems to diagnose faulty electronic equipment. Expert systems have been successfully applied to problems as diverse as medical diagnosis[27] and mineral exploration[28]. In electronics, these systems can be applied to both analog and digital circuits and may operate on information ranging from a detailed description of the circuit to information consisting only of high-level block diagrams.

To understand how expert systems work, it is helpful to understand the distinction between a data base and a knowledge base. Some conventional programming systems consist of simple files of data in which all knowledge about data items and their relationships to one another is contained in the programs. Other systems consist of very complex data bases, and relationships between data items are implied by the structure of the data. For instance, in a data base for a commercial business, specific fields in a record may contain customer name, street address, and date of most recent purchase. In this case, some knowledge is inherent in the structure of the data. In an expert system this trend is taken a step further. Rather than a data base, a *knowledge base* is employed. The knowledge base contains not only the data needed to solve a problem, but also some rules for manipulating the data. A program called the *inference engine* exists which, in theory at least, is independent of the data domain. This inference engine is designed to operate on data in any problem domain and form conclusions based on inference rules contained in the knowledge base. Furthermore, the knowledge base can grow as new knowledge about the problem domain is acquired, without the need for changes to the inference engine.

The Expert System Shell We describe here a system called APEX3[29]. It is an attempt to create an expert system *shell*. A shell is essentially an expert system without a knowledge base; it is an inference engine which can solve problems in different problem domains simply by operating on the data according to rules in the knowledge base. The object is to have a "black box" which becomes an expert system for a specific problem simply by adding specific knowledge. We note here that the effort was not completely successful, in that domain knowledge occasionally had to be incorporated into the inference engine.

The operation of the expert system can be described with the help of Figure 10.11. In this figure, the topmost node represents the most general symptom, for example, an error is detected during system operation. The lowest level nodes represent goals; in this case they define specific faults and the faults in turn define corrective actions that must be performed in order to restore the system to correct behavior.

10.5 ARTIFICIAL INTELLIGENCE METHODS

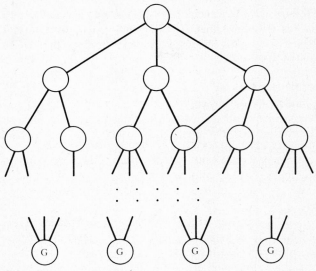

Figure 10.11 Diagnostic tree.

Two methods exist for traversing this tree: forward chaining and backward chaining. In *forward chaining* the system works down the tree toward goals by asking the user questions and making deductions based on the response. In *backward chaining* the system selects a goal and then attempts to determine, again by asking questions, whether it is the correct goal. Forward chaining is attractive in a tree in which every node has an associated question with a yes or no answer. In that case the system reaches a goal in a tree of depth n by asking $n - 1$ questions. Backward chaining becomes more appropriate when there are just a few goals. It also becomes more appropriate when mounting evidence points conclusively to a particular goal.

The APEX3 system actually employs a mixed strategy, since it was determined that this is how test technicians successfully diagnose faulty equipment. The expert first checks causes which are highly probable. For example, he might start by checking whether a fuse has blown. If past history points to a particular component as being responsible for a given symptom, it is checked immediately. After the more obvious possibilities are checked, if the fault is still not diagnosed, then forward chaining is employed in order to gather data about the problem and to traverse the tree in the direction of the goals. During this phase, mounting evidence may again point to a particular goal. The system can then revert to backward chaining in order to try to substantiate the goal with high probability. In general, if there is a high-probability goal, backward chain from it. If not, then forward chain. In addition to a mechanism for deciding whether to forward or backward chain, APEX3 has commands which allow the user to override the strategy currently in effect.

The system may start in the backward chaining mode if initial symptoms point to a high-probability fault, such as the aforementioned blown fuse. The system attempts to verify either that the fuse is blown or that some other similarly obvious fault exists. If no such fault exists, then the system forward chains from the most general fault to more particular faults. To illustrate this, assume that we have a TV set where there is no picture but the audio is okay. We immediately forward chain from defective TV to defective video. Then, a number of causes exist, including "loss of high voltage" at the CRT (cathode-ray tube) or "loss of filament current." Assuming that loss of filament current is the most probable cause on this particular model, the next step is to backchain from the filament current fault and ask questions intended to confirm or deny that the fault is present. If that fault is not found to be present, the next step is to select the loss of high voltage fault and backward chain from that fault to verify that it exists.

If measurements indicate that there is no high voltage at the anode of the CRT, the system may forward chain through the high-voltage transformer and the horizontal output circuits back to the horizontal oscillator where the signal originates for the high-voltage circuit. There may again be an intermediate point, such as the horizontal output circuit, which has a high incidence of failure, making it reasonable to select it as a goal from which to backchain. If it is operating correctly, the high-voltage transformer may next be suspect.

If it is confirmed that both filament current and high voltage are present, then some other goal must be selected; perhaps the wrong voltage is present at the cathode or grid of the CRT. The chaining process may then commence from one of these faults. In any case, if a fault is confirmed it must be specific enough to identify a replaceable unit. Otherwise chaining, forward or backward, must be continued to further isolate the fault.

One objective in a diagnostic system is to minimize the number of questions that must be asked in order to arrive at a diagnosis. Therefore, APEX3 was implemented to permit backchaining from another goal when it becomes more probable than the goal presently being pursued. However, to prevent thrashing between goals with nearly identical probabilities, the probability of another goal must exceed the probability of the present goal by some amount, say 10%, before it can be selected. APEX3 also imposes preconditions on questions at a given goal. Some questions are not asked unless the preconditions are satisfied. For example, in our analysis of the TV set, we investigated the absence of a picture. But, in reality, we would not start immediately checking for anode voltage, filament current, and so on if audio were also absent. The presence of audio is then a precondition for asking questions related to loss of video.

An integral part of APEX3 is the knowledge base construction interface. This allows users to enter knowledge into the user base. Faults are numbered and then given properties, such as:

A description

A mnemonic

An associated question

Prior certainty of the fault

The description is usually a sentence describing the fault, and the mnemonic is some sequence of characters which readily identify a part or component, such as

10.5 ARTIFICIAL INTELLIGENCE METHODS

"U63" to identify an IC. The associated question is one which APEX3 can ask the user in order to acquire more information concerning the fault. The prior certainty of the fault is a number obtained from an expert which represents the prior certainty of the fault in the absence of any other information.

Relationships between faults are also included in the knowledge base. These include:

> Fault A is a cause of fault B.
>
> Fault A has a logical definition in terms of faults B and C.

The system acquires information from the user and this causes the certainty of a particular fault to change. By virtue of the causal relationship, the certainties of other faults then change. These certainties are updated during the diagnostic process to reflect information acquired from the user.

Knowledge Acquisition The acquisition of knowledge is at least as important as the structure which uses the knowledge. In creating a system called CRIB (computer retrieval information bank), the developers examined the question of whether an aggregate of practitioners or a single expert would best serve as a source of knowledge for the knowledge base[30]. Their findings indicate that the knowledge of the nonexperts was factual and additive, whereas the expertise of the experts often resulted not just from their knowledge of the facts, but also from their ability to comprehend relationships between these facts, that is, the structure of the knowledge domain. It was these experts who provided the methods of inference. Factual data could be elicited from several practitioners, but some kind of semantic network had to be provided to detect and eliminate inconsistencies.

One of the decisions made during the development of CRIB was to avoid the use of functional information. This was based in part on the observation that expertise of most field engineers was in module interfaces. Therefore, the development effort concentrated on field-replaceable units and the analysis of faulty systems was accomplished by association of symptoms and faulty subunits. Groups of symptoms were formed to help in the analysis. The symptom groupings reflect analytical skills in the sense that some groupings are more useful than others in diagnosing faults. This parallels the experiences of engineers who can successfully identify the cause of a fault simply by observing particular groups of symptoms.

In order to analyze groups of symptoms, CRIB has a content-addressable file store (CAFS). When several symptoms are observed from a faulty piece of equipment, these symptoms are encoded and sent to the CAFS, where they are matched against groups of symptoms. The CAFS, operating independently of the main processor, returns parts of records which satisfy the match. If the match is exact it is quite simple to determine which fault node should be investigated. If there are several partial matches, then meta-rules are invoked to specify what course of action must be performed. For example, one particular symptom may be contained in several partially matched groups in the knowledge base. In that case, the symptom is selected and analyzed to determine if it exists in the piece of equipment being analyzed. If it is not present, then several potential faults are eliminated. Other, rather pragmatic, rules can be used. For example, a rule may specify: take the symptom with the shortest time factor (the sum of the time it takes to determine whether the symptom is present or not plus the time it takes to investigate the symptom).

CRIB was created with two objectives in mind: reduce training costs for test engineers and increase productivity by reducing average time per fault investigation. Consequently, a user-friendly English language like interface was implemented. Furthermore, the expert system is able to explain its decision processes to the user. This enhances its role as a teacher. However, perhaps more important, the objectives imply that the system must be able to evolve. Experts who use expert systems often suggest that they are either incompetent or, if competent, not able to adapt to new situations. Since industries often track the repair records for their products and redesign either the product or the manufacturing processes when a particular problem occurs frequently, failure probabilities and diagnostic approaches must be modified over time. Opportunities for evolving the system also occur when an expert has to take over and diagnose a problem that could not be successfully diagnosed by the expert system.

The effectiveness of CRIB was checked by having new users attempt to diagnose faults in the ICL 2903 computer through the use of CRIB. It was found that, when problems produced partial matches with symptom groups, CRIB was slower than engineers analyzing the same problem. This was in part due to the fact that the engineers were able to use their understanding of the function of the circuitry to deduce faults despite incomplete knowledge. However, during this validation process symptoms were eliminated when it was found that they were irrelevant or unhelpful to diagnosis. Eventually, after several weeks of fine-tuning, an expert used CRIB to diagnose real faults and he characterized the system as a competent diagnostician. A general assessment of CRIB regards it as useful on faults which have previously been remedied. Since the majority of faults are recurring, the successful diagnosis of those faults may justify the costs of developing and maintaining a system such as CRIB.

10.6 HARDWARE SIMULATION

Simulation is characterized by the fact that it involves enormous numbers of calculations, and most of them are either identical or involve an operation selected from a very small subset of choices. This has led to the observation that simulation could be done by specialized processors. We look at two such processors here. The Yorktown Simulation Engine is composed of many elementary but fast processors operating in parallel. The Metalogician is a data flow machine in which three parts of the simulation algorithm are handled by separate processors. We also look at physical modeling, where an actual component acts as its own model.

10.6.1 The Yorktown Simulation Engine

The Yorktown Simulation Engine (YSE)[31] is a device designed specifically for the purpose of performing simulation. It has a capability of simulating up to 2 million

10.6 HARDWARE SIMULATION

gates and can do over 2 billion simulations per second. An entire mainframe, including CPU, I/O, memory, and channels, can be simulated at the rate of 1000 native instructions per second (IPS). The same system, simulated at the gate level on a computer such as a System 370/168, would simulate at the rate of 0.5 IPS. It is claimed that the instruction sets for some microprocessors could run faster on the YSE than on the original microprocessor.

The YSE is an array of parallel processors. It contains logic processors, array processors, a control processor, and an interprocessor switch. The logic processors are special-purpose gate-level simulation machines. Each logic processor is capable of simulating up to 8K functions and the YSE can be configured with up to 256 logic processors; hence it can simulate a machine of 2 million gate equivalent size. Each gate can compute an arbitrary four-input function using four-valued logic $(0, 1, x, z)$ encoded into 2-bit arguments. The YSE performs unit delay or zero delay simulation. The array processors are used to simulate RAMs and ROMs. The control processor performs initial program load, interrupt handling, and self-test. The processors communicate with each other through the interprocessor switch.

The actual gate simulation mechanism is a table lookup implemented in the logic processor function unit. Each of four inputs must pass through generalized DeMorgan memories (GDMs) before reaching the function RAM. As the 2-bit

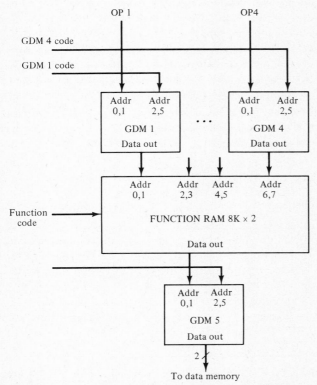

Figure 10.12 Yorktown Simulation Engine (logic processor).

input passes through the GDM, it may be converted to one of the other three possible 2-bit values or it may pass through without change. Whether it changes or not depends on a 4-bit GDM code which is sent to the GDM with the input. The 4-bit code selects one of 16 functions which permits the YSE to map the four input values into any possible combination. For example, one of the 16 functions may map all four values into the 2-bit code 0, 0, corresponding to the value 0. This may be used in an OR gate with fewer than four inputs; the unused inputs would be mapped into 0. For an AND gate all four combinations of an unused input can be mapped into 1. Other useful functions are the identity function, which simply maps each incoming value into itself, and the Invert, which maps each input into its complement. The values at the output of the GDMs are used, in conjunction with a 5-bit function code, to select an output. The function code is the high-order part of the address; it selects one of 32 functions. The eight GDM output bits select the value within that function. At the output of the function RAM another GDM performs a transformation on the output just as the other GDMs performed a transformation on the inputs.

The function unit is part of a logic processor which contains an instruction memory, a data memory, and a program counter. The data memory is actually made up of five data memories, each of which is 16K by 2 bits. Four of the data memories provide the inputs to the function unit. The instruction memory, which is 8K by 128 bits, provides the addresses into the data memory banks for the four input arguments to a given gate as well as the address of the destination. The activities within the YSE are pipelined with a basic clock rate of 80 nanoseconds, so that, in effect, a complete gate level simulation of every element occurs every 80 nanoseconds. Before simulation is started, the circuit to be simulated must be partitioned into subcircuits, none of which can have more than 8K gates. It is possible to partition the circuit into smaller sections and use more processors. This would result in an overall increase in the simulation speed, the amount of increase depending on the number of additional processors added.

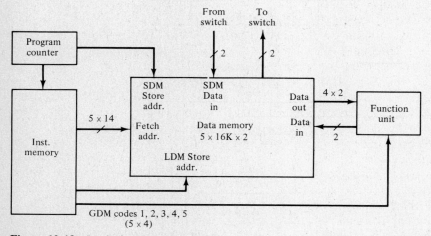

Figure 10.13 Logic function unit.

10.6 HARDWARE SIMULATION

Each of the 8K locations corresponds to one gate in the partition and in most instances a gate communicates only with other gates in its own logic partition. However, when a gate in the actual circuit has a connection to a gate that is represented by a location in some other partition, then the gates must communicate via the switch. Each of the processors makes its output available at each cycle. If processor X needs the results computed by processor Y at time Z, then the switch select memory of X, at time Z, selects the output of processor Y via its switch select memory, and stores it in its switch memory. A potential conflict may occur if two signals are required from another processor on the same clock cycle. This cannot be done, and the software which compiles the circuit must prevent it from occurring. It is claimed that this problem occurs infrequently, and when it does, the compiler handles the problem.

10.6.2 The Metalogician

Although a large number of behavioral and functional or register transfer level simulators have been developed since the mid-1960s, there remains a demand for gate-level simulators. The problem with gate-level simulation, of course, is the fact that it requires simulation of great amounts of minute detail. That, however, is also the attraction of gate- and switch-level simulation. If races and hazards can cause problems in a design, they will likely be uncovered at the detailed level of simulation. Furthermore, it is at the gate level that the simulation most closely approximates the behavior of the physical implementation. A given behavior can be physically implemented in many different ways, some of which are more susceptible to problems than others.

Several hardware simulation accelerators have been built for the event-driven, nominal delay simulator. These include the Zycad LE-1001[32] and the Daisy Metalogician[33]. Others have been proposed[34]. The hardware accelerators are based on the observation that several activities can be performed in parallel. The Metalogician, illustrated in Figure 10.14, separates the simulation task into three activities and assigns them to separate processors. Overall control is exercised by an 80286 microprocessor. The three individual simulation units are the Queue Unit, the State Unit, and the Evaluation Unit, each of which has its own dedicated memory. The Queue Unit is responsible for scheduling elements for future simulation and retrieving elements from the timing queue. The State Unit holds the circuit image as well as the states and strengths of all elements in its dedicated memory. The Evaluation Unit has in its dedicated memory the evaluation routines for the logic elements as well as for any functional elements in the circuit. When the output of an element changes value, the Queue Unit passes the name of that element to the State Unit. The State Unit fetches the elements in the fan-out of the changed element. These elements are evaluated by the Evaluation Unit and passed to the Queue Unit for scheduling. First-in, first-out (FIFO) memories are used to help even out temporary differences in processing speed of the different units.

The Metalogician and similar architectures can provide impressive gains in simulation. The gains are derived from the fact that the simulation task is run in pipeline fashion through three processors and from the fact that the architecture of

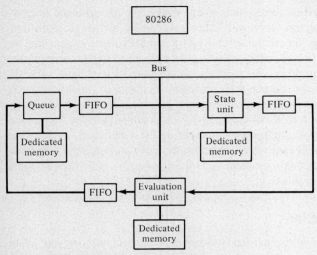

Figure 10.14 Metalogician architecture.

each of these processors is custom-designed for its specific task using high-speed technology. It is claimed that simulation speeds up to 100,000 events per second are possible with this approach as compared to typically about 5000 to 7000 events per second on a general-purpose computer in the performance range of 1 million instructions per second (MIPS). This makes it practical to simulate much larger circuits and makes the gate-level simulator competitive in performance with behavioral level simulators on large circuits. Simulation performance can be further enhanced by using distributed processing which employs two or more of these simulation accelerators[35].

One of the drawbacks of gate-level simulation is the fact that a gate-level circuit description may contain as many as 100,000 gates and their interconnections. These descriptions are much more difficult to read than are functional or behavioral level descriptions. If the circuit responds incorrectly, the amount of detail makes it very difficult to trace through the circuit and isolate the defect. One solution is to derive gate-level models from RTL descriptions. Synthesizers that perform this function behave somewhat like ATPG preprocessors, which access a library of SSI and MSI devices and create a gate-level model from a circuit described as an interconnection of elements in the library. The RTL description is then used to help localize a problem and the gate-level description is used to isolate the specific cause.

10.6.3 The Physical Model

Hardware is being used to assist simulation in yet another way. The problems associated with modeling of components continue to grow. There exist, on a single integrated circuit, 16- and 32-bit microprocessors with features such as instruction prefetch, on-board cache, facilities for instruction retry, facilities for serial I/O, and pipeline architectures which permit several instructions to be in various states of decode at the same time. The list of features incorporated into the chip continues to grow. In addition, the microprocessor families include growing numbers of complex peripheral chips. It is virtually impossible to justify the cost incurred to create simulation models for these devices. An accurate model must predict what signals appear at the microprocessor pins and when they appear. However, even something like an instruction prefetch, which causes the next instruction to be fetched before the current one is complete, can be difficult to model because of the uncertainty as to when the next instruction is fetched. Attempts to model these devices have proved costly and there is no assurance that they are modeled correctly. Furthermore, to be useful a model should be available very soon after the actual IC is available, when many companies are selecting the device for use in new products.

One solution to this modeling problem is the use of physical modeling in conjunction with simulation, as done with physical modeling extension (PMX)[36] and Realchip[37]. These devices permit a component to be physically wired to an engineering workstation through a socket in such a way that the workstation can apply stimuli directly to one or more of these components during simulation and read their response. Because the device operates quite rapidly compared to simulation, it is necessary to provide a table of propagation delays. The simulator reads the logic response from the physical model and uses the propagation delay table to determine the time at which the output response should occur, much as TSETs are used with the high-speed functional tester. When simulating dynamic devices, it is necessary to store up all stimuli applied to the device, and each time there is a stimulus arriving at the device from the simulator, all stimuli must be reapplied.

The use of real devices assumes the availability of a known good chip. However, since propagation delays are taken from a table, and the chip is used only to provide the logic response, the chip can operate at any frequency; for instance, if the designer was planning to use an 8-MHz microprocessor in his design, but could only obtain 4-MHz microprocessor parts for evaluation purposes, they could be used in the modeling extension provided they were not defective.

The modeling extension makes it possible to simulate a printed circuit board and verify that it interfaces correctly with a complex device for which a simulation model does not exist. However, observation of internal registers and flip-flops requires software routines, tailored to the device, capable of causing internal state variables to be read out. Fault simulation with the modeling extension is also somewhat difficult to handle. Without a simulation model, fault coverage cannot be determined within the physical device. Even the task of propagating faults through a physical device is a formidable problem because it may be necessary in a large product to propagate many thousands of faults through the physical device, and these may induce hundreds or thousands of different internal states in the physical device.

10.7 THE SEM NONCONTACT PROBE

When several functional elements are integrated onto a single die, as when a microprocessor encompasses elements such as CPU, cache memory, virtual memory,

memory management units, bus control units, and others, a defect in one of these units can cause an error which does not appear at an I/O pin for several hundred clock cycles. When debugging the design of one of these complex devices, it is imperative that operations inside the device be observable in order to locate the source of errors. Physical probing of the individual die was once possible. Now, however, with shrinking feature sizes and rapidly growing numbers of transistors, physical probing is impractical. The small feature sizes make the die more susceptible to permanent damage and capacitive loading from the probe can distort the signal that is to be observed. Furthermore, the probing process can be extremely time-consuming, tedious, and error-prone because the designer must visually identify an interconnection line that is to be probed from among thousands of such lines that appear nearly identical. The result is that new product design cycles grow longer at an increasing rate.

Noncontact probing can be accomplished by means of the scanning electron microscope (SEM). In this method a die is placed in a vacuum chamber and a focused beam of electrons is directed at the die while the circuits on the die are in operation. The beam is normally blanked (cut off), but is unblanked and allowed to impinge on the die at a time when a voltage sample is desired. When electrons are fired at the die, regions of high voltage attract the electrons while regions of low voltage repel them. A collector captures electrons which are repelled from the surface of the die and the quantity of electrons captured at a given time is used to estimate the voltage at the point on the surface where the beam was aimed. If the SEM and the device are properly synchronized, the SEM can be used to sample voltages at specified points in several consecutive clock cycles.

Capabilities of the SEM include measurement accuracy of 10 millivolts with a time resolution of 100 picoseconds[38]. A beam diameter of 0.8 micrometer can be achieved with a rule of thumb recommending that beam diameter be approximately $W/5$, where W is the width of the interconnections on the die to be investigated[39]. The accelerating voltage of an e-beam must be limited in order to avoid radiation damage to the device being observed. On the order of 1 or 2 kilovolts is usually suggested as a safe limit.

The method of estimating voltage by collecting electrons repelled from the surface, called voltage contrast, can be used to create waveforms or complete images. In the *waveform* mode the electron beam is positioned at a location on the die and the waveform at that point is constructed by periodically strobing while the die is clocked through a number of states. In effect, this mode of operation is quite similar to that of an oscilloscope or logic analyzer. In the *image* mode a picture of the complete die, or some designated part of the die, is constructed by scanning an area of interest. By repeating this operation, several images can be obtained and averaged to minimize the effects of noise and produce a complete image of voltage activity on the top level of the die.

The use of a computer-aided design (CAD) system enhances the efficiency with which an e-beam is used. The CAD system may contain physical information describing the die, including the (x, y) coordinates of the end points of top-level interconnections. This information can be used to locate particular interconnections on a die and can therefore be used to help position the e-beam accurately. This

10.7 THE SEM NONCONTACT PROBE

integration of an e-beam, in the waveform mode, together with CAD and a source of input test vectors then becomes analogous to the printed circuit board tester. The values on a connector are obtained by the e-beam system and can be compared with expected values derived from simulation to determine if the values on the connector are correct.

The e-beam system is not intended to be used as a production tester. It is slow compared to a conventional tester and may need several hours to acquire enough information to diagnose a problem. The logic states provided by the e-beam at the top-level interconnections may not be sufficient to diagnose problems; analog waveforms at components underneath the top level may also be required. To analyze a die that has already been packaged, it is necessary to delid the device, and that is potentially destructive. The e-beam is best used where short, repetitive cycles of operation can be set up. Nevertheless, it has proved successful for such applications as failure analysis and yield enhancement. When excessive numbers of devices fail with similar symptoms, it is reasonable to expect that the same failure mechanism is causing all or most of the failures. The e-beam may help trace those to design or process errors. If a device operates successfully at some clock frequency but fails when the frequency is increased slightly, it may be possible that a single design factor is limiting performance and that identification and correction of that one factor may permit a significant increase in the clock frequency. The e-beam also proves useful as a research tool to characterize technology and circuit properties.

One of the problems encountered when using an e-beam is the fact that it can be difficult to determine which nodes should be probed. If an error is detected at an I/O pin, the fault which caused the error may have occurred many clocks previous to the clock cycle when symptoms were first detected. An approach to solving this problem, called dynamic fault imaging (DFI), uses the image mode to build *fault cubes*[40]. The fault cube (Figure 10.15) is a series of images from successive machine cycles which are stacked on top of each other to show the origin of a fault and the divergence of error signal(s) in subsequent image frames as a result of that fault. The first step in DFI is to construct voltage contrast images for good and faulty dies for several clock cycles. Then the good and faulty device images are differenced to form an image which highlights the areas of the die where

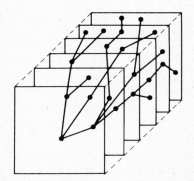

Figure 10.15 Fault cube.

different voltage levels exist. On successive clock cycles the fault effects can then be seen to propagate through the die and affect increasing numbers of other states.

The DFI method is under computer control and employs special image processors. It creates a 512 × 512 image in which each pixel (picture element) is resolved to 8 bits in order to represent a wide range of voltage levels. Pseudocolor lookup tables are used to false color an image so as to enhance visual analysis. As many as 64K images can be averaged to improve resolution. The system has a MOVIE mode in which up to 32 images can be displayed in sequence, either forward or backward in time. A PROBE mode can select the values from the same (x, y) coordinate position of many consecutive images and use these values to construct a waveform corresponding to the voltage at that point on the die. In fact, waveforms corresponding to several (x, y) positions can be created and displayed simultaneously in a logic analyzer format. This kind of integrated design debug system may become routine as more and more complete systems are integrated onto single pieces of silicon.

10.8 SUMMARY

Market forces have had in the past, and will continue to have in the future, a significant influence on test strategies. There are divergent needs within the marketplace that will demand divergent test strategies. Those market forces are related to cost, reliability, and maintainability. The factors assume different degrees of importance to different individuals. Whereas maintainability is important in a large mainframe or complex military electronics, it is not nearly as significant in consumer products, where electronic controls may have a life expectancy of 150 years in a mechanical device with a life expectancy of 10 years. As pointed out at the beginning of this chapter, it is usually sufficient in this case to verify that the device was working when it left the factory.

A major reason for high test costs is the fact that we use highly trained technicians to perform the tedious task of creating test stimuli for digital circuits. This is a task that is best done by computers. However, little recognition has been given to the fact that the collection of gates and wires in a design actually represents a behavioral entity. When the designer is asked to create test stimuli for a circuit that he has designed, he looks at gates and wires only long enough to drive a test to a definable functional border. From there, he uses his knowledge of the circuit function to propagate the test to an observable output. Test algorithms must evolve to a similar plane of reasoning.

REFERENCES

1. Slana, M. F., "Workshop Report: Computer Elements for the 80's," *Computer*, vol. 12, no. 4, April 1979, pp. 98–102.
2. van Cleemput, W. M., "Computer Hardware Description Languages and Their Applications," *Proc. 16th Design Automation Conf.*, June 1979, pp. 554–560.

REFERENCES

3. Barbacci, M. R., "A Comparison of Register Transfer Languages for Describing Computers and Digital Systems," *IEEE Trans. Comput.*, vol. C-24, no. 2, Feb. 1975, pp. 137–150.
4. Hill, F. J., and G. R. Peterson, *Digital Systems: Hardware Organization and Design*, 2nd ed., Wiley, New York, 1978.
5. Chu, Y., "An Algol-like Computer Design Language," *Commun. ACM,* vol. 8, Oct. 1965, pp. 607–615.
6. Duley, J. R., "DDL—A Digital Design Language," Ph.D. dissertation, Department of Electrical Engineering, University of Wisconsin, Madison, June 1967.
7. Hill, F. J., and G. R. Peterson, *Digital Systems: Hardware Organization and Design*, 2nd ed., Wiley, New York, 1978.
8. Thompson, E. W., et al., "The Incorporation of Functional Level Element Routines into an Existing Digital Simulation System," *Proc. 17th Design Automation Conf.,* 1980, pp. 394–401.
9. Menon, P. R., and S. G. Chappel, "Deductive Fault Simulation with Functional Blocks," *IEEE Trans. Comput.*, vol. C-27, no. 8, Aug. 1978, pp. 689–695.
10. Wilcox, P., "Digital Logic Simulation at the Gate and Functional Level," *Proc. 16th Design Automation Conf.*, 1979, pp. 242–248.
11. d'Abreu, M. A., and E. W. Thompson, "An Accurate Functional Level Concurrent Fault Simulator," *Proc. 17th Design Automation Conf.*, 1980, pp. 210–217.
12. Henckels, L. P., et al., "Functional Level, Concurrent Fault Simulation," *Proc. IEEE Int. Test Conf.*, 1980, pp. 479–485.
13. Szygenda, S. A., and A. A. Lekkos, "Integrated Techniques for Functional and Gate-Level Digital Logic Simulation," *Proc. 10th Design Automation Conf.,* 1973, pp. 159–172.
14. Akers, S. B., "Universal Test Sets for Logic Networks," *IEEE Trans. Comput.*, vol. C-22, no. 9, Sept. 1973, pp. 835–839.
15. Thatte, S. M., "Test Generation for Microprocessors" Ph.D. thesis, University of Illinois, Urbana, Ill., May 1979.
16. Thomas, J. J., "Common Misconceptions in Digital Test Generation," *Comput. Des.* Jan. 1977, pp. 89–94.
17. Liaw, C, S. Y. H. Su, and Y.K. Malaiya, "State Diagram Approach for Functional Testing of Control Section," *Proc. IEEE Test Conf.*, 1981, pp. 433–446.
18. Breuer, M. A., and A. D. Friedman, "Functional Level Primitives in Test Generation," *IEEE Trans. Comput.*, vol. C-29, no. 3, March 1980, pp. 223–235.
19. Levendel, Y. H., and P. R. Menon, "Test Generation Algorithms for Computer Hardware Description Languages," *IEEE Trans. Comput.*, vol. C-31, no. 7, July 1982, pp. 577–587.
20. Hill, F. J., and B. M. Huey, "A Design Language Based Approach to Test Sequence Generation," *Computer*, vol. 10, no. 6, June 1977, pp. 28–33.
21. Azema, P., et al., "Petri Nets as a Common Tool for Design Verification and Hardware Simulation," *Proc. 13th Design Automation Conf.*, 1976, pp. 109–116.
22. Torku, K. E., and B. M. Huey, "Petri Net Based Search Directing Heuristics for Test Generation," *Proc. 20th Design Automation Conf.,* 1983, pp. 323–330.
23. Wharton, D. J., "The Hitest Test Generation System—Overview," *Proc. 1983 Int. Test Conf.*, Oct. 1983, pp. 302–310.
24. Robinson, G. D., "Hitest—Intelligent Test Generation," *Proc. 1983 Int. Test Conf.*, Oct. 1983, pp. 311–323.
25. Maunder, C., "Hitest Test Generation System—Interfaces," *Proc. 1983 Int. Test Conf.*, Oct. 1983, pp. 324–332.

26. Barrow, H. G., "Proving the Correctness of Digital Hardware Designs", *VLSI Des.* vol. 5, no. 7, July 1984, pp. 64–77.
27. Shortliffe, E. H., *Computer-based Medical Consultations: MYCIN*, Elsevier Publishing Co., New York, 1976.
28. Duda, R., J. Gaschnig, and P. Hart, "Model Design in the PROSPECTOR Consultant System for Mineral Exploration," in *Expert Systems in the Microelectronic Age*, Edinburgh Univ. Press, Edinburgh, Scotland, 1979.
29. Merry, M., "APEX3: An Expert System Shell for Fault Diagnosis," *GEC J. Res.*, vol. 1, no. 1, 1983, pp. 39–47.
30. Hartley, R. T., "CRIB: Computer Fault-finding Through Knowledge Engineering," *Computer,* vol. 17, no. 3, March 1984, pp. 76–83.
31. Pfister, G. F., "The Yorktown Simulation Engine: Introduction," *Proc. 19th Design Automation Conf.,* 1982, pp. 51–54.
32. Iverson, W. R., "Modular Hardware Simulates Logic," *Electronics*, July 28, 1982, pp. 125–126.
33. Gindraux, L., and G. Catlin, "CAE Station's Simulators Tackle 1 Million Gates," *Electron. Des.*, Nov. 10, 1983, pp. 127–136.
34. Abramovici, M., Y. H. Levendel, and P. R. Menon, "A Logic Simulation Machine," *Proc. 19th Design Automation Conf.*, June 1982, pp. 65–73.
35. Levendel, Y. H., P. R. Menon, and S. H. Patel, "Special Purpose Computer for Logic Simulation Using Distributed Processing," *Bell Syst. Tech. J.*, Dec. 1982, pp. 2873–2909.
36. Daisy Systems Corp., "Megalogician Physical Modeling Extension (PMX) (TM)," Sunnyvale, Calif.
37. Widdoes, L. C., and W. C. Harding, "CAE Station Uses Real Chips to Simulate VLSI Based Systems," *Electron. Des.*, March 22, 1984, pp. 167–176.
38. Goto, Y., et al., "Electron Beam Prober for LSI Testing with 100 PS Time Resolution," *Proc. Int. Test Conf.,* Oct. 1984, pp. 543–549.
39. Kollensperger, P., et al., "Automated Electron Beam Testing of VLSI Circuits," *Proc. Int. Test Conf.,* Oct. 1984, pp. 550–556.
40. May, T. C., et al., "Dynamic Fault Imaging of VLSI Random Logic Devices," *Proc. Int. Rel. Phys. Symp.*, April 1984.

Appendix

Finite Fields: Some Definitions

The definitions, theorems, and examples contained in this appendix provide additional background on the material describing error-correcting codes found in Chaps. 8 and 9. It is not intended to be a rigorous treatment but, rather, an attempt to provide an intuitive understanding of the material.

A *group* G is a set of elements and a binary operator $*$ such that:

1. $a, b, \in G$ implies that $a * b \in G$ (closure)
2. $a, b, c, \in G$ implies that $(a * b) * c = a * (b * c)$ (associativity)
3. There exists $e \in G$ such that
   $a * e = e * a = a$ for all $a \in G$ (identity)
4. For every $a \in G$, there exists
   $a^{-1} \in G$ such that $a * a^{-1} = a^{-1} * a = e$ (inverse)

A group is *commutative,* also called *Abelian,* if

$$\text{for every } a, b \in G, a * b = b * a.$$

EXAMPLE ■

The set $I = \{\ldots, -2, -1, 0, 1, 2, \ldots\}$ and the operator $*$ form a group when $*$ represents the usual addition ($+$) operation. ■ ■

EXAMPLE ■

We define the set $S = \{S_i | 0 \le i \le 3\}$ of squares where S_0 has a notch in the upper left corner and S_i represents a clockwise rotation of S_0 by $i \times 90$ degrees. We define a rotation operator R, $S_i R S_j = S_k$, such that $k = i + j$ (modulo 3). The set S and the operator R satisfy the definition of a group. The element S_k is simply the result of the sum of the rotations of S_i and S_j.
■ ■

Given a group G with n elements and identity 1, then the number of elements in G is called the *order* of G. The order of an element $g \in G$ is the smallest integer e such that $g^e = 1$. It can be shown that e divides n.

A *ring* R is a set of elements on which two binary operators, $+$ and $\times$, are defined, which satisfy the following properties:

1. The set R is an Abelian group under $+$
2. $a, b \in R$ implies that $a \times b \in R$
3. $a, b, c \in R$ implies that $(a \times b) \times c = a \times (b \times c)$
4. $a, b, c \in R$ implies that

$$a \times (b + c) = a \times b + a \times c$$
$$(b + c) \times a = b \times a + c \times a$$

If the set R also satisfies

5. $a \times b = b \times a$

then it is a commutative ring.

EXAMPLE ∎

The set of even integers is a commutative ring. ∎∎

A commutative ring which has a multiplicative identity and a multiplicative inverse for every nonzero element is called a *field*.

EXAMPLE ∎

The set of elements $\{0, 1\}$ in which $+$ is the exclusive-OR and $\times$ is the AND operation satisfies all of the requirements for a field and defines the Galois field GF(2). ∎∎

Given a set of elements V and a field F, with $u, v, w, \in V$ and $a, b, c, d, \in F$, then V is a vector space over F if it satisfies:

1. The product $c \cdot v$ is defined, and $c \cdot v \in V$
2. V is an Abelian group under addition
3. $c \cdot (u + v) = c \cdot u + c \cdot v$
4. $(c + d) \cdot v = c \cdot v + d \cdot v$
5. $(c \cdot d) \cdot v = c \cdot (d \cdot v)$
6. $1 \cdot v = v$ where 1 is the multiplicative identity in F

The field F is called the *coefficient field*. It is GF(2) in this text, but GF(p), for any prime number p, is also a field. The vector space V defined above is a linear associative algebra over F if it also satisfies the following:

7. the product $u \cdot v$ is defined and $u \cdot v \in V$
8. $(u \cdot v) \cdot w = u \cdot (v \cdot w)$

APPENDIX/FINITE FIELDS: SOME DEFINITIONS 407

9. $u \cdot (c \cdot v + d \cdot w) = c \cdot u \cdot v + d \cdot u \cdot w$
$(c \cdot v + d \cdot w) \cdot u = c \cdot v \cdot u + d \cdot w \cdot u$

The *Euclidean division algorithm* states that for every pair of polynomials $S(x)$ and $D(x)$, there is a unique pair of polynomials $Q(x)$ and $R(x)$ such that

$$S(x) = D(x) \cdot Q(x) + R(x)$$

and the degree of $R(x)$ is less than the degree of $D(x)$. The polynomial $Q(x)$ is called the quotient and $R(x)$ is called the remainder. We say that $S(x)$ is equal to $R(x)$ modulo $D(x)$. The set of all polynomials equal to $R(x)$ modulo $D(x)$ forms a *residue class* represented by $R(x)$. If $S(a) = 0$, then a is called a *root* of $S(x)$.

A correspondence exists between vector n-tuples in an algebra and polynomials modulo $G(x)$ of degree n. The vector elements $a_0, a_1, \ldots, a_{n-1}$ correspond to the coefficients of the polynomial

$$b_0 + b_1 g + b_2 g^2 + \cdots + b_{n-1} g^{n-1}$$

The sum of two n-tuples corresponds to the sum of two polynomials and scalar multiplication of n-tuples and polynomials is also similar. In fact, except for multiplication, they are just different ways of representing the algebra. If $F(x) = x^n - 1$, then the vector dot product has its correspondence in polynomial multiplication. When multiplying two polynomials, modulo $F(x)$, the coefficient of the ith term is

$$c_i = a_0 b_i + a_1 b_{i-1} + \cdots + a_i b_0 + a_{i+1} b_{n-1} + a_{i+2} b_{n-2} + \cdots + a_{n-1} b_{i+1}$$

Since $x^n - 1 = 0$, $x^{n+j} = x^j$, and the ith term of the polynomial product corresponds to the dot product of vector a and vector b when the elements of b are in reverse order and shifted $i + 1$ positions to the right.

Theorem A.1. The residue classes of polynomials modulo a polynomial $f(x)$ of degree n form a commutative linear algebra of dimension n over the coefficient field.

A polynomial of degree n which is not divisible by any polynomial of degree less than n but greater than 0 is called *irreducible*.

Theorem A.2. Let $p(x)$ be a polynomial with coefficients in a field F. If $p(x)$ is irreducible in F, then the algebra of polynomials over F modulo $p(x)$ is a field.

The field of numbers $0, 1, \ldots, q - 1$ is called a *ground field*. The field formed by taking polynomials over a field GF(q) modulo an irreducible polynomial of degree m is called an *extension field*; it defines the field GF(q^m). If $z = \{x\}$ is the residue class, then $p(z) = 0$ modulo $p(x)$, therefore $\{x\}$ is a root of $p(x)$.

If $q = p$, where p is a prime number, then the field GF(p^m), modulo an irreducible polynomial $p(x)$ of degree m, is, by Theorem A.2, a vector space of dimension m over GF(p) and therefore has p^m elements. Every finite field is isomorphic to some Galois field GF(p^m).

Theorem A.3. Let $q = p^m$, then the polynomial $x^{q-1} - 1$ has as roots all the $p^m - 1$ nonzero elements of $GF(p^m)$.

Proof. The elements form a multiplicative group. We then use the fact that the order of each element of the group must divide the order of the group. Therefore, each of the $p^m - 1$ elements is a root of the polynomial $x^q - 1$. But the polynomial $x^q - 1$ has, at most, $p^m - 1$ roots. Hence, all the nonzero elements of $GF(p^m)$ are roots of $x^q - 1$.

If $z \in GF(p^m)$ has order $p^m - 1$, then it is *primitive*.

Theorem A.4. Every Galois field $GF(p^m)$ has a primitive element; that is, the multiplicative group of $GF(p^m)$ is cyclic.

EXAMPLE ■

$GF(2^4)$ can be formed modulo $F(x) = x^4 + x^3 + 1$. Let $z = \{x\}$ denote the residue class x; that is, z represents the set of all polynomials which have remainder x, when divided by $F(x)$. Since $F(x) \equiv 0 \bmod F(x)$, x is a root of $F(x)$. Furthermore, x is of order 15. If we divide the powers of x by $F(x)$, the first six division operations give the following remainders:

$$x^0 = 1 \qquad \text{modulo } F(x) \equiv (1, 0, 0, 0)$$
$$x^1 = x \qquad \text{modulo } F(x) \equiv (0, 1, 0, 0)$$
$$x^2 = x^2 \qquad \text{modulo } F(x) \equiv (0, 0, 1, 0)$$
$$x^3 = x^3 \qquad \text{modulo } F(x) \equiv (0, 0, 0, 1)$$
$$x^4 = 1 + x^3 \qquad \text{modulo } F(x) \equiv (1, 0, 0, 1)$$
$$x^5 = 1 + x + x^3 \text{ modulo } F(x) \equiv (1, 1, 0, 1)$$

The interested reader can complete the table by dividing each power of x by $F(x)$. With careful calculations, he or she should find that $x^{15} = 1 \bmod F(x)$ but that no lower power of x equals $1 \bmod F(x)$. Furthermore, when dividing x^i by $F(x)$, the coefficients are cyclic; that is, if the polynomials are represented in vector form, then each vector will appear in all of its cyclic shifts. ■■

INDEX

AALG, 60
Abbreviated descripter cell, 164
Activity vector, 38
Acyclic, 69
Adaptive experiments, 103
Addressable registers, 281
Adjacent pin shorts, 200
AHPL (A Hardware Programming Language), 5, 360
All-0s test, 292
All-1s test, 292
Ambiguity, 104
Ambiguity delay, 138
Ambiguity region, 147
Arbitrary value, 44
Asynchronous, 69
ATLAS (Abbreviated Test Language for All Systems), 252

Backdrive, 248
Back-propagation, 60
Backward chaining, 391
Bare-board testing, 229
Behaviorally equivalent circuits, 213
Behavioral model, 119
BILBO (Built-In Logic Block Observer), 330
Bistable, 70
Blocked signal path, 208

Block-oriented analysis, 150
Boolean differences, 50
Branch address hashing, 345
Branch-and-bound, 47
Bridging faults, 199
Bulletins, 180
Burst errors, 347
Bus activity conflict, 154
Bus state conflict, 154

CAFS (content addressable file store), 393
Carriers, 358
Causative link, 147
CDL (Computer Design Language), 5
Chain rule, 54
Checkerboard test, 293
Checking sequence, 107
Checkpoint arcs, 192
Checkpoint faults, 192, 363
Circuit level model, 121
Cirrus Circuit Language, 381
Cirrus Waveform Language, 381
Clock rate, 236
Clock rate tester, 243
CMOS (Complementary Metal Oxide Semiconductor) faults, 198
COMET, 315
Common ambiguity, 147
Compaction, test pattern, 206

409

Compiled simulator, 125
Comprehension, fault, 205
Concurrent fault simulation, 160
Cone, 134, 208
Conflict, 26
Connected, two nets are, 38
Consistency operation, 25
Consistent, two singular cubes are, 30
Contains, singular cube, 30
Controllability, 256
Controllability equations, 265
Controllability relation, 310
Converged fault, 161
COP (controllability/observability analysis), 270
Cover, 31
Cover of F, 31
CRC (cyclic redundancy check), 337
CRIB (Computer Retrieval Information Bank), 393
Critical path, 43, 88
Critical race, 79
Critical value, 44
Cross-coupled latch, 69
Crosspoint fault, 201
Cyclic, 69

D-algorithm, 28
D-chain, 38, 157
D-frontier, 38, 157
D-intersection, 38
D-list, 157
D-notation, 28
DA (design automation), 5
DDL (Digital Design Language), 5
Decision table, 26
Deductive fault simulation, 159
Delay faults, 194
Delta-t loop, 139
Deracing, 146
Descriptor cell, 142, 187
Destination registers D (Ij), 319
Detectable, 24
Difference operator, 51
Display formatter, 180
Distinguishing sequence, 104
Diverged fault, 161
DL (defect level), 9, 122
Dominates, 191
Don't care, 25

Double latch design, 277
DRBACK, 62
DROPIT, 62
Dual clock serial scan, 280
Dummy model, 216
DUT (device under test), 2
Dynamic compaction, 207
Dynamic hazard, 79
Dynamic tester, 243

ECL (Emitter Coupled Logic) faults, 197
Economics of testing, 8
EDAC (Error Detection and Correction) codes, 289
Edge sensitive flip flop, 71
Edge set E(Ij), 319
Effective test rate, 236
Electronic knife, 236, 239
Elementary function, 200
Elementary gate, 200
ELIST (emission list), 164
ENF (equivalent normal form), 58, 326
Equivalent faults, 191
Error, 2
Error correcting codes, 289, 297
Evaluation techniques, 144
Evaluation unit, 397
Event, 134
Event-driven simulator, 134
Event notice, 139
Exercise sequence, 93
Expert system shell, 390
External fault, 367
Extremal, 31

Fall time, 138
FAN, 64
Fault, 2
Fault collapsing, 191
Fault cube, 401
Fault dictionary, 210
Fault dropping, 134, 211
Faulted machine, 28
Fault effect, 161
Fault list, 26
Fault model, 15
Fault origin, 161
Fault parameter, 55
Fault secure, 346
Fault simulation, 122, 156

INDEX

Field faults, 202
Finite state machine, 69, 76
Fire codes, 347
First degree hardcore, 310
Flexible signals, 60
Floaters, 329
Flush test, 280
Forcing value, 44
Forward chaining, 391
Functional fault model, 361
Functional fault simulation, 361
Functional model, 120
Function hazard, 80
FUNTAP (FUNctional Testability Analysis Program), 270

Galloping diagonal, 293
Galois field, 406
Galpat memory test, 292
Gate, 19
Geometric level model, 121
Glitch, 79
Golden board, 233
Good machine, 28
Graph model of microprocessor, 315
Guard circuit, 248
Guided probe, 236

Hamming codes, 299
Hamming distance, 300
Hamming weight, 300
Hardcore, 310
Hardware design languages (HDL), 357
Hazard, 79
Hazard detection, 129, 136
High speed functional tester, 243
Hitest, 379
Hold time, 73
Homing sequence, 104
Huffman model, 68

Image mode, 400
Implicant, 32
Implication, 26, 204
In-circuit tester, 229, 247
Independent, a function is, 52
Indeterminate value, 74
Indistinguishable, 311
Inertial delay, 138

Inference engine, 390
Initialization problem, 76
Initialization sequence, 93
Input-bridging fault, 200
Input fault origin, 161
Internal fault, 367
Interpretive simulator, 125
Intersection of singular cubes, 30
Irredundant circuit, 194
ISL (Integrated Schottky Logic) faults, 195
Iterated codes, 305
Iterative test generator, 82
ITTAP (ITT Testability Analysis Program), 270

Justification, 25

Knowledge acquisition, 393
Knowledge base, 390
Known-good-board (KGB), 233

LASAR (Logic Automated Stimulus And Response), 43
Level sensitive, 71, 276
LFSR (linear feedback shift register), 324
Literal propositions, 56
Logic faults, 77
Logic hazard, 80
Logic level model, 121
Logic symbols defined, 19
Loop unrolling, 169
LSSD (level sensitive scan design), 276

Macroblocks, 310
Maintenance processor, 334
Manufacturing faults, 202
Manufacturing management system, 231
March test, 293
Marker, 378
Master/slave flip flop, 72
Maximum likelihood decoding, 301
Media delay, 138
Memory address test, 294
Memory fault types, 291
Merging test sequences, 206
Metalogician, 397
Microblocks, 310
Microprocessor matrix, 314
Min-max timing, 146
MISR (multiple input shift register), 331

Mode control, 273
MOS (Metal Oxide Semiconductor) faults, 197
Moving inversions memory test, 294
MTTR (Mean Time To Repair), 350
Multi-layer board, 229

Necessary value, 44
Net, 20
Net oriented description, 187
Nine-value iterative test generator, 87
Nine-value simulation, 136
Node, 20
Nominal delay simulator, 135
Nonintegral event timing, 140
Nonprocedural language, 359

Observability, 256
Observability equations, 267
Observability relation, 311
1-controllability, 265
1-point, 31
One-step diagnosis, 340
Operators, 358
Opportunistic test generator, 383
Ordering relation, 309
Ordering tree, 311
Order of a polynomial, 347
Oscillations, 75
Outline waveform, 383
Output fault origin, 161

Parallel fault simulation, 131
Parametric faults, 77
Parity bit, 343
Parity matrix, 301
Partially symmetric function, 191
Pass-fail vector, 210
Path enumeration, 149
Pattern rate, 236
PDCF (primitive D-cube of failure), 32
Performance monitoring, 343
Period of a net, 84
Petri net, 378
Physical modeling, 398
Ping-pong test, 292
Places, 378
PMS (Processors, Memories, Switches), 4
Poage's method, 55
PODEM (Path Oriented Decision Making), 47

Potential bus activity conflict, 154
Potential bus state conflict, 154
Potentially detectable, 181
Potentially detected, 76
Power margining, 336
Power-of-ten rule, 12
Predecessor, 83
Preset experiment, 103
PRG (pseudo random generator), 330
Primary input, 20
Primary output, 20
Prime cube, 31
Prime implicant, 32
Primitive, 32
Primitive polynomial, 330
Probable tested faults, 75
Problem interpreter, 383
Problem solver, 383
Procedural language, 359
Propagation, 24
Propagation D-cubes, 35
Propagation delay, 138
Propagation search, 374
Prototype, 119
Psuedo input, 82
Pseudo output, 82

Queue unit, 397

Race, 78
Race detection, 145
Random patterns, 202, 329
READ array, 127
Read(Ri), 319
Receive list (RLIST), 164
Reconvergent path, 28
Redundant faults, 194
Reference tester, 233
Register transfer level languages, 357
Register transfer level model, 120
Repairable memories, 295
Resolution, fault, 205
RETAIN (Remote Terminal Access Information Network), 336
Rise time, 138
Root of a polynomial, 347
Rule-of-ten, 228

SALT (Sequential Automated Logic Test), 114
Scan-in, 273

INDEX

Scanning electron microscope, 399
Scan-out, 273
Scan path, 273
Scan/set, 283
Scheduler, the, 138
SCIRTSS (Sequential CIRcuit Test Search System), 373
SCOAP (Sandia Control./Observ. Analysis Program), 28, 265, 338
SEC/DED code, 303
Second degree hardcore, 310
Selective trace, 270
Self-checking, 346
Self-test, 323
Sensitization search, 374
Sensitized path of a fault, 24
Sequential Path Sensitizer, 90, 193
Sequentially t-fault diagnosable, 341
Serial fault simulation, 131
Seshu's heuristics, 80
Setup time, 73
SFP (single fault propagation), 167
Shadow register, 283
Shared resource tester, 244
Shift register latch (SRL), 276, 339
SIFT (software implemented fault tolerance), 351
Signature analysis, 323
Signatured instruction stream, 344
Single-fault assumption, 22
Singular cube, 30
Slack, 150
Slew rate, 243
Sliding diagonal test, 293
Soft errors, 289
SOFTG (Software Oriented Fault Test Generator), 167
Source registers $S(Ij)$, 318
State planner, 383
States applied analysis, 168
State table, 69
State unit, 397
Static hazard, 79
Static tester, 243
Stimulus bypass, 133
Stored response tester, 234
Stored state variable, 70
Strongly connected component (SCC), 83
Structural model, 122
Stuck-at-0, 21
Stuck-at-1, 21

Stuck-open, 198
Subscripted D-algorithm, 60
Successor, 83
Super flip-flop, 91
Super output block, 91
Surround read disturb test, 293
Surround write disturb test, 294
Switch level model, 120
Symmetric function, 191
Synchronizing sequence, 104
Synchronous, 69
Syndrome, 301
Systematic code, 300

t-fault diagnosis, 340
Table driven simulation, 125
Terminal, 20
Ternary algebra, 129
Test cube, 38
Test cube $c(T,F)$, 43
Test links, 340
Test measure effectiveness, 271
Test pattern languages, 219
Test patterns, 2
Test vectors, 2
Testdetect, 156
Tester per pin architecture, 244
Three-valued simulation, 127
Timing generators, 244
Timing verification, 148
Timing wheel, 139, 173
Token, 378
Totally self-checking, 346
Transfer path $T(Ij)$, 319
Transition, 378
Transmission path, 193
Trapped fault, 374
Triple modular redundancy, 350
TSET (timing set), 245

Unate function, 191
Unit delay simulation, 135, 148

Vector spaces, 297
Vertex, 30
Visible fault, 161

Wafer, 9
Walking pattern, 293
Wave formatter, 244
Waveform mode, 400

Well known signal, 384
WRITE array, 127
Write recovery test, 294
Write (Ri), 319

x-generator, 215
x-point, 31

Yield, 9
Yorktown Simulation Engine, 394

0-controllability, 265
Zero delay simulator, 135
0-point, 31
Zoom table, 145, 168, 178